ACS SYMPOSIUM SERIES 307

Excited States and Reactive Intermediates

Photochemistry, Photophysics, and Electrochemistry

A. B. P. Lever, EDITOR
York University

Developed from a symposium sponsored by
the Divisions of Inorganic Chemistry
of both the American Chemical Society
and the Chemical Institute of Canada
at the 1985 Biennial Inorganic
Chemical Symposium,
Toronto, Ontario,
June 6–9, 1985

American Chemical Society, Washington, DC 1986

Library of Congress Cataloging-in-Publication Data

Excited states and reactive intermediates.
(ACS symposium series, ISSN 0097-6156; 307)

"Developed from a symposium sponsored by the American Chemical Society and the Chemical Institute of Canada at the 1985 Biennial Inorganic Chemical Symposium, Toronto, Ontario, June 6-9, 1985."

Bibliography: p.
Includes indexes.

1. Excited state chemistry—Congresses. 2. Chemistry, Physical organic—Congresses.

I. Lever, A. B. P. (Alfred Beverly Philip)
II. American Chemical Society. III. Chemical Institute of Canada. IV. Series.

QD461.5.E914 1986 541.3 86-7908
ISBN 0-8412-0971-5

PRINTED IN THE UNITED STATES OF AMERICA

ACS Symposium Series

FOREWORD

The ACS SYMPOSIUM SERIES was founded in 1974 to provide a medium for publishing symposia quickly in book form. The format of the Series parallels that of the continuing ADVANCES IN CHEMISTRY SERIES except that, in order to save time, the papers are not typeset but are reproduced as they are submitted by the authors in camera-ready form. Papers are reviewed under the supervision of the Editors with the assistance of the Series Advisory Board and are selected to maintain the integrity of the symposia; however, verbatim reproductions of previously published papers are not accepted. Both reviews and reports of research are acceptable, because symposia may embrace both types of presentation.

CONTENTS

PREFACE

MOLECULES IN THEIR EXCITED STATES and molecules of transient existence generated by photochemical stimulation or by other processes, such as electrochemistry, are rapidly drawing considerable interest and gaining importance. The excited state of a molecule is, in many ways, a new species different chemically from the ground state molecule and endowed with additional energy; it is often capable of chemical processes that are not possible in the ground state. The ability to do "test tube" experiments with such short-lived species is currently under intensive development.

The conference from which this book was developed addressed many of the techniques that may be used to probe these systems and dealt with the new chemistry that is being learned.

Approximately 160 participants took part in the Biennial Inorganic Chemical Symposium 1985. The participants came primarily from Canada and the United States; however, some came from as far as Japan, England, Belgium, Italy, East and West Germany, and The Netherlands, thus generating an international atmosphere. They heard the latest ideas in excited state photochemistry and photophysics, species at electrode surfaces, resonance Raman spectroscopy, electrochemiluminescence, photochemistry of organometallic and cluster species, and gas phase organometallic chemistry to name a few topics. Some fifty posters were also presented and the Chemical Abstracts Services displayed the latest in on-line searching.

Funding for the conference came from the divisions of inorganic chemistry of the American Chemical Society (ACS) and the Chemical Institute of Canada, the Petroleum Research Fund (ACS), the Natural Sciences and Engineering Research Council, York University, and eight industrial companies: Strem Chemicals Inc.; Merck Frosst Canada Inc.; Union Carbide Canada Limited; EG&G Canada Limited; Xerox Research Centre of Canada; Tasman Scientific Inc.; Guided Wave, Inc.; and Lumonics Inc.

I hope this volume will stimulate the readers to consider how they might also contribute to this rapidly growing area.

A. B. P. LEVER
York University
Toronto Ontario
Canada M3J 1P3

January 1986

INTRODUCTION

CHEMISTS HAVE BEEN CONCERNED predominantly with the chemistry of species that exist in their molecular ground state, often stable for an indefinite period. Structures can, in principle, be obtained by X-ray crystallographic methods, and physical data such as NMR, IR, and UV/VIS spectra can be obtained with conventional spectrometers.

The advent of lasers and electronic devices that can record extremely fast events has led to growing interest in the study of molecules in excited states, usually, though not exclusively, the lowest excited state. The detailed study of excited states is a field that is now growing rapidly and promises to deliver a fascinating new view of chemistry in the future.

Excited States: Characteristics

A molecule in its first excited state is, in a very real sense, a different molecule from the ground state of the species. It possesses additional energy and probably has a different structure, at least in respect to small changes in bond lengths and angles, and indeed may have a totally different stereochemistry. It has different electronic and vibrational spectra and clearly has a different chemistry. Such chemistry is referred to as *photochemistry* because it is accessed by a light absorption event. These excited state molecules commonly exist for time intervals ranging from picoseconds to microseconds, rarely longer, except when solids at cryogenic temperatures are being studied. Nevertheless, methods of analysis are available to probe the photophysics and photochemistry of these species even on such a short time frame.

Light absorption will not generally occur to the lowest excited state, but rather a series of excited states may become populated. These will generally decay rapidly (in picoseconds) to the lowest excited state. One can anticipate the even richer chemistry of these higher excited states, but their lifetime will usually—though not exclusively—be so short that this chemistry has no time to be expressed. Nevertheless, light emission and photochemistry may sometimes be observed from these higher energy levels. Special techniques such as ultra-short laser pulses (measured in femtoseconds) are becoming available to probe this chemistry (for example, on the time scale of bond breaking).

In general the lowest excited state will decay back to the ground state by one or more pathways, including radiationless *deactivation* (loss of excited state energy as heat to the surroundings), one or more photochemical reactions, or by *luminescence* (fluorescence or phosphorescence). The study of these processes leads to a better understanding of the electronic structure of the excited state.

In the short term, the value of such studies must lie in what we can learn about how chemistry changes when the quantum mechanical state of a molecule changes, and how the additional energy, distributed over the molecule, modifies its chemistry. In the long term, new industrially important processes may depend upon the use of excited state molecules.

With the exception of a few highly studied states, such as, for example, the redox active lowest metal-to-ligand charge transfer excited state (MLCT) in the $[Ru(bipyridine)_3]^{2+}$ cation, or the ligand field active ^2E state of Cr(III), we know relatively little about these excited states. An enormous body of knowledge is waiting to be explored.

Excited State Quenching Reactions

An important facet of the chemistry of excited states is that additional energy confers upon the state both greater oxidizing power and greater reducing power, relative to the ground state. The greater reducing power originates in the higher energy electron that has been excited, while the greater oxidizing power resides in the hole created by the excitation of an electron. Electron transfer reactions may be observed by reaction of the excited state with an electron donor or acceptor.

Where the excited state luminesces, redox reactions with various species can be monitored by observing the quenching of excited state luminescence, or reduction in excited state lifetime, as a function of the concentration of quenching species (Stern–Volmer plot). In this fashion one can determine the rates of chemical reaction between excited state and quencher, and, using models such as those developed by Marcus or Angmon and Levine, determine various parameters such as free energies of activation and reorganization energies.

Because of the much greater driving forces potentially available in reactions between substrates and excited state molecules, difficult—but valuable—electron transfer reactions, such as the oxidation of water or chloride ion, may be accessed through excited state photochemistry. The question of how to separate hole-electron pairs generated in a quenching reaction, how to provide kinetic pathways to lead these two highly reactive species far apart from each another, and how to couple in some useful chemistry are currently of interest.

Of related interest is the problem of how far apart, in a fixed sense, an excited state and a quencher can be, and yet have electron transfer take place. Thus one can have quenching via a collision process, or by overlap of donor and acceptor orbitals at long distances. If the excited state and quencher are linked by a long conjugated pathway, quenching might be expected to take place. However, what about nonconjugated and through-space pathways, both of which can also lead to quenching? Such studies can lead to geometric information about proteins and impurity sites in crystal.

Some extremely fascinating oxidative addition-type chemistry is possible at cryogenic temperatures when metal atoms are irradiated in the presence of C-H and C-O bonds. Metal atoms are inserted, presumably via an excited state. This may well have significance for the activation of alkanes.

Some of the more interesting and valuable redox processes are multi-electron in nature, suggesting the utility of coupling a two- or many-electron event into an excited state process. The study of the excited state photochemistry and photophysics of binuclear and polynuclear (cluster) molecules is thus becoming of importance, and two-electron reactions are being identified.

The sensitization of semiconductors is a special example of electron transfer quenching and may prove to be very important. A photoexcited electron may, for example, be injected with high quantum yield into the semiconductor conduction band, to produce a photovoltaic device. The "hole" that is "left behind" may then perform some useful oxidation process.

Excited state quenching is not restricted to electron transfer processes, but may also occur by atom abstraction (for example, hydrogen atom abstraction), or by energy transfer to another species. In addition, the excited state energy may be used along a reaction coordinate leading ultimately to ligand loss or ligand exchange. New molecules may be formed by shining light upon the old. Organometallic photochemistry is particularly rich in providing unusual molecules after stimulation by light. The mechanisms and dynamics of such reactions are areas of serious study.

Indeed elucidation and understanding of the many processes that can occur upon light stimulation, and the chemical dynamics associated therewith, are major goals of current excited state chemistry.

Photoexcitation of biological molecules, proteins, and enzymes also has interest (such as watching a carbonyl group photodissociate from carbonyl heme and studying the chemistry of the resulting products).

Spectroscopy

As one might expect, various spectroscopies, especially electronic spectroscopy and resonance Raman spectroscopy, can provide detailed information about the electronic and vibrational nature of an excited state. Conventional electronic spectroscopy, absorption and emission, can provide information about the geometry and bond distances in excited states, while resonance Raman reveals the nature of the coupling between an electronic state and vibrational modes of the molecule.

Transient absorption spectroscopy, wherein one measures the electronic absorption spectrum of a molecule in an excited state, is still in its infancy, but the growing availability of ultra-high-speed, rapid-scan spectrometers augurs well for this area of spectroscopy. Thus one may, in the future, routinely probe excited state absorption spectra as well as ground state absorption spectra. The former can be expected to be as valuable in obtaining information about the excited state as is the latter for the ground state.

Time-resolved spectroscopies of various kinds have proven useful in probing the life of an excited state. As an excited state decays, perhaps through a chain of species, time-resolved spectroscopy (e.g., luminescence, excitation, resonance Raman) can provide data for these various steps. Such studies have led, for example, to the view that the first MLCT excited state in $[Ru(bipyridine)_3]^{2+}$, is localized in one bipyridine ring rather than delocalized over all three rings.

Electrochemistry

Electrochemically generated chemiluminescence provides an unusual method for studying excited state energies. Thus, for example, an oxidant and a reductant can be generated at the same electrode (with alternating polarization), or at two closely spaced electrodes. Given appropriate energetics, the oxidant and reductant quench one another to generate an excited state rather than a ground state product, and luminescence may be observed.

Excited States of Mononuclear and Dinuclear Chromium(III) Complexes

Hans U. Güdel

Department of Chemistry, University of Bern, Freiestrasse 3, CH-3000 Bern 9, Switzerland

Excited states of Cr^{3+} complexes were explored by single crystal spectroscopy at low temperatures. In the dimeric $[a_4Cr(OH)_2Cra_4]^{4+}$ the sharp ^2E single excitations were used to determine orbital exchange parameters. Out-of-plane interactions are dominant. The complex $CrCl_6^{3-}$ was studied in two exactly octahedral crystal environments. Broad-band $^4T_{2g} \longrightarrow {}^4A_{2g}$ luminescence with a great deal of fine structure was observed. The equilibrium geometry of the luminescent $^4T_{2g}$ state is a distorted octahedron with an equatorial Cr-X elongation of 0.1 Å and a small axial compression.

Luminescence from Cr^{3+} complexes, both in the solid state and in solution, is a widespread phenomenon. The great majority belong to type a) in Figure 1, where the luminescent state is ^2E and the optical transitions are sharp. The well-known ruby emission is a prototype for this situation. In a weaker ligand field the situation b) in Figure 1 is approached, the $^4T_2$ state becomes competitive with ^2E as the luminescent state. The $^4T_2$ emission, corresponding to a spin-allowed d-d transition, is vibronically broadened. Pure $^4T_2$ luminescence from Cr^{3+} has been observed in halide and oxide coordinations (1). Intermediate situations with both ^2E and $^4T_2$ emissions are also known.

The ^2E and $^4A_2$ states both derive from the $(t_2)^3$ electron configuration. The two states have approximately the same chemical bonding and thus the same equilibrium geometry. The resulting sharpness of the corresponding optical transitions in absorption and emission at low temperatures provides a great deal of information about the nature of the excited ^2E state. In the case of dinuclear Cr^{3+} complexes very useful information about the exchange coupling can be obtained from a detailed study of the singly and doubly excited ^2E dimer states (2). The reason lies in the intraconfigurational nature of the ^2E excitations, which greatly simplifies the theoretical ap-

0097-6156/86/0307-0001$06.00/0

proach to the problem. It is possible to deduce the dominant orbital contributions to the net exchange from an analysis of the energy splittings in the excited states which result from exchange interactions. This information about the mechanisms of exchange is not accessible by studying the ground-state properties alone. We will give an illustration of this type of study in the first part of this paper.

The $^4T_2$ state derives from the $(t_2)^2(e)^1$ electron configuration and is therefore displaced with respect to $^4A_2$ in the diagram of Figure 1b. Exploring the nature of the $^4T_2$ state, in particular its equilibrium geometry with respect to the ground state, is chemically and physically relevant. In the low-field situations of Figure 1b $^4T_2$ is the excited state with the longest physical lifetime. Its photochemical relevance is thus enhanced. The broad-band $^4T_2 \longrightarrow {}^4A_2$ luminescence lies in the near infrared (NIR). Solid state materials with broad NIR luminescence are of current interest as possible candidates for tunable lasers (1) as well as "solar concentrators" (3). In the second part of this paper the potential of crystal luminescence spectroscopy to investigate excited state properties will be illustrated for $CrCl_6^{3-}$.

Superexchange in bis(μ-hydroxo)-bridged chromium (III) dimers

Optical spectroscopy is a valuable complement to magnetochemical techniques for the study of exchange effects in polynuclear paramagnetic complexes. A recent review was given in Ref. (3). Dinuclear Cr^{3+} complexes have received a great deal of attention, mainly because of the sharpness and thus the high information content of their $^2E \longleftrightarrow {}^4A_2$ transitions. The complexes $[a_4Cr(OH)_2Cra_4]^{4+}$, where a= NH_3 or a_2 = en, have been studied in great detail(5,6). They will serve as illustrative examples of the types of effects which are observed and the conclusions which can be drawn.

When both ions of a Cr^{3+} dimer are in the electronic ground state, the exchange interaction can be represented by the well-known Heisenberg Hamiltonian

$$\hat{H}_{ex,g} = -2J_{ab}\,\vec{S}_a \cdot \vec{S}_b \qquad (1)$$

One of the main aims of magnetochemical studies is a determination or estimate of the exchange parameter J_{ab}.

When one of the ions is in the ^2E state we have a singly excited dimer state, and the exchange interactions can be represented by

$$\hat{H}_{ex,e} = -2\sum_{i,j} J_{a_i b_j}\,(\vec{s}_{a_i} \cdot \vec{s}_{b_j}) \qquad (2)$$

where i and j number the singly occupied t_2 orbitals. $J_{a_i b_j}$ are orbital exchange parameters, which are related to J_{ab} by

$$J_{ab} = \frac{1}{9} \sum_{i,j} J_{a_i b_j} \tag{3}$$

The simplicity of equation (2) results from the fact that both the $^4A_2$ and the ^2E state correspond to a half-filled t_2 shell in the strong-field limit. As a consequence the ^2E excitations are pure spin-flip transitions. More complicated expressions for the exchange Hamiltonian result even for the case of orbital changes within the $(t_2)^3$ configuration ($^2T_1$, $^2T_2$ excitations). On the basis of equation (2) it is possible to determine the individual orbital parameters from a knowledge of the experimental energy splittings in the singly excited ^2E dimer state. The dominant orbital pathways of the exchange coupling can thus be derived. In the title complexes we distinguish between in-plane and out-of-plane orbital interactions, as shown schematically in Figure 2. With a Cr-Cr separation of approximately 3.0 Å and a CrOCr angle of 100° it is not a priori clear which of the two is dominant. The question was recently resolved by a spectroscopic determination of the exchange splittings (5).

Figure 3 shows that, as a result of exchange splittings, there is a great deal of fine structure in the ^2E, $^2T_1$ region of the 6K single crystal absorption spectrum of $[(NH_3)_4Cr(OH)_2Cr(NH_3)_4]Br_4 \cdot 4H_2O$ (abbreviated $[NH_3]Br_4 \cdot 4H_2O$). The temperature dependence of the low-energy part of the spectrum is shown for $[(en)_2Cr(OH)_2Cr(en)_2]Br_4 \cdot 2H_2O$ (abbreviated $[en]Br_4 \cdot 2H_2O$) in Figure 4a. On the basis of the observed polarizations and temperature dependencies the dimer transitions can be assigned. The exchange splitting pattern of the ground and excited state emerges (Figure 4b). Under very high spectral resolution the dimer transitions in $[NH_3]Br_4 \cdot 4H_2O$ were found to consist of several sharp lines. This "zero field splitting" is illustrated in Figure 5. It provides an additional handle for the assignment of the dimer states.

As elaborated in detail in Ref. (5) there are two principal intensity mechanisms for dimer excitations. The single-ion mechanism is based on the combined action of spin-orbit coupling and an odd-parity ligand field potential at the Cr center. It is by this mechanism that spin-forbidden transitions obtain their intensity in mononuclear complexes. The pair mechanism, on the other hand, is restricted to exchange-coupled systems. It leads to the selection rules $\Delta S = 0$, $\Delta M_S = 0$, where S, M_S characterize the dimer states. The single-ion mechanism also allows transitions with $\Delta S = \pm 1$ and $\Delta M_S = \pm 1$. In addition, both mechanisms lead to orbital, i.e. symmetry selection rules. In order to make full use of the latter in the assignment of dimer states it is essential to have single crystal data.

The fact that the crystals used in this study are primitive triclinic proved to be an advantage rather than a disadvantage. All the dinuclear molecules are lined up parallel in the crystal, and it was found that the crystal extinction directions in the relevant wavelength range more or less coincide with the molecular symmetry axes (6). In higher symmetry crystal systems one often encounters the problem that it is difficult or impossible to extract molecular po-

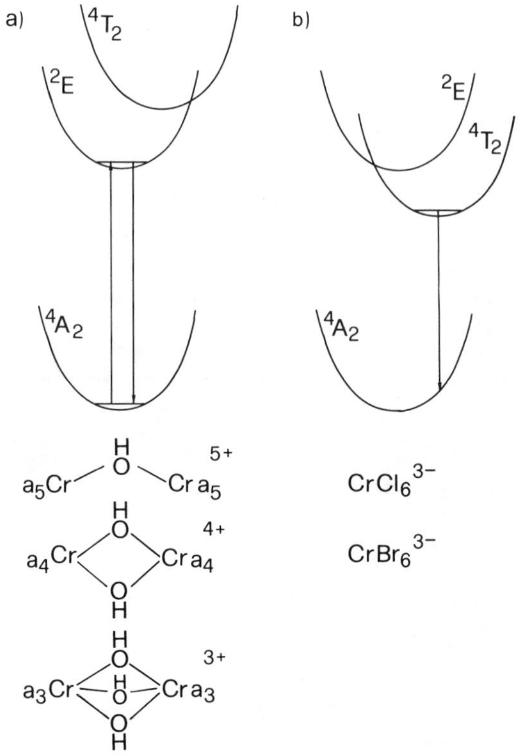

Figure 1. Luminescent states of Cr^{3+} complexes.

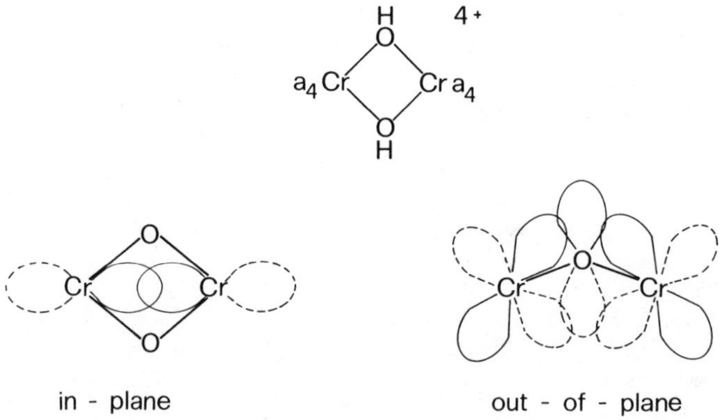

Figure 2. Exchange interaction pathways in $[a_4Cr(OH)_2Cra_4]^{4+}$ complexes.

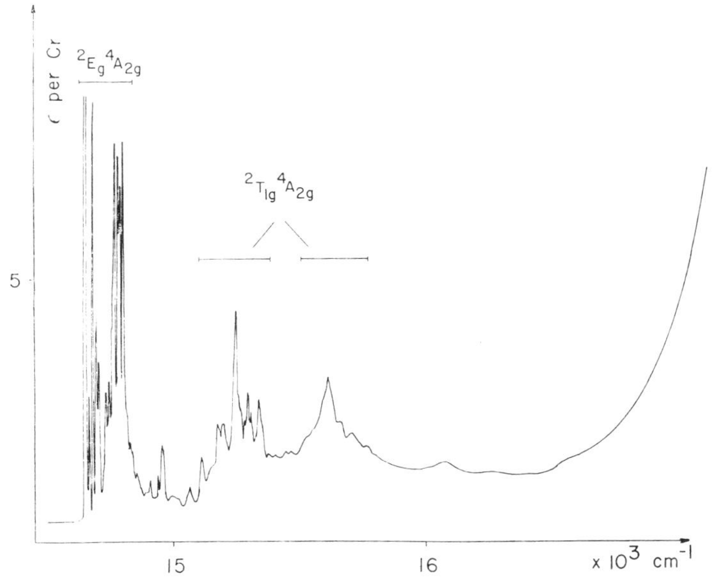

Figure 3. Single crystal absorption spectrum of $[NH_3]Br_4 \cdot 4H_2O$ at 6K in the $^2E/^2T_1$ region. "Reproduced from Ref. 5 - Copyright 1982, American Chemical Society".

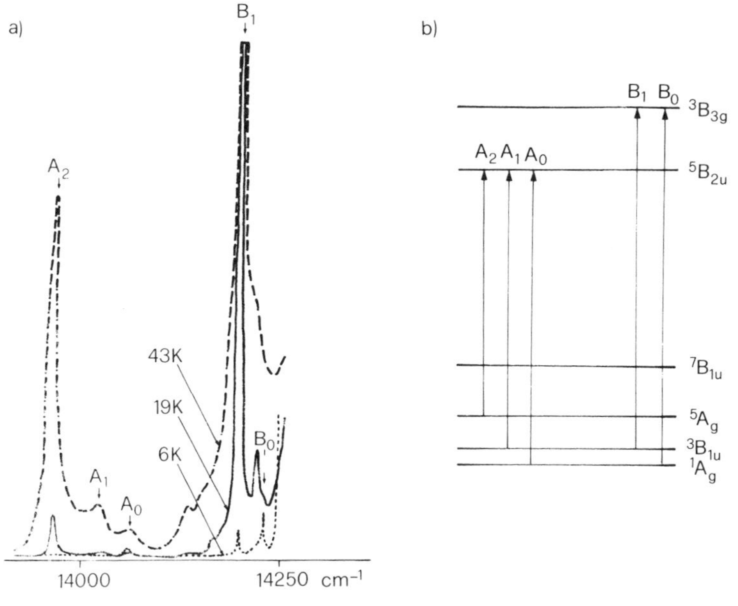

Figure 4. Low-energy dimer excitations of $[en]Br_4 \cdot 2H_2O$. "Reproduced with permission from Ref. 4. Copyright 1985, Reidel Publishing Company."

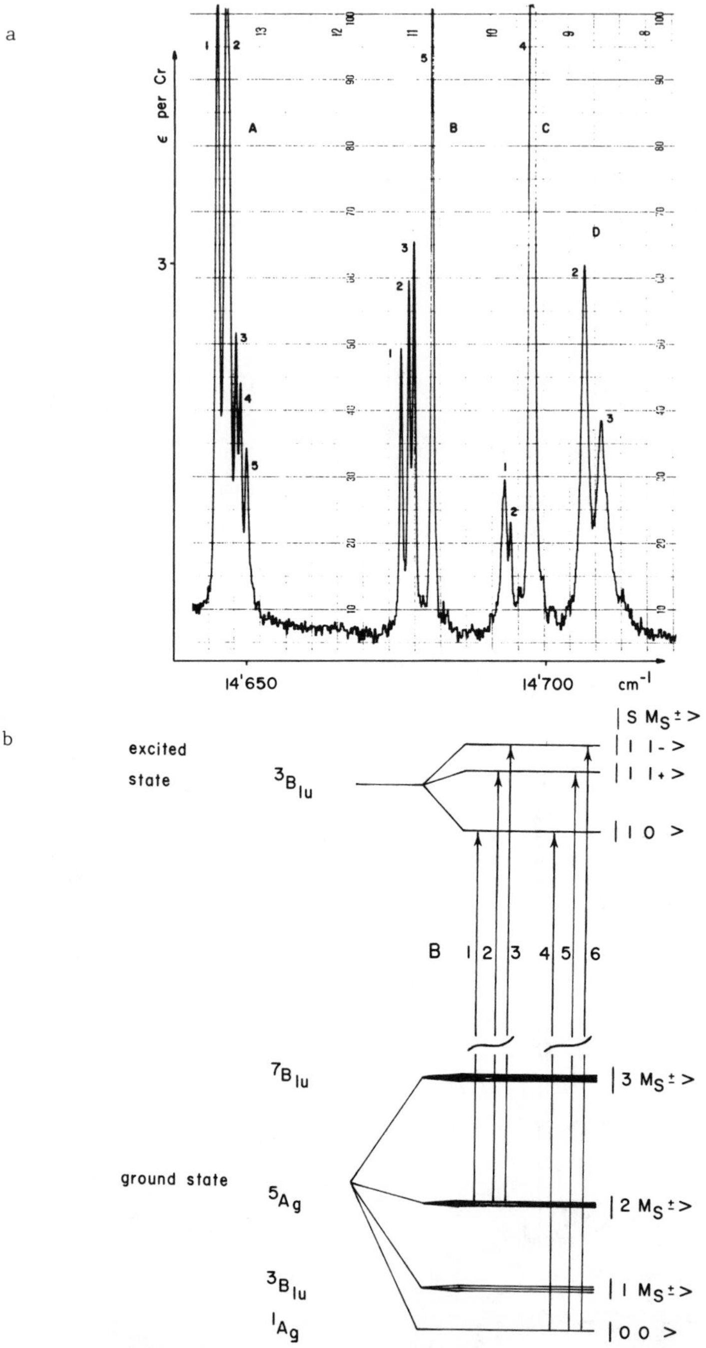

Figure 5. High-resolution absorption spectrum of $[NH_3]Br_4 \cdot 4H_2O$ in the region of ^2E excitations at 6K (a) Assignment of the components of band B (b). Reproduced from Ref. 5. Copyright 1982, American Chemical Society.

larisations from the polarized crystal spectra. The results obtained from a full analysis of the very extensive experimental data on the [en] and $[NH_3]$ complexes are summarized in Table I, which lists the antiferromagnetic contributions to the total J_{ab}. The out-of-plane interaction is found to be dominant. Its magnitude and, as a result, the magnitude of the total J_{ab}, is correlated with the position of the H atom in the bridge. In $[en]Br_4 \cdot 2H_2O$ the H atom lies more or less in the CrOCr plane. As a consequence the p(π) orbital at the oxygen is fully available for out-of-plane superexchange. In the $[NH_3]$ salts the H atom lies outside the CrOCr plane, and J (out-of-plane) is reduced accordingly. We have neglected ferromagnetic orbital contributions to J_{ab} in this summary discussion. As shown in Ref. (5) they can also be deduced from the optical spectroscopic data.

Table I. Structural and antiferromagnetic orbital exchange parameters

	Cr-Cr (Å)	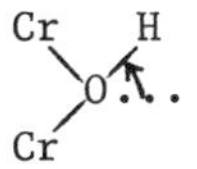 deg	J (in plane) (cm^{-1})	J (out of plane) (cm^{-1})	J_{ab} (cm^{-1})
$[en]Br_4 \cdot 2H_2O$	3.038(4)	6(2)	-5	-145	-16
$[NH_3]Cl_4 \cdot 4H_2O$	3.041	50(3)	-8	- 65	-0.9
$[NH_3]Br_4 \cdot 4H_2O$	~3.04	~50	-8	- 55	-0.4

We conclude that the high information content of the intraconfigurational ${}^4A_2 \leftrightarrow {}^2E$ transitions has enabled us to derive a very accurate picture of the singly excited state in the dimeric complexes $[a_4Cr(OH)_2Cra_4]^{4+}$. By using single crystals, high resolution spectroscopy in the temperature range 1.5K - 300K and a straightforward theoretical treatment, numerical values for the orbital exchange parameters were obtained. Antiferromagnetic orbital parameters can be compared with the results of theoretical calculations. The parameters $J_{a_i b_j}$ are connected with one-electron transfer integrals $A_{a_i b_j}$ as follows:

$$J_{a_i b_j} = \frac{A_{a_i b_j}^2}{2U} \qquad (4)$$

where U is the electron transfer energy (7). The transfer integrals can be related to energy differences between molecular orbitals, which are obtained by an approximate method like, e.g. an Extended Hückel calculation. For the $[a_4Cr(OH)_2Cra_4]^{4+}$ complexes the dominance of out-of-plane superexchange (Figure 2b) is nicely reproduced by such a calculation (7).

A large number of dimeric Cr^{3+} systems, both natural dinuclear complexes and pairs obtained by doping Cr^{3+} into a suitable host lattice, have been investigated by optical spectroscopy (8) (9). Examples of exchange-coupled 3d complexes other than Cr^{3+}, whose excited states were studied in detail, are given in Refs. (10) - (13).

Nature of the luminescent state in $CrCl_6^{3-}$

Cr^{3+} can be doped into the cubic elpasolite lattices $Cs_2NaInCl_6$ and Cs_2NaYCl_6. The Cr^{3+} site symmetry is exactly octahedral, which makes these elpasolite systems particularly attractive. With Cr^{3+} doping levels of approximately 2% broad-band luminescence corresponding to the $^4T_{2g} \longrightarrow {}^4A_{2g}$ transition is observed. Figure 6 shows the 6K emission spectra. They exhibit a great deal of fine structure, much more than in any spin-allowed d-d band ever measured in absorption. We also notice some differences in the intensity distribution between the two lattices. They result from a difference of approximately 0.1 Å in the M^{3+} - Cl distance in the host lattices.

A vibrational analysis of the spectra is straightforward. Details for the $Cs_2NaInCl_6$:Cr system were given in Ref. (14). In both spectra weak electronic origin lines on the high-energy side of the broad band can be identified as magnetic dipole transitions from the lowest-energy spin-orbit component E" of $^4T_{2g}$ to the ground state U'($^4A_{2g}$). They are followed by stronger electric dipole false origins involving t_{2u} and t_{1u} vibrations of the $CrCl_6^{3-}$ unit as enabling modes. The remainder of the fine structure consists of progressions in a_{1g} and e_g, based on the vibronic origins. Information about the nature of the luminescent $^4T_{2g}$ state is obtained from the rich fine structure and the highly resolved electronic origins.

The observation of an e_g progression is a clear indication of a Jahn-Teller effect (15). An orbital electronic T_{2g} state coupling to an e_g vibration is a classical Jahn-Teller situation, which can be treated theoretically (14/15). It leads to the well-known picture of three potentials displaced along the e_g coordinates. The $^4T_{2g}$ state in our host lattices thus shows a tetragonal e_g distortion in addition to the normal distortion along the a_{1g} coordinate. Its equilibrium geometry is no longer fully octahedral. This symmetry reduction leads to a partial quenching of orbital angular momentum, which is usually called a Ham effect (16). Designating the orbital components of $^4T_{2g}$ as $\psi_i = |\xi\rangle, |\eta\rangle$ and $|\zeta\rangle$ we have the following nonzero matrix elements of orbital angular momentum:

$$\langle\xi|\underline{L}_y|\zeta\rangle = \langle\zeta|\underline{L}_x|\eta\rangle = \langle\eta|\underline{L}_z|\xi\rangle = i\hbar \quad (5)$$

In the basis of Born-Oppenheimer product states $\Psi_i = \psi_i \cdot \chi_i$ we get

$$\langle\Psi_1|\underline{L}_y|\Psi_3\rangle = \langle\Psi_3|\underline{L}_x|\Psi_2\rangle = \langle\Psi_2|\underline{L}_z|\Psi_1\rangle = i\hbar\gamma \quad (6)$$

where

$$\gamma = \langle \chi_1 \mid \chi_2 \rangle$$

γ, the Ham quenching factor, corresponds to the vibrational overlap of the displaced Jahn-Teller potentials. Assuming Harmonic potentials, γ is given by

$$\gamma = \exp\{-3\, E_{J.T.}/(2\hbar\omega)\} \qquad (7)$$

where $E_{J.T.}$ is the Jahn-Teller stabilisation energy.

All the off-diagonal matrix elements of the spin-orbit coupling in the $|\xi\rangle$, $|\eta\rangle$, $|\zeta\rangle$ basis are thus reduced by the factor γ, and we use the experimentally observed quenching to calculate $E_{J.T.}$ and the corresponding geometrical distortion (14). In the Cs_2NaYCl_6 host lattice the total spread of the four spin-orbit components of $^4T_{2g}$ is 32 cm^{-1}, whereas crystal field theory without considering a Jahn-Teller effect predicts a total spread of approximately 107 cm^{-1}.

Another source of information about the $^4T_{2g}$ distortions, both along a_{1g} and e_g, lies in the distribution of intensity within the respective progressions in the luminescence spectrum. The procedures, by which the displacements ΔQ_i with respect to the ground state equilibrium geometry along the coordinates i are obtained from the experimental data have been given in detail in Refs. (14) and (17). The results for our systems are summarized in Table II.

Table II. Distortion of the $^4T_{2g}$ state of Cr^{3+} with respect to the ground state $^4A_{2g}$. The parameters are defined in the text.

host lattice	$\Delta Q_{a_{1g}}$ (Å)	ΔQ_{e_g} (Ham) (Å)	ΔQ_{e_g} (progr.) (Å)	ΔQ_{e_g} (Å)	$\Delta(Cr-X)_{eq}$ (Å)	$\Delta(Cr-X)_{ax}$ (Å)
$Cs_2NaInCl_6$	0.150	-0.121	-0.111	-0.116	0.095	-0.006
Cs_2NaYCl_6	0.154	-0.132	-0.140	-0.136	0.102	-0.016

We find that the e_g distortions derived from the Ham quenching and the intensity distribution in the progression differ by less than ten per cent, thus confirming the soundness of our analytical procedure. In order to get the actual displacements in Cr-X bond lengths, $\Delta(Cr-X)_{eq}$ and $\Delta(Cr-X)_{ax}$, for the equilibrium geometry of the luminescent $^4T_{2g}$ state, the ΔQ_i values have to be linearly transformed (17). For the $|\zeta\rangle$ component of $^4T_{2g}$ the values in the last two columns of Table II are obtained. The result for the Cs_2NaYCl_6 lattice is visualized in Figure 7.

The highly resolved fine structure of the low-temperature broad-

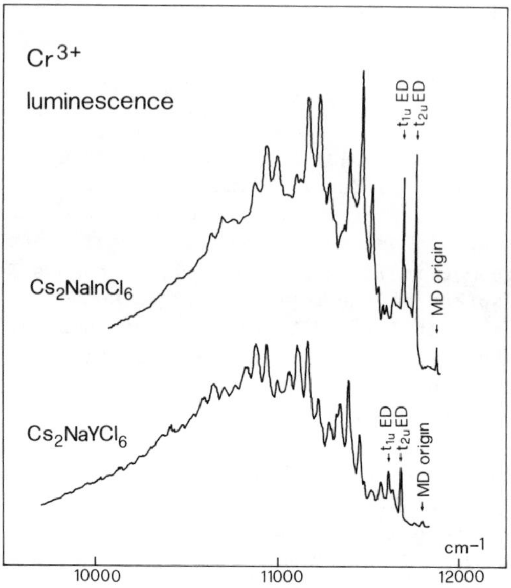

Figure 6. Luminescence spectra of Cr^{3+} at 6K.

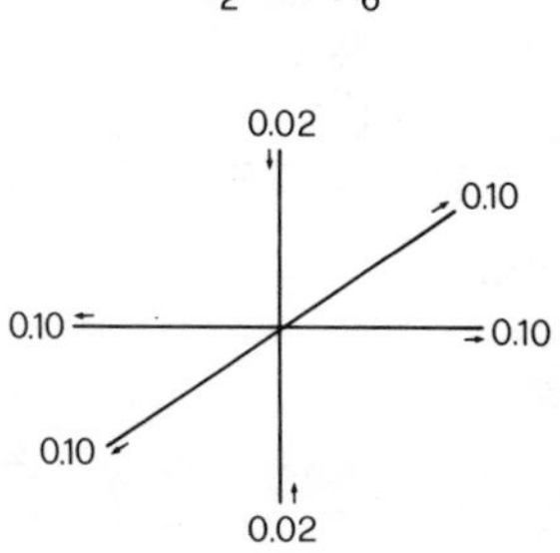

Figure 7. Equilibrium geometry of luminescent state of Cr^{3+} in Cs_2NaYCl_6.

band luminescence of Cr^{3+} in the $Cs_2NaY(In)Cl_6$ host lattices has enabled us to derive a very accurate picture of the emitting $^4T_{2g}$ state. Luminescence spectroscopy, whenever applicable, has the advantage that the low-temperature spectrum consists of only one electronic transition, whereas in absorption the transitions to the four spin-orbit components of $^4T_{2g}$ are superimposed. As a result the absorption spectrum is not as well resolved.

In the $CrCl_6^{3-}$ units the combined distortions along the a_{1g} and e_g coordinates lead to a compressed octahedron, with equatorial Cr-Cl distances approximately 0.1 Å larger than in the ground state and a much smaller axial compression. Similar distortions were deduced from optical absorption spectroscopy for the $^4T_{2g}$ state in $Cr(NH_3)_6^{3+}$ (18) as well as the $^1T_{1g}$ and $^3T_{1g}$ states in $Co(NH_3)_6^{3+}$ (17). There are important photochemical implications of these distortions. It is intuitively clear that the chemical bonding is quite different in the excited state. Bond strengths and force constants in the equatorial plane are reduced and, consequently, ligand substitution is facilitated. Such photochemical processes are, of course, much more likely to occur in solution than in the elpasolite systems studied here. But high-resolution crystal spectroscopy is a very powerful technique to study the physical properties of the relevant states.

Literature Cited

1. Kenyon, P.T.; Andrews, L.; McCollum, B.; Lempicki, A. IEEE J. Quant. Electronics, 1982, QE-18, 1189.
2. Güdel, H.U. Comments Inorg. Chem., 1984, 3, 189.
3. Hermann, A.M. Sol. Energy, 1982, 29, 323.
4. Güdel, H.U. in "Magneto-Structural Correlations in Exchange Coupled Systems"; Willett, R.D., Ed.; Reidel, 1985; p. 297.
5. Decurtins, S.; Güdel, H.U. Inorg. Chem., 1982, 21, 3598.
6. Decurtins, S.; Güdel, H.U.; Pfeuti, A. Inorg. Chem., 1982, 21, 1101.
7. Leuenberger, B.; Güdel, H.U. Inorg. Chem., submitted.
8. References 7, 8, 10 - 15 quoted in Ref. 4.
9. Riesen, H.; Güdel, H.U.; Chaudhuri, P.; Wieghardt, K. Chem. Phys. Lett., 1984, 110, 552.
10. References 24 - 28 quoted in Ref. 3.
11. McCarthy, P.J.; Güdel, H.U. Inorg. Chem., 1984, 23, 880.
12. Güdel, H.U. Inorg. Chem., 1983, 22, 3812.
13. Riesen, H.; Güdel, H.U. Inorg. Chem., 1984, 23, 1880.
14. Güdel, H.U.; Snellgrove, T.R. Inorg. Chem., 1978, 17, 1617.
15. Sturge, M.D. Solid State Phys., 1967, 20, 91.
16. Ham, F.S. Phys. Rev. A, 1965, 138, 1727.
17. Wilson, R.B.; Solomon, E.I. J. Amer. Chem. Soc., 1980, 102, 4085.
18. Wilson, R.B.; Solomon, E.I. Inorg. Chem., 1978, 17, 1729.

RECEIVED November 8, 1985

2

Ab Initio Analysis of Charge Transfer Excitations

The $Cr(CN)_6^{3-}$ Complex

L. G. Vanquickenborne, L. Haspeslagh, and M. Hendrickx

Department of Chemistry, University of Leuven, Celestijnenlaan 200F, 3030 Leuven, Belgium

For the $Cr(CN)_6^{3-}$ complex, a number of charge transfer transitions (both LMCT and MLCT) have been analyzed at the SCF-level of approximation. Apart from the formal electron transfer taking place in the orbitals that change their occupation numbers, significant density shifts take place in the other orbitals as well. The energy spectrum of the charge transfer transitions is found to be considerably simplified by a level clustering, strongly reminiscent of the related d^{n+1} or d^{n-1} ligand field spectrum.

In a previous paper (1), ab initio SCF-calculations using a very large basis set have been reported for the ground state and the ligand field excited states of $Cr(CN)_6^{3-}$. Figure 1 shows a partial molecular orbital diagram, based on the SCF-orbitals for the d^3-configuration average. All orbitals are either predominantly ligand based, or predominantly metal-based. In fact, this predominancy is very pronounced, since all the orbitals are 85 or 90 per cent metal or ligand. In this respect, the Hartree-Fock calculations are closer to the pure ligand field picture than to certain extended Hückel calculations (2) where one obtained considerable metal-ligand mixture. Another striking difference between SCF and EH calculations is that the metal orbitals in Figure 1 are obviously not the frontier orbitals. Although this fact is quite well known and has been reported for several other transition metal complexes (3-6), it is particularly relevant in the study of charge transfer transitions.

The main reason why the t_{2g}^3-configuration of Figure 1 has a lower energy than any other configuration, including t_{2g}^4 coupled to a hole in any one of the topmost orbitals, is connected to the just mentioned dichotomy between ligand orbitals and metal orbitals. Indeed, the orbitals characterized by predominant ligand (L) character are highly delocalized, whereas the $2t_{2g}(3d\pi)$-orbital is almost entirely localized on the chromium metal (M) atom. As a consequence, adding a fourth electron into the $3d\pi$-shell will lead to an increase of the interelectronic repulsion energy, which is much larger than the decrease resulting from the removal of an electron from a ligand

0097-6156/86/0307-0012$06.00/0

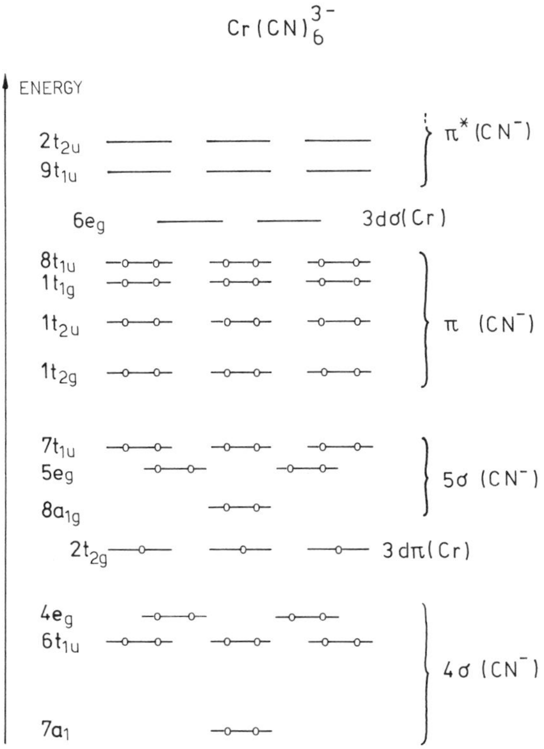

Figure 1. Partial and qualitative molecular orbital diagram of the $Cr(CN)_6^{3-}$-molecule. Circles indicate electron occupation. The relative order is based on the solutions of the Hartree-Fock equations for the average of all d^3-states (1).

orbital. Numerically, the dd-Coulomb integral $J(2t_{2g},2t_{2g}) = 0.790$ hartree, whereas typical ligand Coulomb integrals, or typical metal-ligand Coulomb integrals are of the order of 0.2 hartree.

Therefore transitions of the type $2t_{2g}^3 8t_{1u}^6 \rightarrow 2t_{2g}^4 8t_{1u}^5$ do <u>not</u> lead to a stabilization; on the contrary, they correspond to charge transfer (CT) excitations, characterized by a rather high transition energy. SCF-calculations are not sufficiently reliable to predict quantitative values for these transition energies. Rather, they should be used to answer some very general questions, such as : are the dd bands lower than the CT bands also at the Hartree-Fock level of approximation ? Do the SCF results reproduce the qualitative features of ligand field theory ? Is there a pattern in the energy of the CT levels ?

<u>Ligand Field Considerations</u>

Although the main emphasis of this work is on CT transitions, it is useful to consider first the intraconfigurational energy splittings within the t_{2g}^3-configuration, as shown in Figure 2. In the framework of ligand field theory(purely atomic d-orbitals), the ^2E_g and $^2T_{1g}$ states would be degenerate, and the remaining gaps $(^2E_g,^2T_{1g} - {}^4A_{2g})/(^2T_{2g} - {}^4A_{2g})$ are determined by the Racah B, C parameters; they are predicted to be in the ratio 3/5, which is reasonable if compared with the experimental ratio of about 0.7.

Figure 2 goes beyond ligand field theory and is based on the assumption that the t_{2g}-orbitals are (predominantly metal-centered) molecular orbitals. At the left hand side of Figure 2, the four states are described by one single set of molecular orbitals. Let us suppose that these orbitals are solutions of the Hartree-Fock equations for the $^4A_{2g}$ ground state. In order to describe the interelectronic repulsion within these orbitals, it is not possible to work with the two Racah parameters, which are based on the special rotational properties of the atomic d-orbitals. Instead, one must recur to the Griffith parameters (in this case a, b and j). At the left hand side of Figure 2 (frozen orbital approximation) the energy splittings are functions of these 3 parameters. Conceptually and qualitatively, the frozen orbital approximation is very similar to the pure ligand field picture.

The right hand side of Figure 2 shows the result of a complete SCF-calculation, where the Hartree-Fock equations have been solved for each state separately. Therefore, each state is now characterized by its own set of optimal orbitals. Since a Hartree-Fock calculation is basically a variational treatment, the relaxation of the orbitals to their optimal shape causes the energy to drop somewhat in going from the left to the right on the diagram (except for $^4A_{2g}$). One consequence is that the transition energies can no longer be described by means of simple expressions as in the frozen orbital scheme : each state is now characterized by its own set of Griffith parameters. But also the other orbitals, more specifically the ligand orbitals, and even the core orbitals are slightly modified, and contribute to the transition energy : all simplicity appears to be lost, at least formally. The main simplification that is left however, is that the relaxation energy is - numerically - very small indeed, being of the order of a few hundreds of cm^{-1}. Therefore, the global

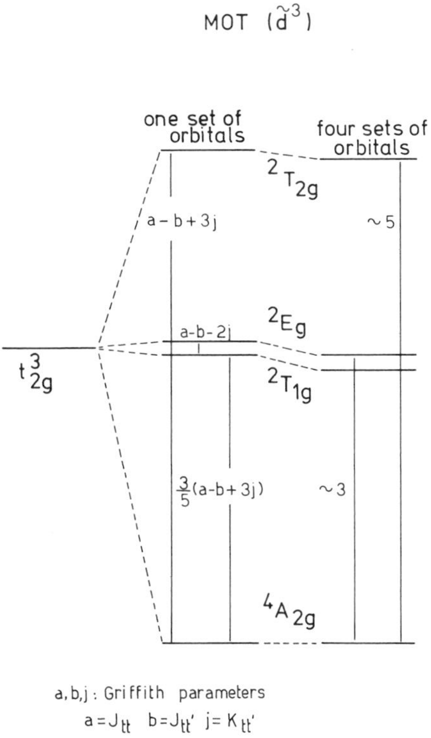

Figure 2. Relaxation of t_{2g}^3 states. Left : energies are based on the frozen orbitals of the $^4A_{2g}$ ground state. Right : results of full-scale SCF calculations. The energies are not drawn to scale.

pattern, and the relative energies at the right hand side of Figure 2 are virtually identical to the left hand side.

Yet, the relaxation process is important from another point of view. In the frozen orbital approximation, the doublet states have higher repulsion energies than the ground state. Since the orbitals are only optimal for the ground state situation, the orbital shape will not be ideally adapted to this increased repulsion. Therefore, the essence of the relaxation process can be described as a reshaping of the orbitals so as to restore the balance. One of the more important factors in this reshaping process is an expansion of the d-orbitals. Since a Hartree-Fock calculation yields an energy minimum, a small expansion means next to nothing for the total energy, but it means very much for the energy components V (potential energy) and T (kinetic energy).Table I shows the results for the interelectronic repulsion energy C (V = C + L, where L is the nuclear-electron attraction energy).

Table I : Interelectronic repulsion energy in the frozen orbital approximation (C) and in the SCF approximation (C')(1). All energies are relative to the ground state repulsion.

State	C (cm^{-1})	C'(cm^{-1})
$^2T_{2g}$	33 502	- 91 047
^2E_g	20 111	- 53 187
$^2T_{1g}$	20 095	- 52 821
$^4A_{2g}$	0	0

Obviously, the expansion (while not affecting the total energy to any significant extent) is so important that it overcompensates the original change in interelectronic repulsion : C' is smaller in the excited states than in the ground state. This rather thorough modification of the conceptual framework of ligand field theory (and also of atomic multiplet theory) is discussed more fully in other publications (1,7). Apparently, the conventional models of transition metal chemistry yield reasonable predictions of transition energies on the basis of physically unsound assumptions. If one is only interested in the energy pattern, ligand field theory remains a reliable guide. If one is interested in the reason why a certain energy pattern emerges, ligand field theory should be abandoned (1,7).

Charge Transfer Transitions : General Considerations

The CT-excited state which was discussed in the introduction, is of the ligand to metal type (LMCT) : $8t_{1u} \rightarrow 2t_{2g}$. If its energy is calculated by using the ground state orbitals, one obtains a very high value (more than 110 000 cm^{-1}); the experimental CT bands start at much lower energy ($\sim$ 40 000 cm^{-1}) (8). If a more complete SCF calculation is carried out (by solving also the SCF equations for the CT excited state and using the same basis set as before (1), the transition energy drops to 68 000 cm^{-1}. The relaxation energy, which was negligible from one ligand field state to another one, amounts to more than 40 000 cm^{-1} from the ground state to a CT state.

The main reason is that the transfer of an electron from the outside of the molecule toward the center induces important secondary shifts in the other orbitals as well. A first indication of these shifts can be obtained from a standard Mulliken population analysis, as shown in Table II. In each case, the actual shifts are much smaller than the formal transfer of one charge unit, indicating that the relaxation effects are characterized by a significant backflow of electrons. The relaxation is seen to be more important for the MLCT transitions, leading in one case ($2t_{2g} \rightarrow 9t_{1u}$) to the counterintuitive result shown in Table II, where the formal charge on the metal increases. This suggests that relaxation effects in this particular case overcompensate the originally induced change. Since we use a very large basis set, containing a number of rather diffuse functions, this result might be an artefact of the Mulliken population analysis.

Table II. Population shifts upon CT-excitation

L → M	Cr	6 CN
hypothetical	+ 1	- 1
$8t_{1u} \rightarrow 2t_{2g}$	+ 0.60	- 0.60
$1t_{2u} \rightarrow 2t_{2g}$	+ 0.50	- 0.50
M → L	- 1	+ 1
$2t_{2g} \rightarrow 9t_{1u}$	+ 0.22	- 0.22
$2t_{2g} \rightarrow 2t_{2u}$	- 0.13	+ 0.13

Density difference plots are perhaps more informative, as can be seen in Figure 3 for one of the LMCT states. The figure shows the difference between the two orbital densities for which the population has been changed. The figure corresponds to the simple picture one would get from the classical ideas : the ligand π-zone is depopulated and the metal dπ-zone is populated. From a classical point of view, one might be inclined to expect that this figure describes the essence of the charge transfer phenomenon. Yet, the total density difference in Figure 4 reveals a completely different picture : it shows the global density shifts upon excitation. $\Delta\rho$ is made up of two contributions : first the formal orbital density difference (Figure 3), and next the density shifts, associated with the relaxation taking place in the other orbitals. Obviously, figure 4 retains some of the

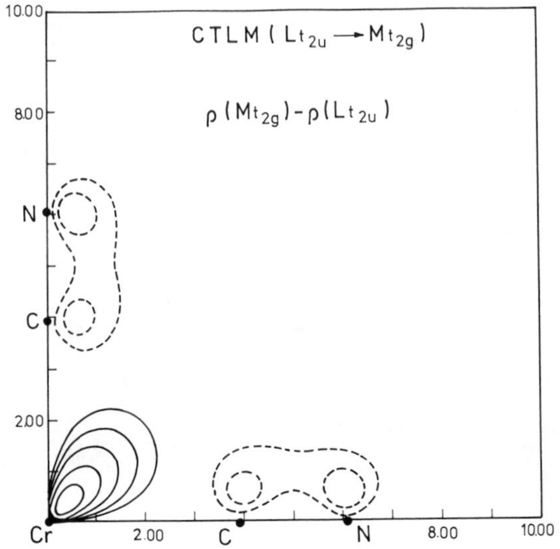

Figure 3. Orbital density difference plot Δρ in one of the coordinate planes. Full lines : Δρ > 0; dashed lines : Δρ < 0; dotted lines : Δρ = 0. Δρ at the different isodensity lines equals ± 0.16, ± 0.08, ± 0.04, ... ± 0.0025.

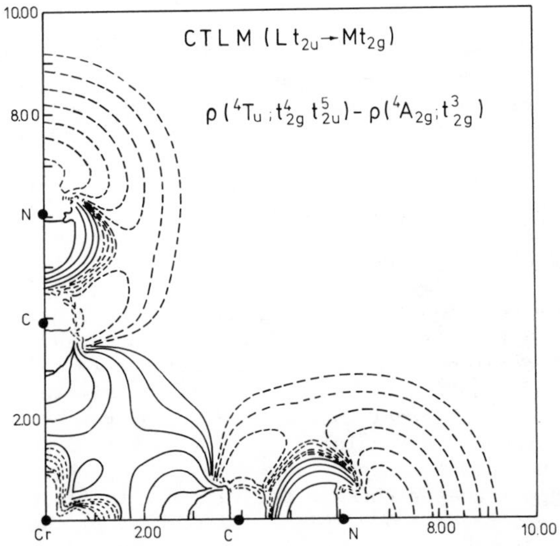

Figure 4. Total density difference plot; the conventions are as in Figure 3.

characteristics of Figure 3, but many other shifts are complicating the picture.

Similar diagrams can be obtained for MLCT-states. In the $2t_{2g} \rightarrow 9t_{1u}$ case, the formal orbital density difference does correspond to the classical picture; as a matter of fact, it is quite similar to Figure 3, except for the reversal between dotted lines and full lines, indicating a depletion of the metal dπ-orbitals and a population increase in the ligand π-region. In the total density difference map, remnants of the metal dπ-depletion can still be observed, but one also finds a depopulation of the outer ligand region, which seems to point in the same direction as the surprising results of the population analysis. In all cases, the classical picture is only confirmed, if we restrict ourselves to the density differences between the orbitals directly involved in the transition. If we consider the total density differences, the relaxation effects in the other orbitals are always very important and always correspond to density shifts of opposite direction. The backflow of electrons does not seem to follow any simple pattern, for instance by being concentrated in one particular set of molecular orbitals. It is as if each nucleus tends - by whatever channels available - to maintain roughly the same charge in its immediate neighborhood.

Energy Pattern of Charge Transfer Transitions

As indicated before, SCF theory has no problem in situating the CT transitions well above the ligand field transitions, and in this sense, it fits in completely with the traditional ideas. As for the relative position of different CT transitions, many problems remain to be solved. Consider for example the MLCT excited configuration corresponding to $2t_{2g} \rightarrow 9t_{1u}$. Table III shows the 15 states resulting from the single $t_{2g}^2 t_{1u}$ configuration. In an attempt to discover a pattern in this multitude of states, a simplifying model has been

Table III. Different states corresponding to the $2t_{2g}9t_{1u}^1$ configuration

t_{2g}^2	$^3T_{1g}$	x	$^2T_{1u}$	$^{2,4}T_{1u}$, $^{2,4}T_{2u}$, $^{2,4}A_{1u}$, $^{2,4}E_u$
	^1E_g			$^2T_{1u}$, $^2T_{2u}$
	$^1T_{2g}$			$^2A_{2u}$, ^2E_u, $^2T_{1u}$, $^2T_{2u}$
	$^1A_{1g}$			$^2T_{1u}$

introduced (9,10), based on the distinction between weak and strong interactions. The t_{2g}-electrons are localized on the metal and therefore their repulsion should be relatively large. The t_{1u}-electron is situated on the ligands, at the outside of the complex and relatively far from the metal. Therefore, it was believed that this electron should be coupled only weakly to the metal π-electrons. If this hypothesis holds true, the fifteen states can be expected to exhibit a very simple pattern, where we have large energy separations between the four metal-centered states, and where the final doublets and quartets deviate only slightly from their parent metal state. In other words, in the MLCT state of a d^3 system, we should recognize the structure of the ligand field spectrum of a d^2-complex. This model has been applied to a whole series of examples (9,10) and a more recent and complete review has been given by Lever (11).

From the present point of view, it would be interesting to verify if the model is confirmed by the Hartree-Fock calculations. In the original papers (9,10), the applicability of the model was limited by the observability of the expected energy pattern;the observability had to be verified in each particular case by assuming an electric dipole transition mechanism. In carrying out SCF-calculations however, one is freed from these limitations, and one can verify the model for an arbitrary case, including the case of Table III. Figure 5 shows an SCF energy diagram for the CT transitions under consideration. And indeed, it is gratifying to see that one finds exactly the pattern predicted by the weak coupling model. One has three groups of closely spaced levels, separated by relatively large energy gaps. The closely spaced levels have the same metal parentage. The relative position of the parentage averages corresponds to the ligand field states of a two-electron system, satisfying the same qualitative relationships. The ^1E and $^1T_2$ state are nearly, albeit not completely degenerate. The ratio between the two larger energy gaps is 0.41 instead of the ligand field value of 0.4.

In a previous Section, it has been shown how the ligand field energy pattern was confirmed at the SCF level, but how the underlying physical reason was thoroughly modified. The same situation is found to hold true for the CT-transitions : the states with $^3T_1$ parentage have the lowest total energy, but are characterized by the largest interelectronic repulsion (Table IV). Another rather striking example of the role of interelectronic repulsion energy is provided by Table V, where two different CT configurations are compared. The t^3e^1 (^5E_g) parentage corresponds to a high spin state while the t^4 ($^3T_{1g}$) parentage corresponds to a low spin state. The SCF excitation energies are quite close to each other, but, contrary to the classical expectations, the low spin state is characterized by the lowest repulsion energy.

Table IV. Total SCF energy (E') and repulsion energy (C') of the parentage averages of the $2t_{2g}^5 9t_{1u}^1$ CT states. All energies are relative to the $^3T_{1g}$-values

Parentage average	E'(cm^{-1})	C'(cm^{-1})
$^1A_{1g}$	34 903	- 23 268
^1E_g	14 372	- 8 685
$^1T_{2g}$	14 358	- 8 566
$^3T_{1g}$	0	0

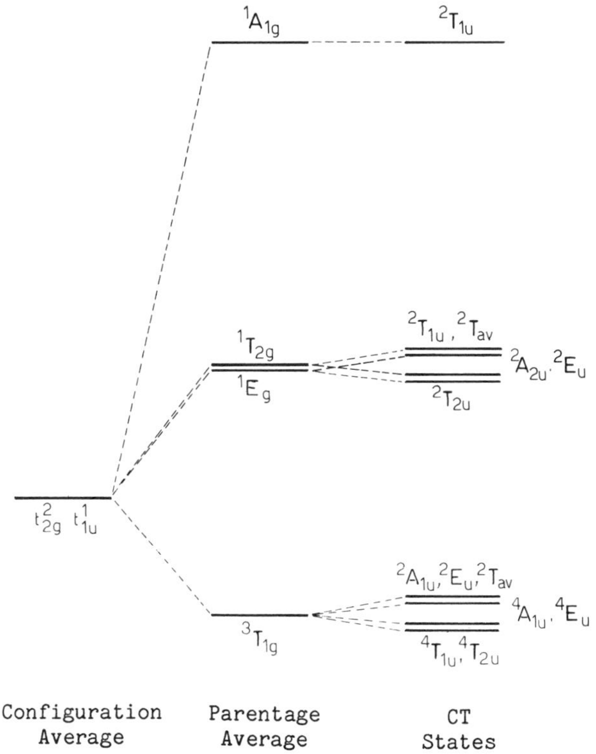

Figure 5. SCF Energy level diagram of the different CT states corresponding to the $t_{2g}^2 t_{1u}$-configuration. The energies are not drawn to scale. The symbol $^2T_{av}$ refers to the average energy of $^2T_{1u}$ and $^2T_{2u}$; Roothaan's open shell formalism does not allow these states to be calculated separately (1); only their average energy is accessible. The same remark holds for one of the entries in Table V. The intraconfigurational energy gaps are some 30 times smaller than the interconfigurational energy gaps.

Table V. SCF excitation energy $\Delta E'$ and repulsion $\Delta C'$ for two LMCT states; $^4T_{av}$ refers to the average of $^4T_{1u}$ and $^4T_{2u}$

Configuration and state	$\Delta E'(cm^{-1})$	$\Delta C'(cm^{-1})$
$^4T_{2u}$ $[t_{2g}^3 e_g^1(^5E_g);t_{2u}^5]$	75 201	1 036 845
$^4T_{av}$ $[t_{2g}^4 e_g^o(^3T_{1g});t_{2u}^5]$	75 994	837 016

As a conclusion, Hartree-Fock calculations are seen to be qualitatively compatible with the simple models (ligand field theory for dd-transitions, and the weak coupling model for CT-transitions). However, the ab initio work strongly suggests that the results be situated in a different conceptual framework.

Literature Cited

1. Vanquickenborne, L.G.; Haspeslagh, L.; Hendrickx, M.; Verhulst, J. Inorg. Chem. 1984, 23, 1677-84
2. Alexander, J.J.; Gray, H.B. Coord. Chem. Rev. 1967, 2, 29-43
3. Demuynck, J.; Veillard, A.; Vinot, G. Chem. Phys. Lett. 1971, 10, 522-5.
4. Wachters, A.H.J.; Nieuwpoort, W.C. Phys. Rev. B, 1972, 5, 4291-301.
5. Sano, M.; Yamatera, H.; Hatano, Y. Chem. Phys. Lett. 1979, 60, 257-60.
6. Sano, M.; Kashiwagi, H.; Yamatera, H. Inorg. Chem. 1982, 21, 3837-41.
7. Vanquickenborne, L.G.; Haspeslagh, L. Inorg. Chem. 1982, 21, 2448-54.
8. Alexander, J.J.; Gray, H.B. J. Am. Chem. Soc. 1968, 90, 4260-71.
9. Vanquickenborne, L.G.; Verdonck, E. Inorg. Chem. 1976, 15, 454-61.
10. Verdonck, E.; Vanquickenborne, L.G. Inorg. Chim. Acta 1977, 23, 67-76.
11. Lever, A.B.P. "Inorganic Electronic Spectroscopy"; Elsevier Science Publishers B.V. : Amsterdam, 1984.

RECEIVED November 8, 1985

Excited State Geometries of Coordination Compounds

Obtained from Vibronic Spectra and Photon Flux Fluctuation Measured by Time Resolved Spectroscopy

Hans-Herbert Schmidtke

Institut für Theorestische Chemie der Universität, D-4000 Düsseldorf 1, Universitätsstraße 1, Federal Republic of Germany

The intensity distributions of well resolved vibronic spectra recorded in absorption and emission at low temperature are used to determine the geometric distortions of the electronically excited states of coordination compounds. In particular for complexes of lower symmetry, band analysis is necessary leading to results with which bond distance changes can be calculated. For spectra exhibiting no vibrational fine structure, a new technique is proposed which uses time resolved methods, considering deviations from the Poisson distribution of photons by recording time intervals between two successively emitted photons.

Photochemical reactivity primarily results from electron distributions which are different to those in the ground state. With a change of electron density the geometry of the excited molecule may be distorted from that of the ground state molecule. These excited states, provided with large amounts of excess energy, are short lived species which are difficult to characterize. Since quantum chemistry is not able to calculate molecules of the size we are interested in (it also cannot satisfactorily consider cooperative effects resulting from interaction with the environment) one is restricted to experimental investigation. In particular some spectroscopic methods are fast enough to follow the physical conversions taking place in the molecule, by detecting the excited state species and by measuring at least some of its properties.

Vibronic spectra reflect changes in the electronic and vibrational state of a molecule at the same time. It is possible to calculate the geometry of the excited species and the potential hypersurface close to the equilibrium state. For this, a spectrum is required with sufficiently well resolved vibronic structure to carry

0097-6156/86/0307-0023$06.00/0

out a band analysis. This should allow for decomposition of the bands into different electronic transitions and a further resolution of the vibrational fine structure. Usually, the electronic spectra of transition metal coordination compounds in solution, or in the solid state, exhibit relatively broad overlapping bands, which do not show any sign of vibrational fine structure. However, there are many cases known where distinct vibrational structure is detected although often, the structure is not completely resolved. This has led some authors, on the basis of such spectra, to explain their results, in an incorrect fashion. To avoid misinterpretation one is urged to consider only those spectra with optimal resolution. This is, for instance, achieved when the vibrational quanta measured from a band progression in a luminescence spectrum agree with the fundamental modes of a vibrational spectrum taken from the ground state (Raman or IR). The MIME effect (missing mode effect), i.e. the absence of normal modes in the vibrational progression intervals (1,2), may exist on various occasions where the damping in the dissipative system is too large to detect separate modes. However, one cannot exclude the possibility that an unusual progressional frequency is simulated by incomplete resolution imposed on the system by insufficiencies in the experiment.

To obtaining high quality, well resolved absorption or emission spectra, various techniques have been applied.

e.g. (1) decreasing the temperature of the probe, if necessary below the λ-point of liquid He (~2 K)

(2) investigating single crystals

(3) considering doped chromophore materials with (inert) host crystals

(4) diluting the chromophore to be investigated with appropriate counter ions or looking at double salts
e.g. $[Co(NH_3)_6][Ir(CN)_6]$ (3) or
$[Cr(NH_3)_6](ClO_4)_2Cl \cdot KCl$ (4)

(5) using polarized light

(6) improving the apparatus to be used (monochromators, detection systems, photon counting, micro-optics etc.).

However, in many cases, these efforts may not lead to any resolution of the vibrational fine structure. This obviously has a physical reason. Since condensed systems are investigated, interaction with the environment is involved in the transition. The chromophore is an open system which dissipates vibrational energy into the surrounding medium by irreversible processes. This phenomenon can be used for detecting fine structure from the time resolved measurements of photon events, by monitoring the correlations between successively emitted photons. This new technique will be reported in the second part of this article.

Conventional Vibronic Spectra

Here we will discuss the visible and UV absorption and emission spectra of some selected transition metal and main group coordination compounds. The spectra are due to electronic transitions and exhibit extensive vibrational fine structure in long band progressions superimposed on each other. The transition metal d-d transitions become allowed by a complicated coupling mechanism, which mixes levels of different parity by a vibronic coupling operator, and different spin states by a spin-orbit coupling operator. These intermixings prepare the initial and final states of the transition for appropriate symmetry selection rules. Vibrational progressions and broad band spectra (in cases where resolution is not achieved) are explained, in general, by shifts of the potential energy curves along specific nuclear coordinates, along which large changes of bond properties may occur by excitation of the molecule to a higher state (Figure 1). These states are likely to be photoactive.

The theoretical analysis of the spectrum to obtaining information about the excited state structure, must start from a definite assignment of band components and a reliable band analysis carried out by resolving the band progression into gaussian or lorentzian profiles. The importance of good band analyses should not be underestimated if the intensity distribution of the measured spectra is compared with the theoretical band profile function. A band system which remains undetected under the main progression may lead to different results when calculating geometry distortions.

Theoretical band profiles are usually obtained from Franck-Condon factors which are adjusted to the band peaks of vibrational components in the progresssion. In our band analyses, we are using distribution functions $I_j\ (m,n;\Delta,\beta)$ which collect Franck-Condon integrals and Herzberg-Teller factors into a comprehensive function which has more general applicability than earlier band analysis procedures. The method can be used for j-fold degenerate vibrations, where the vibrational quanta may be different from the ground state $\beta = \nu_e/\nu_g \neq 1$ and for vibrational excitation in the initial state $m>0$, by which temperature dpendence will be introduced into the band profile function (5-7). The parameter $\Delta = (\omega M/\hbar)^{1/2}\Delta Q$ supplies the shift of the potential curve minimum of the excited state compared to the ground state. We shall apply these functions to some of the spectra discussed below.

Emission Spectra. The chance to obtain vibrational fine structure is usually higher for luminescence than for absorption spectra since in emission, absolute light intensities with high sensitivities are measured rather than intensity differences. However, few compounds exhibit luminescence intense enough to measure a reliable

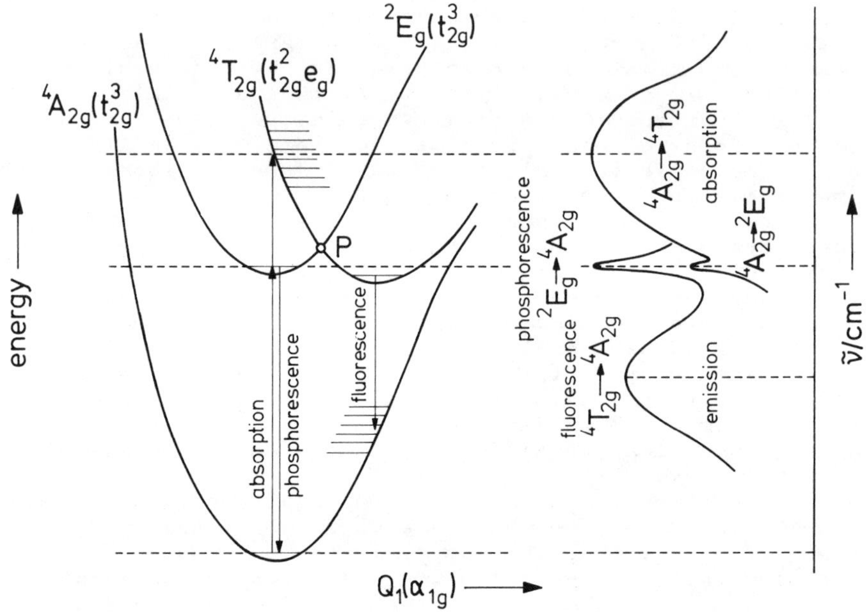

Figure 1. Potential energy curves in the totally symmetric subspace Q_1 in the case of octahedral d^3 systems and resulting band shapes of optical spectra.

spectrum. Also, in many cases, only emission from the lowest electronic state can be observed, due to radiationless deactivation from higher excited states.

tr[Rh py$_4$Cl$_2$] Cl: The powder emission spectrum of this compound at 2 K (immersed in liquid Helium) has been recorded by Crosby and coworkers (8). A single progression in 350 cm^{-1} has been found, which does not correspond to any of the vibrational frequencies obtained from a vibrational spectrum of the ground state. Franck-Condon analysis of the spectrum, carried out by these authors on the basis of an elaborate theory including anharmonic effects, was about to be revised when a better resolved spectrum was obtained which exhibited three superimposed progressions with equal vibrational quanta (9). The series of band peaks (Figure 2) follow the formula

$$\nu_n^{(i)} = \nu_{00} + \nu_{ui} + n\nu_g \qquad i = 1,2,3$$

in which ν_{ui} are ungerade promoting modes inducing the d-d transitions which by vibronic coupling become electric dipole allowed. The intervals $\nu_{u1} = 370$ cm^{-1}, $\nu_{u2} = 251$ cm^{-1}, and $\nu_{u3} = 176$ cm^{-1} now agree with quanta obtained from the far infrared spectrum. Also the progressional interval $\nu_g = 295$ cm^{-1} compares well with the Raman band $\nu_1(a_{1g}) = 289$ cm^{-1} (at room temperature). With these data, the vibrational structure of the spectrum can be well understood.

K_2PtCl_6: At 1.9 K the emission spectrum of cubic single crystals (Figure 3) exhibits three progressions due to promoting modes $\nu_3(t_{1u})$, $\nu_4(t_{1u})$ and $\nu_6(t_{2u})$ with band intervals agreeing with infrared (t_{1u}) and theoretical data (t_{2u}) (9). Progressional intervals are in each case $\nu_g = 323$ cm^{-1} corresponding to a $\nu_2(e_g) = 320$ cm^{-1} fundamental vibration reported from a Raman measurement (10). Emission occurs from the lowest excited electronic Γ_3 state which is one of the $^3T_{1g}(t_{2g}{}^5e_g)$ levels split by spin-orbit coupling (11). This level is degenerate and subject to a Jahn-Teller effect distorting the molecule either by e_g or t_{2g} vibrations. Since a progression in e_g is observed, we can conclude that this vibrational mode is predominantly Jahn-Teller active, Analysis of the band profile by fitting the spectrum to the theoretical function for the transition rate (band profile function) in the zero-temperature limit (m = 0) (5-7) yields appropriate fitting parameters Δ and β from which a distortion of $\Delta z = 0.19$ Å and $\Delta x = \Delta y = -0.095$ Å is calculated for one of the possible Γ_3 perturbations (9). The frequency effect is $\beta \approx 1$ indicating no large changes of force constants for e_g distortions compared to the ground state. The zero phonon line extrapolated from band analysis is at 19 500 cm^{-1}.

Cs_2SeCl_6 and Cs_2SeBr_6: These compounds have central ions with $d^{10}s^2$ electron systems. Their emission spectra (cf. Figure 4) are well resolved, furnishing extremely

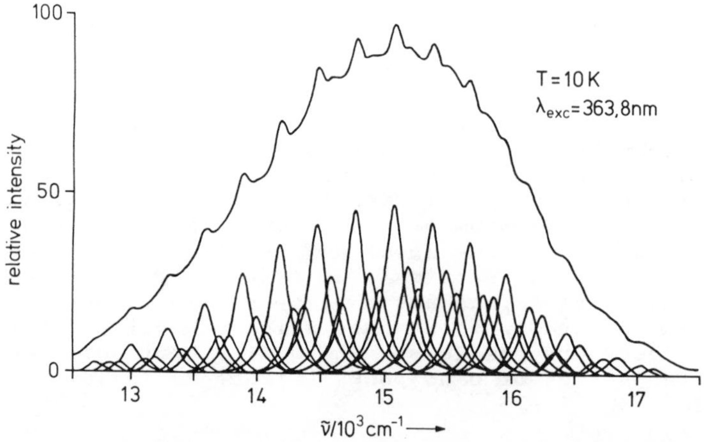

Figure 2. Powder emission spectrum of tr-[Rh py_4Cl_2]Cl and a lorentzian band analysis yielding three superimposed progressions. Reproduced with permission from Ref. 9. Copyright 1981, VCH-Verlag.

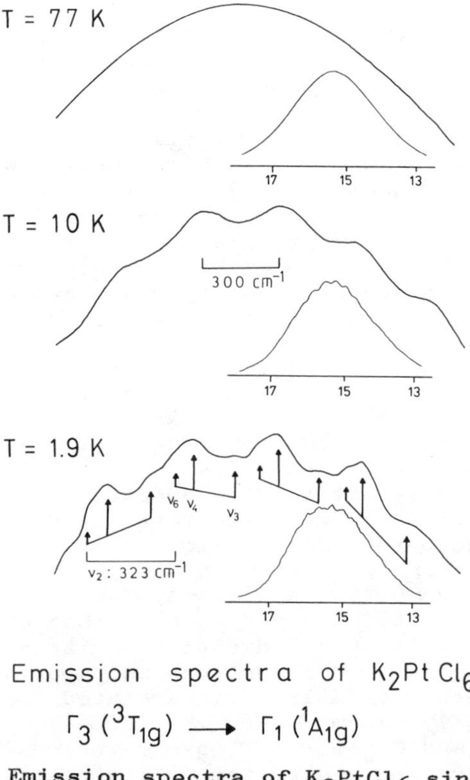

Figure 3. Emission spectra of K_2PtCl_6 single crystals showing the evolution of vibrational structure with decreasing temperature. Excitation wave length λ_{exc} = 454.5 nm from an argon ion laser.

long vibrational progressions (more than 25 members can be detected at 10 K) with almost equal quanta throughout, indicating harmonic conditions up to high vibrational level numbers. From other s^2 systems (Tl^+, Sn^{2+}) it is well known that the octahedral excited state configuration $a_{1g}t_{1u}$ (sp) has the energy level systems $\Gamma_1^-(^3T_{1u}) < \Gamma_4^-(^3T_{1u}) < \Gamma_5^-(^3T_{1u}) < \Gamma_3^-(^3T_{1u}) < \Gamma_4^-(^1T_{1u})$. Since transition from Γ_1^- is strongly forbidden (which cannot be circumvented by vibronic coupling effects since an octahedral molecule has no t_{1g}-type vibrations necessary for this coupling), the spectrum is due to a $\Gamma_4^-(^3T_{1u}) \rightarrow \Gamma_1^+(^1A_{1g})$ transition which is symmetry and parity allowed for an electric dipole transition mechanism; the spin selection rule is suspended by the large spin orbit coupling of about $\zeta_p \cong 2\,400\ cm^{-1}$ (12). Since analysis of the progressions yields, for both compounds, only e_g quanta (240 cm^{-1} for Cl, 154 cm^{-1} for Br) the excited Γ_4^- state must be distorted by Jahn-Teller forces yielding a mainly tetragonal species with a C_4 axis collinear with one of the corresponding axes of the octahedron. Due to coupling of the three degenerate states, i.e. two Γ_4^- and one Γ_5^-, by spin-orbit and vibronic coupling, the excited state hypersurfaces in the e_g space are scrambled in a particular way; in Figure 5 one of these potential surfaces is shown, calculated with a set of parameters relevant to this compound, i.e. spin-orbit, vibronic coupling and electron repulsion parameter. The band profile analysis yields Jahn-Teller distortions in the excited states of $\Delta z = -2\Delta x = -2\Delta y = 0.2$ Å for one of the three Γ_4^- states i.e. expansion along z- and a compression in the other directions. The other component states yield equivalent distortions into x- and y-directions. The frequency effect is $\beta = 1.4$ indicating increased slopes of the potential curve compared to the ground state for nuclear displacements close to the X-minima, which are all equivalent and give rise to only one type of emission.

Absorption Spectra. Since fundamental vibrations of excited electronic states are usually different from those in the ground state (frequency effect $\beta \neq 1$) the vibrational intervals obtained from the fine structure of an absorption spectrum cannot be expected to be equal to the vibrational spectra (IR and Raman) of the ground state. Using the identity of normal frequencies measured in both types of spectra, as a criterion for the degree of band resolution, as taken for analysing emission spectra, is only a useful guide when applied to absorption spectra. However, if the vibronic spacing in the progression agrees with the vibrational quantum calculated from the frequency effect β, obtained from fitting the measured curves to the intramolecular distribution $I_j(m,n;\Delta,\beta)$, the spectrum is well enough resolved to determine the potential curve at least for that part of the nuclear coordinate hyperspace which refers to geometry changes in the excited state.

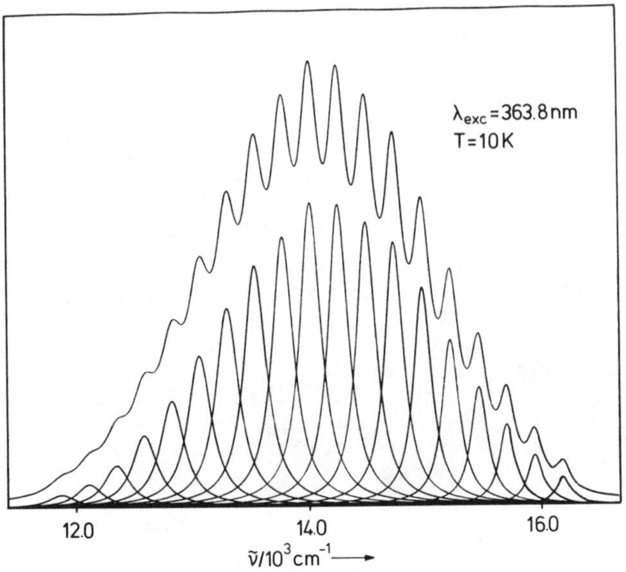

Figure 4. Powder emission spectra of Cs_2SeCl_6 and a band resolution in terms of lorentzian functions.

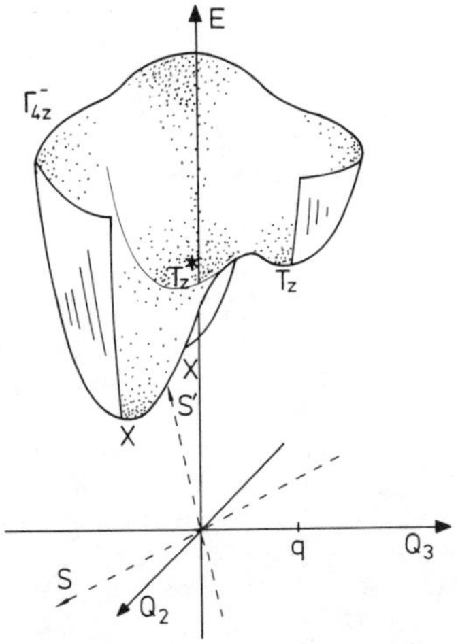

Figure 5. One of the adiabatic energy surfaces along the Q_2, Q_3 coordinates belonging to the Jahn-Teller active e_g mode calculated for a parameter set appropriate to $SeCl_6^{2-}$. Reproduced with permission from Ref. 12. Copyright 1980, North-Holland Publishing Company.

tr-[$Co(NH_3)_4(CN)_2$]Cl and tr-[Co(en)$_2$(CN)$_2$](ClO_4): Low temperature (4.2 K) polarized absorption spectra from these crystals, lead to well resolved spectra belonging to the first excited state (14,15). This example shows that with carefully performed band profile analyses, excited state geometries can also be calculated for compounds of lower symmetry. The low energy spectrum has two ligand field transitions into $^1A_{2g}$ and $^1E_g(D_{4h})$; the first shows vibrational structure which can be further resolved into a vibrational progression that in itself is well structured due to a series of vibronic origins which, together with the zero phonon band, are denoted by "fundamental spectral pattern". The band analysis, e.g. for the deuterated compound given in Figure 6, shows that this fundamental pattern is repeated many times with increasing wavenumber of equal spacing, weighted by varying intensity factors. The intensity distribution in this progression, as analyzed by using the band profile function with $I_1(0,n;\Delta,\beta)$ functions, yields an elongation of 0.096 Å for the Co-NH_3 bond lengths with no changes for the Co-CN distances. For the ethylenediamine complex an expansion of 0.06 Å for the C-N bond is calculated, with the Co-CN bond left unchanged. In this case, because of the many vibrational degrees of freedom for the chelate ring, a normal-coordinate analysis had to be carried out (15). This identified the accepting mode giving rise to the progression of 236 cm^{-1}, as due to almost equal contributions from Co-N, N-C stretching, and NCoN angular vibrations.

Time Resolved Emission Spectra

Due to the damping effect arising from interaction with the medium, the vibrational fine structure of electronic transitions in most cases cannot be resolved by conventional spectroscopy even when elaborate procedures are used. Therefore we have tried to apply time resolved techniques, looking at time intervals for photon events emitted by a continuously (cw-) irradiated probe (16,17). With an experimental device which uses some parts of the TCSP (time correlated single photon) technique, i.e. start and stop inputs of a TAC (time to amplitude conventer) and recording with a MCA (multi channel analyzer), the time difference t between two successively recorded photons is monitored many times (typically 10^6 measurements taken each minute). The relative frequency f(t) for the occurrence of the time interval t monitoring these photon pairs, supplies an exponential function $f(t) = b\exp(-mt)$ with decreasing slope since short time photon pairs are counted more often and those with larger time intervals are discriminated by being neglected ("photon pile up") (18). Photon statistics applied to these events, provide a Poisson distribution $p_i(\mu)=\frac{\mu^i}{i!}e^{-\mu}$ ($\mu = \langle I \rangle \cdot t$) if all of these photon pairs are non-

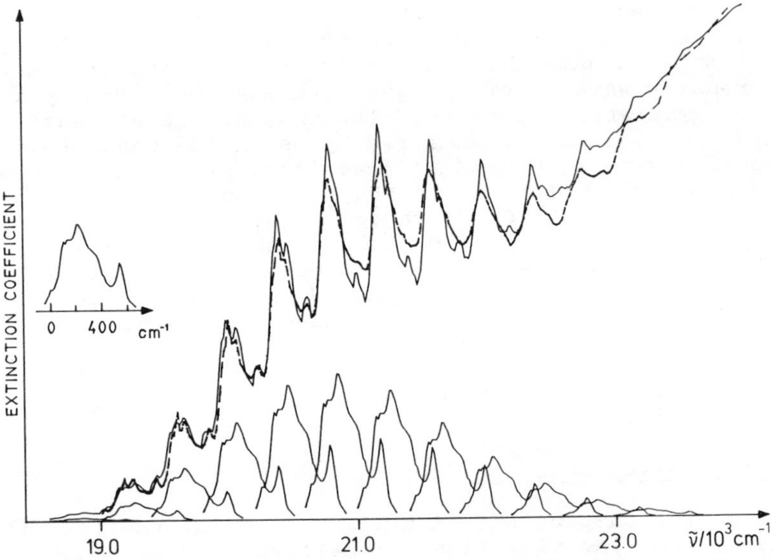

Figure 6. Vibrational progression of the $^1A_{1g} \rightarrow {}^1A_{2g}$ transition in the absorption spectrum of tr-[$Co(ND_3)_4(CN)_2$]Cl single crystals at 4.2 K (---- experimental, —— simulated by the theoretical band profile function). The insert illustrates the fundamental spectral pattern (see text). Reproduced with permission from Ref. 14. Copyright 1982, North-Holland Publishing Company.

correlated, i.e. are completely independent of each other. In this case the f(t) plot is strictly exponential and the parameters b and m are proportional to the mean intensity obtained for a large time interval $\langle I \rangle = \lim_{T\to\infty} \frac{1}{T}\int_0^T I(t)dt$. We found, however, e.g. for a Cs_2TeBr_6 powder example, that a plot of m or b against photon energy does not have such a simple dependence, it shows distinct fine structure. Photons of different energy deviate differently from the usual Poisson behavior. For uncorrelated photons an m and b plot with frequency should vary in the same way as the mean intensity $\langle I \rangle$, i.e. it should not exhibit any fine structure if an $\langle I \rangle$ plot is not structured.

This variation (fluctuation) becomes more evident from Poisson distribution if we look at photon pairs which arrive within relatively small time intervals. If we define cumulative frequencies for some representative time intervals equal or less to a certain time limit Δt

$$c(\Delta t) = \int_0^{\Delta t} f(t)dt$$

the fluctuation from the mean intensity $\langle I \rangle$ becomes more evident since for small time intervals the errors introduced (if $\langle I \rangle$ is not small enough) by the partial neglect of photon pairs with large time intervals become less important. The $c(\Delta t)$ plot (Figure 7) (20) with photon energy exhibits a band structure even at ~70 K. If the time limit Δt, at a certain temperature (10 K), is increased to infinity, the curve would represent the mean intensity $\langle I \rangle$ corresponding to a conventional emission measurement showing no vibrational structure for this sample at that particular temperature. From this progression an energy interval of about 150 cm^{-1} is calculated, the average agreeing with the e_g vibrational mode obtained from Raman measurements (19). The analysis indicates a Jahn-Teller effect in the excited Γ_4^- state as also found for corresponding Se compounds (s. above).

Such fluctuations of the photon flux, emitted from a molecule, have been predicted to be due to cooperative effects (21,22). The theory is based on an idea of Prigogine and coworkers (s. e.g.(23)) who treated the irreversible part of a physical process by transforming the wavefunctions of a dissipative system into another space using a "dynamical" non-unitary representation $D = \exp(-iV\mathcal{T}/\hbar)$ with a "star-Hermitian" time operator $\mathcal{T}$ and V describing the interaction of a relevant local system H_0, e.g. the complex chromophore, and the total system H, i.e. our crystal. In the new representation $\tilde{\psi} = D\psi$ no additional time dependence is introduced, $dD/dt = 0$, any expectation value of an operator $\tilde{M} = DMD^{-1}$ should be unchanged $\langle M \rangle = \langle \tilde{M} \rangle$ and the total Hamiltonian is transformed by $\tilde{H} = DHD^{-1} = H_0$ to the local system Hamiltonian (21,22). To describe the time development in the new representation, the electron density $\rho = |\psi\rangle\langle\psi|$

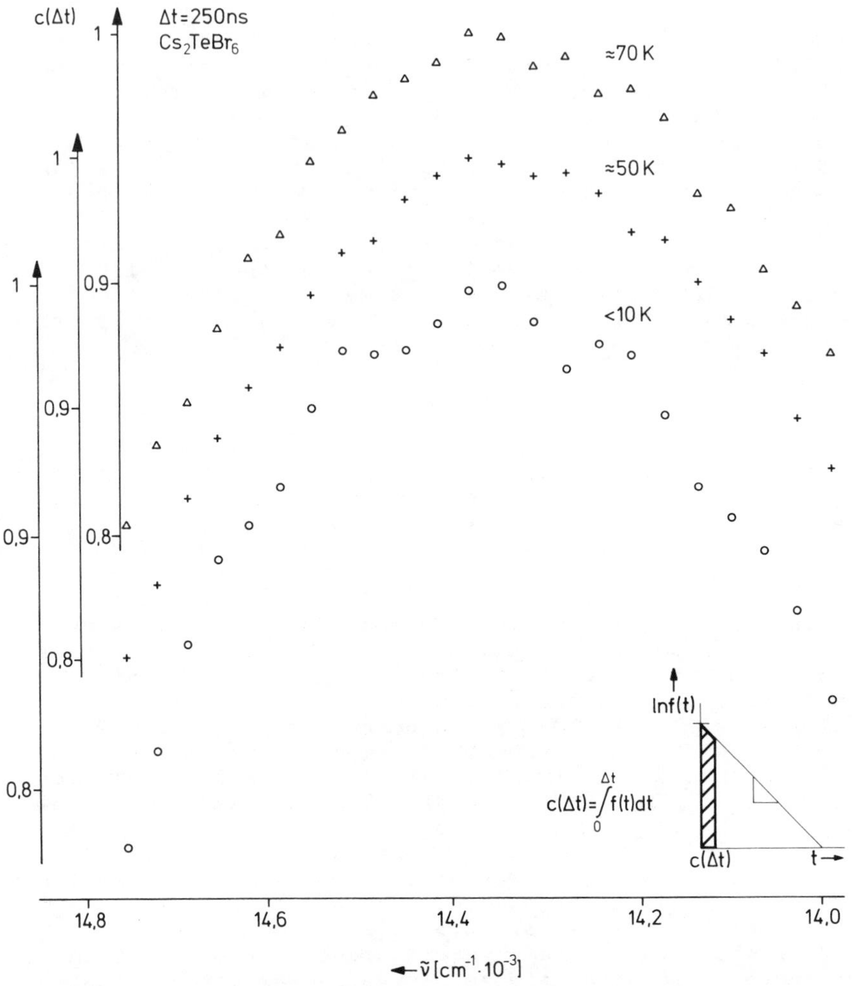

Figure 7. Cumulative relative frequencies c(Δt) up to the time limit Δt for the occurrence of the time interval t between successively recorded photons plotted against photon energies for a Cs_2TeBr_6 powder sample. The mean error is about 0.5 %, i. e. the error bars have the size of the symbols used for the data points.

satisfying the von Neumann-Liouville equation $i\partial\varrho/\partial t = L\varrho$ ($L = [H,..]/h$) is transformed by $D(+L)$ and all other operators by $D(-L)$, leading to time operators $\mathcal{T}(+L)$ and $\mathcal{T}(-L)$ which, due to the $Lt\mathcal{T}$ symmetry, were shown to be substituted by $\mathcal{T}(\pm L) = \pm i\mathcal{T}^0$ where $\mathcal{T}^0$ is the odd functional part $\mathcal{T}^0(-L) = -\mathcal{T}^0(+L)$ of $\mathcal{T}$. With the "intrinsic time" $\mathcal{T}^0$, the total time evolution of the system has two contributions

$$i\frac{\partial}{\partial t} = i\widetilde{\left(\frac{\partial}{\partial t}\right)} + \frac{\partial}{\partial \mathcal{T}^0} \quad \text{with} \quad \widetilde{\left(\frac{\partial}{\partial t}\right)} = D\frac{\partial}{\partial t}D^{-1}$$

One respresents evolution with respect to universal time controlled by the local system Liouvillian L_0, the other part describes evolution with respect to intrinsic time being the irreversible component of the process determined by the interaction Liouvillian L_V. The average $\langle \partial\tilde{\varrho}(t)/\partial\mathcal{T}^0 \rangle$ then corresponds to any deviation of the usual density changes due to the population or depopulation (decay) of the states. For a stationary luminescence experiment, this term therefore describes the photon fluctuations from a vibronic state of the local system H_0, caused by environmental interactions.

These photon flux fluctuations are also observed in the time resolved spectra of other systems. Two more examples refer to Sn^{2+} doped in KI (which is also an s^2 electronic system) and $[Ru(bipy)_3]Cl_2$ in aqueous solution (Ru^{2+} to bipyridine charge transfer transition) (20). Figures 8 and 9 show $c(\Delta t)$ plots for these systems exhibiting distinct band structures for small time limits. Assignments to superimposed progressions and to several emitting states are possible; the obtained resolution does not allow, however, a detailed analysis. With the equipment available to us, the method is limited to compounds which exhibit a relatively large emission intensity. Therefore the zero phonon regions which eventually show the most pronounced fine structure, could not be investigated. Since transition group elements are rather low emitters only few measurements (e.g. for K_2PtCl_6) were successful until now, supplying better resolved vibrational structures than those obtained by usual emission or absorption spectroscopy.

Acknowledgment

I am grateful to Dr. A. B. P. Lever for his editorial assistance to prepare the manuscript.

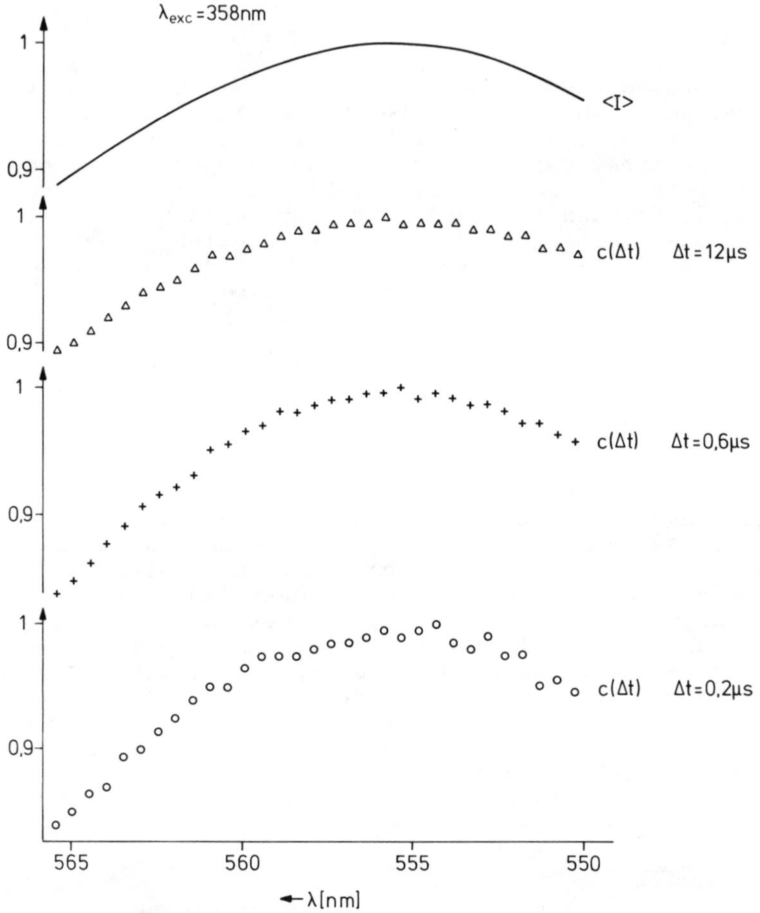

Figure 8. Corresponding plots as in Figure 7 for KI:Sn^{2+} single crystals for various given time limits Δt compared to the emission spectrum ⟨I⟩.

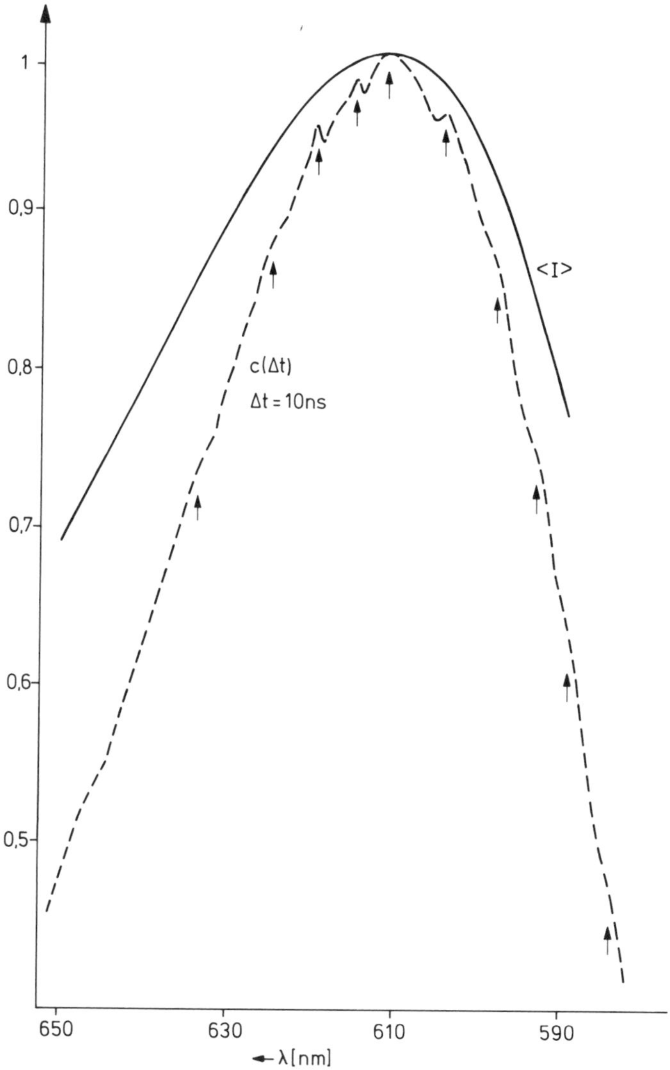

Figure 9. Emission intensity ⟨I⟩ and c(Δt) plots for $[Ru(bipy)_3]Cl_2$ in aqueous solution at normal temperature. Excitation wave length λ_{exc} = 358 nm.

Literature Cited

1. Tutt, L.; Tanner, D.; Heller, E. J.; Zink, J. I. Inorg. Chem. 1982, 21, 3858.
2. Tutt, L.; Tanner, D.; Schindler, J.; Heller, E. J.; Zink, J. I. J. Phys. Chem. 1983, 87, 3017.
3. Komi, Y.; Urushiyama, A. Bull. Chem. Soc. Jpn. 1980, 53, 979.
4. Wilson, R. B.; Solomon, E. I. Inorg. Chem. 1978, 17, 1729.
5. Kupka, H. Mol. Phys. 1978, 36, 685.
6. Kupka, H.; Enßlin, W.; Wernicke, R.; Schmidtke, H.-H. Mol. Phys. 1979, 37, 1693.
7. Kupka, H.; Schmidtke, H.-H. Mol. Phys. 1981, 43, 451.
8. Hipps, K. W.; Merrell, G. A.; Crosby, G. A. J. Phys. Chem. 1976, 80, 2232.
9. Eyring, G.; Schmidtke, H.-H. Ber. Bunsenges. Phys. Chem. 1981, 85, 597.
10. Woodward, L. A.; Creighton, J. A. Spectrochim. Acta 1961, 17, 594.
11. Eyring, G.; Schönherr, T.; Schmidtke, H.-H. Theor. Chim. Acta 1983, 64, 83.
12. Wernicke, R.; Kupka, H.; Enßlin, W.; Schmidtke, H.-H. Chem. Phys. 1980, 47, 235.
13. Fukuda, A. Phys. Rev. B 1970, 1, 4161.
14. Urushiyama, A.; Kupka, H.; Degen, J.; Schmidtke, H.-H. Chem. Phys. 1982, 67, 65.
15. Hakamata, K.; Urushiyama, A.; Degen, J.; Kupka, H.; Schmidtke, H.-H. Inorg. Chem. 1983, 22, 3519.
16. Degen, J.; Schmidtke, H.-H. Theor. Chim. Acta 1985, 67, 33.
17. Degen, J.; Schmidtke, H.-H.; Chatzidimitriou-Dreismann, C. A. Theor. Chim. Acta 1985, 67, 37.
18. Oliver, C. J. In "Photon Correlations and Light Beating Spectroscopy"; Cummins, H. Z.; Pike, E. R., Ed.; Plenum Press: New York, 1974; p. 151.
19. Stufkens, D. J. Rec. Trav. Chim. Pays Bas 1970, 89, 1185.
20. Degen, J. Ph.D. Thesis, University of Düsseldorf, 1985.
21. Chatzidimitriou-Dreismann, C. A. Int. J. Qu. Chem. Symp. 1982, 16, 195.
22. Chatzidimitriou-Dreismann, C. A. Int. J. Qu. Chem. 1983, 23, 1505.
23. Prigogine, I. "From Being to Becoming-Time and Complexity in Physical Science"; W. H. Freeman: New York, 1980.

RECEIVED November 8, 1985

4

Excited State Distortions Determined by Electronic and Raman Spectroscopy

Jeffrey I. Zink, Lee Tutt, and Y. Y. Yang

Department of Chemistry, University of California, Los Angeles, CA 90024

The geometry changes which transition metal complexes undergo when excited electronic states are populated are determined by using a combination of electronic emission and absorption spectroscopy, pre-resonance Raman spectroscopy, excited state Raman spectroscopy, and time-dependent theory of molecular spectroscopy. Bond length changes in the lowest d-d excited states of $W(CO)_5$pyridine and $W(CO)_5$piperidine are calculated. The results are consistent with the predictions of the ligand field theory and with the observed photochemical reactions. Bond bending distortions are observed in metal-nitrosyl compounds. The photochemical reactions of $Co(CO)_3NO$ and the photohydrogenation catalyzed by $Rh(PPh_3)_3NO$ provide indirect support for the bending. Excited state Raman spectroscopy of $Fe(CN)_5NO^{2-}$ provides direct support for the bending distortion.

The geometry changes which a molecule undergoes when it is electronically excited are important in determining its spectroscopic and photochemical properties. These geometrical distortions can include changes in both the metal-ligand bond lengths and bond angles. Changes in the formal charges may result from these distortions.

A simple type of distortion is metal-ligand bond lengthening along one totally symmetric normal coordinate. In a highly symmetrical complex containing a small number of atoms (e.g. $PtCl_4^{2-}$), there is only one totally symmetric normal mode and the electronic spectrum shows a well defined vibronic progression in that one mode. The distortion can be readily determined by using standard Franck-Condon theory (1).

Multiple distortions along more than one totally symmetric normal mode are common in complexes containing many atoms. These distortions are symmetry preserving. The point group of the molecule is the same in both the ground and the distorted excited states. The problem which frequently arises with large molecules in condensed media is that vibronic structure, when

0097–6156/86/0307–0039$06.00/0

present at all, is not well enough resolved to reveal the individual components of each of the distorted modes. Typical examples to be discussed below are $W(CO)_5L$,L=pyridine and piperidine. New theoretical and experimental methods to determine the magnitudes of the distortions along all of the distorted normal modes in the excited state potential hypersurface are the subjects of the first part of this paper.

Distortions along non-totally symmetric modes may occur in certain excited states. These distortions are non-symmetry preserving; the point group of the molecule changes in the excited state. The specific examples in this paper are the linear to bent geometry changes of metal nitrosyls (e.g., from C_{4v} to C_s in $[Fe(CN)_5NO]^{2+}$.)

The analyses of distortions and their spectroscopic and photochemical consequences are the subjects of this paper. First, multi-mode symmetry-preserving distortions arising from d-d excitation of monosubstituted tungsten carbonyls will be calculated by using a combination of time dependent theory, pre-resonance Raman spectroscopy, and electronic spectroscopy. The points of connection between the ligand field theory of transition metal photochemistry, the measured photochemical reactions, and the excited state distortions will be discussed. Secondly, non-symmetry preserving distortions arising from charge transfer excitation of the complexes containing the MNO group are studied. Both indirect evidence for the distortions from photochemical reactions and direct evidence from excited state Raman spectroscopy are discussed.

Excited State Distortions of W(CO)5pyridine and W(CO)5piperidine from Time-Dependent Theory, Pre-resonance Raman Spectroscopy, and Electronic Spectroscopy

The emission spectrum of $W(CO)_5py$, shown in figure 1, is typical of the spectra of large organometallic molecules taken in condensed media at low temperatures (2). This spectrum, taken from a single crystal at 10 K, shows a long regularly spaced progression with a peak to peak separation of 550 ± 10 cm^{-1}. Instrumental resolution is two orders of magnitude higher. The emission has been assigned to the ^3E to $^1A_1$, d_z2 to (d_{xz},d_{yz}) transition (3,4). In a luminescence spectrum of a small molecule, a regularly spaced progression is almost always caused by a distortion along a totally symmetric normal mode whose frequency is equal to the vibrational frequency of that mode in the ground electronic state. Surprisingly, there are no totally symmetric modes with a frequency of 550 cm^{-1} in $W(CO)_5py$. Thus, a calculation of the distortion along one 550 cm^{-1} mode could not be correct. Unfortunately, a meaningful calculation of the excited state distortions is impossible from this spectrum alone. The emission spectrum of a single crystal of $W(CO)_5$piperidine is shown in figure 2. The assignment is the same as that for the pyridine complex. The major progression is regularly spaced with a peak separation of 520 ± 15 cm^{-1}. The major difference between this spectrum and that of the pyridine complex is that the progression is significantly longer. The

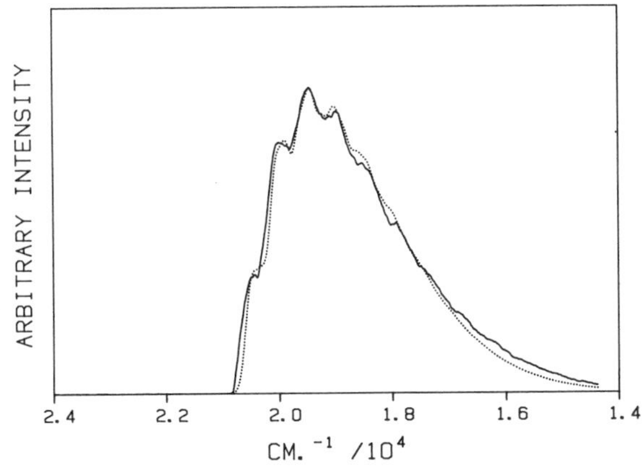

Figure 1. Luminescence spectrum of a single crystal of $W(CO)_5$pyridine at 10 K. The solid line is the experimental spectrum. The dotted line is the spectrum calculated from the pre-resonance Raman intensities as discussed in the text.

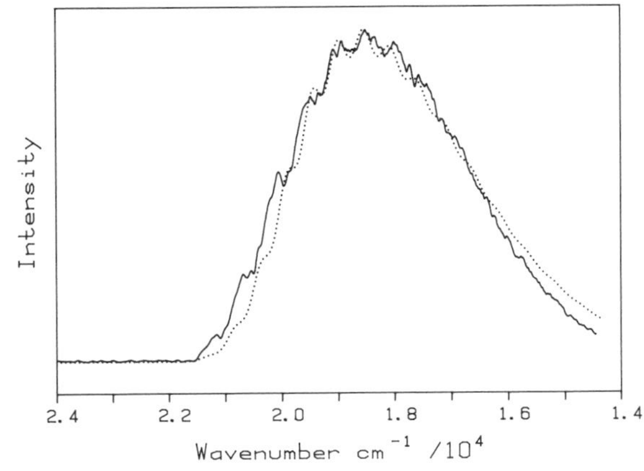

Figure 2. Luminescence spectrum of a single crystal of $W(CO)_5$piperidine at 10 K. The solid line is the experimental spectrum. The dotted line is the spectrum calculated as discussed in the text.

peak maximum occurs at the sixth vibrational quantum instead of the third. Again there is no ground state normal vibrational mode with an energy near 520 cm^{-1} and a meaningful determination of the excited state distortions is not possible from this spectrum alone. The detailed calculation of the multi-mode distortions is possible by using pre-resonance Raman spectroscopy, the emission spectra, and time-dependent theory of molecular spectroscopy. Time dependent theory provides both a quantitative and an intuitive physical picture of the inter-relationship between the excited state distortions, pre-resonance Raman spectra, and the emission spectra (5). The theory is briefly discussed in the following section and then used to determine the details of the excited state distortions.

Electronic Absorption and Emission Spectroscopy. The electronic absorption process begins when the ground state vibrational wavepacket undergoes a vertical transition and is placed on the excited state potential surface (5). The initial vibrational wavepacket, ϕ, propagates on the excited state potential surface which in general is displaced by Δ_k relative to the excited state surface. The displaced wavepacket is not a stationary state and evolves according to the time-dependent Schrodinger equation. The quantity of interest is the overlap of the initial wavepacket with the time-dependent wavepacket, $\langle\phi|\phi(t)\rangle$. The overlap is a maximum at t=0 and decreases as the wavepacket moves away from its initial position. At some later time t, the wavepacket may return to its initial position giving rise to a recurrence of the overlap. A plot of the overlap as a function of time in the time domain shows the initial overlap decreasing and then recurring at a time t when the wavepacket (and thus all of the atoms in the molecule) returns to its original position. This pattern is repetitive. In the simple case of harmonic potential surfaces and no change in vibrational frequencies between the ground and excited electronic states, the overlap is

$$\langle\phi_k|\phi_k(t)\rangle = \exp\,[-\sum_k(\Delta_k/2(1-e^{i\omega_k t} - i\omega_k t/2)] \qquad (1)$$

where E_o is the energy difference between the minima of the two surfaces, Γ is a phenomenological damping factor (vide infra), and ω_k and Δ_k are the frequency and the displacement of the kth normal mode.

The electronic absorption spectrum in the frequency domain is the fourier transform of the overlap in the time domain. The spectrum is given by

$$I(\omega) = C\omega \int_{-\infty}^{\infty} e^{i\omega t} \langle\phi_k|\phi_k(t)\rangle \, dt \qquad (2)$$

where C is a constant, ω is the frequency of incident radiation, and the other quantities are those defined above. Thus, the absorption spectrum can be calculated when the frequencies ω_k and the displacements Δ_k of the normal modes are known.

The time-dependent theoretical treatment of the electronic emission spectrum is very similar to that of the absorption spectrum. The principal difference is that the initial wavepacket starts on the upper (excited state) electronic surface and propagates on the ground electronic state surface. The overlap of the initial wavepacket with the time-dependent wavepacket is given by equation 1. The emission spectrum is given by

$$I(\omega) = C\omega^3 \int_{-\infty}^{\infty} e^{i\omega t} \langle \phi_k | \phi_k(t) \rangle \, dt \qquad (3)$$

where all of the symbols are the same as those in eq. 2. Note that for emission, the intensity is proportional to the cube of the frequency times the Fourier transform of the time dependent overlap.

In most transition metal and organometallic compounds, many normal modes are displaced. The expressions discussed above describe the case where one specific normal mode (the kth) is displaced. In the usual case of many displaced normal modes, the total overlap is given by

$$\langle \phi | \phi(t) \rangle = \prod_k \langle \phi_k | \phi_k(t) \rangle \exp\left(iE_0 t/\hbar - \Gamma^2 t^2\right) \qquad (4)$$

where the shift of the electronic energy between the minima of the two surfaces E_0 and a Gaussian damping Γ have been included. The complete overlap is thus

$$\langle \phi | \phi(t) \rangle = \exp\left\{-\sum_k \left[(\Delta_k^2/2)(1-e^{-i\omega_k t}) - i\omega_k t/2\right] - iE_0 t/\hbar - \Gamma^2 t^2\right\} \qquad (5)$$

This expression for the complete overlap is Fourier transformed to give the electronic emission spectrum. In order to carry out the calculation it is necessary to know the frequencies ω_k and the displacements Δ_k for all of the displaced normal modes. In addition, the energy difference between the minima of the two potential surfaces E_0 and the damping Γ must be known. As will be discussed below, the frequencies and displacements can be experimentally determined from pre-resonance Raman spectroscopy, and the energy difference between the ground and excited states and the damping can be obtained from the electronic absorption spectrum and/or emission spectrum.

The damping factor Γ arises because of relaxation into other modes, the "bath", etc. It is instructive to examine the meaning of the damping factor in the time domain. At one extreme, it will have a value of zero if every atom in the molecule returns to exactly its starting position at the same time during the time evolution of the system. This condition means that each of the wavepackets in eq. 4 must have a recurrence in the overlap at time t which is equal to the overlap at t=0. If even one atom of the molecule does not return to exactly its starting position at that time t, the total overlap in eq. 4 will be smaller at time t than at time t=0. The effect of a non-zero damping factor is to decrease the value of the recurrence of the overlaps in the time domain. The effect on the spectrum in the frequency domain is to decrease the resolution, i.e. to "fill in" the spectrum. In the case where the damping is large enough to prevent any recurrence, the spectrum in the frequency domain will consist of only the envelope with no vibronic structure.

Raman Spectroscopy The time-dependent picture of Raman spectroscopy is similar to that of electronic spectroscopy (6). Again the initial wavepacket propagates on the upper excited electronic state potential surface. However, the quantity of interest is the overlap of the time-dependent wavepacket with the final Raman state ϕ_f, i.e. $\langle\phi_f|\phi(t)\rangle$. Here ϕ_f corresponds to the vibrational wavefunction with one quantum of excitation. The Raman scattering amplitude in the frequency domain is the half Fourier transform of the overlap in the frequency domain,

$$\alpha_{fi}(\omega_i) = \int_0^\infty e^{i\omega_i t - \Gamma t} \langle\phi_f|\phi_i(t)\rangle \, dt \tag{6}$$

It is very difficult to experimentally obtain the values of the scattering cross section. However, it is relatively easy to obtain the intensity of given normal mode k relative to that of another mode k'. A simple expression relating the relative intensities has been derived for the special conditions of harmonic oscillators, no Duschinsky rotation, no change in normal mode frequencies, and pre-resonance (short time) condition spectra. Under these conditions the relative intensities of two modes is given by

$$\frac{I_k}{I_{k'}} = \frac{\Delta_k^2\omega_k^2}{\Delta_{k'}^2\omega_{k'}^2} \tag{7}$$

The important experimental condition which must be fulfilled is the short time, pre-resonance condition. When the incident frequency is slightly off resonance with the excited electronic state of interest, the propagation time of the wavepacket on the upper potential surface is governed by the time-energy uncertainty principle $\Delta\omega\Delta t \sim 1$ where $\Delta\omega$ is the frequency mismatch. Under short time conditions the wavepacket moves in a region localized near the equilibrium geometry of the ground electronic state, i.e., the Franck-Condon region. The pre-resonance Raman intensity is dominated by the slope of the potential surface in this region. The greater the slope, the greater the motion of the wavepacket, the greater the overlap with the final state and the greater the intensity.

Calculation of Excited State Distortions and Electronic Spectra from Raman Intensities

The dynamics of the wavepacket on the upper potential surface determines both the absorption spectrum and the Raman spectrum. The emission spectrum is determined by the dynamics on the ground state potential surface with the same displacements as those which determine the absorption and Raman. In the short time limit, the intensities in the Raman spectrum are related to the displacements by eq. 7. In the short time limit, the absorption spectrum becomes

$$I(\omega) = C\omega \exp[-(\omega-E)^2/2\sigma^2] \tag{8}$$

The quantity $2\sigma^2$ is the width of the electronic absorption spectrum at 1/e of the height. This quantity is also related to the displacements

$$2\sigma^2 = \Sigma\Delta_k^2\omega_k^2 \tag{9}$$

Thus $2\sigma^2$ is experimentally found from the electronic spectrum, ratios of the Δ's are found from the Raman spectrum, and the Δ_k's are calculated (except for sign) by pairwise comparison of the Raman intensities. Once the Δ_k's are calculated, the electronic spectra are calculated by using eq. 2 or 3.

Calculation of Excited State Distortions of $W(CO)_5L$. The emission spectrum discussed earlier, the theory discussed above and pre-resonance Raman data will now be used in concert to calculate the multi-mode distortions. The relative intensities of the peaks in the pre-resonance Raman spectra were determined by integrating the peaks. All of the peaks in the experimental spectrum having intensities greater than three percent of that of the most intense peak were measured and used in the calculations.

The electronic spectrum is calculated by using equations 3 and 5. The distortions used in these equations are determined from the pre-resonance Raman intensities by using equations 7 and 9. Both the vibrational frequencies of the normal modes and the displacements of the excited state potential surfaces along these normal modes are obtained from the pre-resonance Raman spectrum.

The major uncertainty encountered in applying the theory to large molecules is the unknown effect of nearby excited states. The luminescence spectrum predominantly originates from a single spin-orbit state, but the Raman and absorption spectra are influenced by other spin-orbit states and by nearby charge transfer states. The results reported here are based on pre-resonance Roman data obtained as close to the experimentally determined orgin of the lowest excited state as possible. Attempts to obtain full Raman enhancement profiles to probe the possible effects of other states are continuing, but they have been hindered by the high photosensitivity of the compounds.

Once the displacements of the excited state surface along the normal modes are determined, the wavepacket is propagated on the multidimensional hypersurface and the overlap $\langle\phi|\phi(t)\rangle$ is calculated from equation 5. The overlap in the time domain is then Fourier transformed (eq. 3) to give the calculated electronic spectrum. When good agreement with experiment is found, the agreement indicates that the simplifying assumptions discussed previously are met and that the distortions which are calculated are meaningful.

The emission spectrum of $W(CO)_5py$ calculated as discussed above is shown in figure 1 (2). In this calculation exactly the displacements and frequencies obtained from the experimental

Raman data were used. Excellent agreement between the experimental spectrum and the theoretical spectrum calculated from the 18 dimensional excited state potential surface is obtained. Interpretation of these results will be discussed below.

The calculated emission spectrum of $W(CO)_5pip$ is shown superimposed on the experimentally determined spectrum in figure 2. The agreement between the calculated and experimental spectra is not as good as that in figure 1. Two factors are probably involved. First, the excited state is significantly more distorted than that in $W(CO)_5py$. Thus more vibrational quanta are involved and anharmonicity, which was not included in the calculation, will play a larger role. Secondly, the W-N stretching mode was not observed in the Raman spectrum. Even if a low frequency mode has a significant distortion, the observed relative intensity will be small because of the inverse frequency squared term in eq. 7. The major effect of a very low frequency mode on the emission spectrum can be included in the damping factor (vide infra). These considerations explain why a larger damping factor is required for piperidine than for pyridine.

The excited state geometries of the $W(CO)_5py$ and $W(CO)_5pip$ complexes in their lowest excited electronic states are calculated by converting the relative displacements from eq. 7 to bond length changes in Å units. First, the relative displacements are converted to absolute displacements by using eqs. 7 and 9. The value of $2\sigma^2$ is obtained from the absorption spectrum. The resulting displacements, Δ_K, are converted from dimensionless normal coordinates to lengths and angles by transforming to the desired units. A complete calculation requires a complete normal coordinate analysis. A good approximation is achieved by assuming that the normal coordinates are uncoupled and that the masses appropriate to a specific normal coordinate can be used. The latter calculation is reported here because a complete normal coordinate analysis is not available.

The calculated changes in the bond lengths (in Å) for $W(CO)_5py$ show that the most highly elongated bond is the W-N bond. The W-C bond trans to the pyridine is also highly elongated. The lengths of the W-C bonds cis to the pyridine are only slightly changed. The excited state bond length changes are W-N, 0.18 Å, trans W-C, 0.12 Å, and cis W-C, 0.04 Å. All of the WCO bond angles of the carbonyls cis to the pyridine are changed from 180°. A determination of the angle change in degrees requires a normal coordinate analysis. The meaning of the bond length changes in terms of the bonding changes in the excited state and the connections of these bond lengthenings to the photochemical reactivity of the molecule are discussed below.

The bond length changes in the $W(CO)_5pip$ complex are larger than those in the pyridine complex. The W-C bond trans to the piperidine is lengthened by 0.25 Å and the cis W-C bonds are lengthened by 0.05 Å. The bond length change of the W-N bond has an upper limit of 0.3 Å.

Orbital Characteristics of the Lowest Excited State. The lowest energy excited state of C_{4v} $W(CO)_5L$ compounds, (where L is a ligand lower in ligand field strength than CO), has been assigned by Wrighton, et al., to the (d_{xz},d_{yz}) to d_{z^2} transition (3,7). This assignment has been confirmed by detailed spectroscopic studies including MCD spectroscopy (8). Both the pyridine and piperidine complexes have ^3E lowest energy excited states.

The orbital components of the lowest energy excited state in C_{4v} d^6 complexes are given by equation 10 (9). Because of the proximity of other states of the same symmetry, mixing occurs and some $d_{x^2-y^2}$ orbital character is involved.

$$\phi_E = \frac{1}{\sqrt{1+\lambda^2}}\left[\frac{\sqrt{3}+\lambda}{2}(d_{xy}{}^2d_{xz}{}^2d_{yz}{}^1d_{z^2}{}^1)+\frac{1-\sqrt{3}\lambda}{2}(d_{xy}{}^2d_{xz}{}^2d_{yz}{}^1d_{x^2-y^2}{}^1)\right] \quad (10)$$

The mixing coefficient λ is related to the difference between the ligand field strengths of the carbonyl ligand and the unique ligand (9). For an octahedron (where L=CO), $\lambda = 0$ and the sigma interactions between the metal and all six ligands are equal. When the unique ligand L is weaker than carbonyl as is the case for pyridine and piperidine, $\lambda > 0$ and the increase in d_{z^2} character over $d_{x^2-y^2}$ character causes increased sigma antibonding in the z direction compared to that in the xy plane. The relationship between λ and the percent d_{z^2} character has been discussed in detail (9). The most important conclusions from these orbital considerations are 1) that in the one-electron ligand field picture of $W(CO)_5py$ and $W(CO)_5pip$, the d_{z^2} orbital lies lower in energy than the $d_{x^2-y^2}$ orbital and 2) that the lowest energy excited state wavefunction is dominated by a large d_{z^2} character.

The excited state distortions can be predicted from the orbital characteristics (10). The d_{z^2} orbital and the $d_{x^2-y^2}$ orbital are both sigma antibonding molecular orbitals. Populating the d_{z^2} orbital is thus expected to weaken sigma bonding primarily in the z direction. Populating the $d_{x^2-y^2}$ orbital is expected to weaken sigma bonding in the xy plane. The larger the value of λ in equation 10, the greater the bond weakening in the z direction.

The largest distortion expected for $W(CO)_5py$ and $W(CO)_5pip$, based on the above considerations, is metal-ligand bond lengthening along the z axis, the axis containing the unique ligand. Smaller but non-zero distortions are also to be expected in the xy plane. Although metal-ligand bond lengthenings are predicted to be the biggest distortions, small changes in bond lengths in the ligands themselves are also expected. For example, lengthening a W-CO bond should reduce back bonding to the CO thus strengthening the CO bond and decreasing the CO bond length. Small changes are also expected in the pyridine ring.

Trends in the magnitudes of the major distortions can also be predicted from equation 10. The piperidine ligand is a weaker ligand in the spectrochemical series than pyridine.

(The energy of the absorption maximum is 22,100 cm^{-1} in $W(CO)_5pip$ and 22,400 cm^{-1} in $W(CO)_5py$.) Thus λ is larger for the piperidine complex than that for the pyridine complex and the distortions along the z axis should be greater.

The experimental results provide the first spectroscopic substantiation of the ligand field based bonding predictions. The largest experimentally determined distortions occur along the z axis and smaller distortions occur along in the xy plane. The piperidine complex is more highly distorted than the pyridine complex. Small distortions of the bond lengths within the ligands are also observed. The predicted bonding changes which have been used to predict photochemical reactivity are verified by the combination of pre-resonance Raman spectroscopy, electronic spectroscopy, and time-dependent theory.

Correlations between Excited State Distortions and Photochemical Reactivity

The ligand field theory of transition metal photochemistry is based on the idea that the bonding changes in excited electronic states are correlated with ligand photolabilization (11-13). Populating the d_{z^2} or $d_{x^2-y^2}$ orbitals increases sigma antibonding in the z and xy directions respectively. Depopulating d orbitals which have pi symmetry simultaneously change the pi bond order with a directionality determined by which d orbital is depopulated. Sigma and pi bond weakening along a given metal-ligand bond is correlated with the photochemical ligand labilization of that bond.

In cases such as those studied here where two different ligands are on the same molecular axis such as the z axis, the more complicated question arises of which of the two ligands experiences the greatest antibonding. Three approaches to answering this question have been used. The first successful approach used molecular orbital theory, specifically overlap populations, to calculate the distribution of antibonding along a given axis (14). This approach is predictive, but it requires a complete calculation for each compound of interest. A second approach uses ligand field theory. When the original theory is rewritten in terms of angular overlap parameters, contributions from each ligand can be apportioned (13). This approach is also predictive, but it is not useful for many metals because the required parameters have not or cannot be determined. The third approach, used specifically for tungsten carbonyls, is an empirical analysis based on infrared data (4). It is to some extent predictive for compounds far from the empirical dividing lines. The above three approaches are indirect methods of inferring antibonding character in a given excited state.

The bond length changes determined from pre-resonance Raman spectra, electronic spectra and time dependent theory provide a detailed picture of the results of bonding changes caused by populating excited electronic states. There is a direct but not linear correlation between bond length changes and the photochemical labilizations of the ligands. In the pyridine complex, the most highly distorted metal-ligand bond, the W-N bond, is lengthened by 0.18 Å. The quantum yield for pyridine loss is 0.22. The much less distorted metal-carbon bond is much less reactive; the quantum yield for CO loss is less than 0.01. It is interesting to note that the piperidine complex has a larger distortion and a larger quantum yield for unique ligand loss, 0.58, than the pyridine complex. Further work is needed to determine whether or not this type of correlation is general for $W(CO)_5L$ complexes.

The Missing Mode Effect (MIME) Both of the compounds $W(CO)_5py$ and $W(CO)_5pip$ exhibit the "Missing Mode Effect" (MIME), a regularly spaced vibronic progression in the luminescence spectrum which does not correspond to any ground state normal mode vibration (2). In the luminescence spectrum of $W(CO)_5py$, the MIME spacing is 550 cm^{-1}, and in the luminescence spectrum of $W(CO)_5pip$ the MIME spacing is 520 cm^{-1}. No totally symmetric vibrational modes of the frequencies are found in the vibrational spectra of these molecules.

The MIME effect is easily understood from the viewpoint of time dependent theory. In the time domain, the most important characteristics of the overlap are the rapid decrease near $t=0$, the partial recurrence in the overlap near $t_m=2\pi/\omega_m$ (where ω_m is the frequency spacing of the observed progression in the frequency domain,) and the quenching of further recurrences in the time domain due to the damping factor Γ and to the presence of displacements in several different modes. The partial recurrence at $t=t_m$ is responsible for the appearance of the regularly spaced progression at frequency $\omega_m=2\pi/t_m$. Two or more displaced modes can conspire to give such a partial recurrence which is not expected of any mode alone. In the simplest pedagogical example, a two mode case, the total overlap is

$$\langle\phi|\phi(t)\rangle = \langle\phi_1|\phi_1(t)\rangle\langle\phi_2|\phi_2(t)\rangle \exp(-iE_0t/\hbar - \Gamma^2t^2) \qquad (11)$$

If $\langle\phi_1|\phi_1(t)\rangle$ and $\langle\phi_2|\phi_2(t)\rangle$ peak at different times, the product $\langle\phi|\phi(t)\rangle$ may peak at some intermediate time. The compromise recurrence time t_m is not just the average of t_1 and t_2. The MIME frequency may be smaller than any of the individual frequencies. It is usually between the highest and lowest frequencies. It cannot be larger than the highest frequency.

Each of the displaced modes in the molecule can contribute to the MIME frequency. Each of these modes k has a time dependence whose magnitude is given by eq. 1. The larger the displacement Δ_k in the kth mode, the sharper the peaks in $\langle\phi_k|\phi_k(t)\rangle$. The total overlap $\langle\phi|\phi(t)\rangle$ is the product (eq. 10) of the individual modes' overlaps and t_m will tend to be closest to $t_k=2\Pi/\omega_k$ for that mode with the largest Δ_k.

The 550 cm^{-1} MIME frequency of $W(CO)_5py$ was calculated by using the Raman-determined distortions and frequencies. The normal modes which give a recurrence in the time domain at $t=t_m$

(i.e., a MIME frequency w_m=2-/t_m=550 cm^{-1}) are predominantly the W-C stretches in the 400-500 cm^{-1} region, the WCO bend at 636 cm^{-1} and the W-N stretch at 195 cm^{-1}. Although all of the modes including the high frequency CO stretching modes contribute to the MIME frequency, the primary effect on the luminescence spectrum of these modes with small displacements is to "fill in" the red end of the spectrum.

The 520 cm^{-1} MIME frequency of $W(CO)_5pip$ was calculated by using the same procedures as those used for $W(CO)_5py$. The major contributing modes to the MIME frequency are the W-C stretches in the 400-500 cm^{-1} region and the WCO bending mode at 596 cm^{-1}.

The emission spectra of $W(CO)_5py$ and $W(CO)_5pip$ are typical of the spectra obtained from perturbed polyatomic molecules. The spectra show structure on a scale of five hundred wavenumbers although the instrumental resolution is two orders of magnitude higher. The natural tendency to interpret the regularly spaced progression in terms of one displaced mode is far from correct. Instead, eighteen displaced modes contribute to the observed MIME progression. The origin of the MIME effect is readily explained by using time-dependent theory. In addition, displacements of the modes, (which are hidden in the emission spectrum) are determined.

<u>Excited State Bending Distortions of the MNO Group and Their Photochemical and Spectroscopic Consequences</u>

Non-symmetry preserving geometry changes are a second important type of excited state distortion. The specific example to be discussed here is the bending of the MNO unit after MLCT excitation of a linear MNO unit. Because of the symmetry change, the bending cannot be as easily treated as a symmetry preserving distortion by the time-dependent theory discussed above. The spectroscopic goal of the studies described here is to measure the excited state MNO distortion by excited state Raman spectroscopy. First, the simple theoretical ideas which motivated the studies are presented. Three types of photochemical reactions which provide indirect evidence for MNO bending are then discussed. Finally, the excited state Raman studies of the bending are discussed.

The theoretical motivating ideas for the photochemical and spectroscopic studies are illustrated by the simplified energy level diagram drawn below.

linear $M^- - NO^+$ }

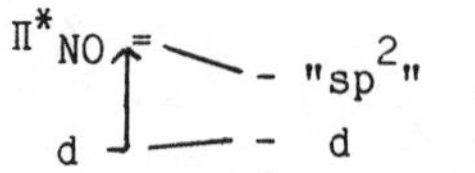

{ bent $M^+ -N^{O-}$

The HOMO, which is mainly metal d in character, is more stable in the linear geometry than in the bent geometry. The LUMO, which is the totally antibonding pi orbital of the MNO unit, has a component which is stabilized by the bending. Part of the driving force for the stabilization occurs because this component becomes an sp^2-like bonding orbital. Populating the metal to nitrosyl charge transfer excited state (which corresponds to the one electron transition shown by the arrow) could cause the linear MNO bond to bend. In a limiting valence bond description, the linear ground state contains a "NO^+" bonded to a metal "M^{-1}" while the bent excited state corresponds to a "NO^-" bonded to a "M^{+1}". In this VB description, the bending causes a two electron oxidation of the metal. In the remainder of this discussion, formal oxidation states will be used for illustrative purposes with the realization that the MO description of this highly covalent unit is more accurate (15).

Photochemistry Three types of photochemical reactions support the idea that excited state bending occurs. The first of these is the gas phase reaction of $Co(CO)_3NO$ with HCl (16). This reaction was chosen for study because it can indirectly probe the geometry and charge changes. If the bending idea is correct, the H^+ of HCl should interact with the bent NO^- and the Cl^- should interact with the Co^+. On the other hand, if photolysis merely activates the linear species, then Cl^-could interact with NO^+ to produce NOCl. The experimental results are shown in eq. 12. The system is photoactive (with negligible thermal reactivity on the time

$$Co^+\text{-}N^{O-} + HCl \rightarrow Cl^-Co^+\text{-}N^{O-}\text{+} \rightarrow [CoCl] + [HNO] \qquad (12)$$

scale of the experiment.) H^+ and NO^- do interact, ultimately disproportionating to produce N_2O and H_2O. Co^+ and Cl^- interact, ultimately undergoing subsequent reactions to produce non-gaseous $CoCl_2$ which forms a powder. No NOCl was detected.

The second photochemical reaction which was studied was the reaction of $Co(CO)_3NO$ with Lewis base ligands L (16). The observed solution phase photochemical reaction is carbonyl photosubstitution. This result initially did not appear to be related to the proposed excited state bending. Further reflection led to the idea that the bent molecule in the excited state is formally a 16 electron coordinatively unsaturated species which could readily undergo Lewis base ligand association. Thus, an associative mechanism would support the hypothesis. Detailed mechanistic studies were carried out. The quantum yield of the reaction is dependent on both the concentration of L and the type of L which was used, supporting an associative mechanism. Quantitative studies showed that plots of $1/\phi$ vs. $1/[L]$ were linear supporting the mechanism where associative attack of L is followed by loss of either L or CO to produce the product. These studies support the hypothesis that the MNO bending causes a formal increase in the metal oxidation state.

The third type of photochemical reaction, photocatalytic hydrogenation of olefins, was pursued because of the possibility that the bent, four coordinate, formally 16 electron excited state of $Rh(PPh_3)_3NO$ could act in a manner similar to

Wilkinson's catalyst. Irradiation at 366 nm of 0.001 M $Rh(PPh_3)_3NO$ and 1 M cyclohexene in o-dichlorobenzene was carried out under 1 atm H_2 at room temperature. The hydrogen uptake was monitored using a mercury manometer attached to the reaction flask. Hydrogen was added periodically in order to maintain 1 atm pressure in the system. The solvent and olefin were distilled twice and degassed by three freeze-pump-thaw cycles before use. A 1000 watt Hg lamp filtered with a glass filter to isolate the 366 nm Hg line was used for all photolysis experiments. The light intensity, measured by ferrioxalate actinometry, was 1.0×10^{-6} einsteins/min.

Figure 3 shows the results of a typical catalysis experiment (17). After mixing the solvent and catalyst in the dark, one atmosphere of H_2 is introduced into the system. A thousand-fold excess of cyclohexene is then added. Initially, hydrogenation is negligible on the scale shown. When the solution is exposed to light there is a short induction period which is followed by hydrogen uptake proceeding roughly linearly with the number of photons absorbed. After irradiation ceases the hydrogenation gradually diminishes, although after 15 hours in the dark a significant thermal reaction is still present. When irradiation is continued, the hydrogenation rate rises to approximately the same value as observed in the previous irradiation period. This rate is limited by the photon flux, and the turnovers achieved to date are limited by the length of the experiment. In a total reaction time of 28 hours (20 thermal, 8 photochemical) 15 turnovers with respect to moles of rhodium were observed. One turnover per hour was observed with a photon flux of 10^6 einsteins/minute.

The quantum yield can be described in two ways. The amount of cyclohexene produced (determined by hydrogen uptake) was measured during the initial irradiation period giving an average value of ϕ=0.75. The net quantum yield includes the thermal hydrogenation that occurs in the dark as a result of the photogenerated catalyst. Since the number of moles of cyclohexane produced is greater than the number of einsteins of photons put into the system, the net quantum yield will be greater than one.

Several possible mechanisms for the catalysis are being studied. The possibility which motivated the study is the production of a coordinatively unsaturated species by the excited state RhNO bending. This excited state could then oxidatively add H_2 and follow a pathway similar to that of Wilkinson's catalyst. Alternatively, the dihydride formed as above could reductively eliminate HNO yielding $HRh(PPh_3)_3$ which itself should be a good catalyst. Another possibility is that heterolytic cleavage of H_2 occurs with H^+ interacting with NO^- as was observed in the reaction with HCl discussed above. Loss of HNO would produce an active metal hydride. Studies are in progress to differentiate between these possibilities.

Excited State Raman Spectroscopy A direct method of studying the geometry of an excited state is to measure the vibrational spectrum of the molecule while it is in the excited state. The direct vibrational spectroscopy complements the indirect photochemical evidence for MNO bending. Raman spectra of metal complexes in excited electronic states have been obtained by using either pulsed or CW lasers to produce a near saturation yield of excited states and to simultaneously provide the probe beam for Raman scattering from the excited molecule (18-22). The pioneering studies of Woodruff, et al., showed that the method could probe electronic changes in the MLCT excited state of $Ru(bipy)^{2+}$(18). In the experiments reported here, excited state Raman spectra were obtained by exciting and probing with 406 nm, 9 nsec pulses at a 40 Hz repetition rate from an excimer pumped dye laser (Lambda Physik EMG 102E and FL2001.) The absorption band at 400 nm has been assigned to the 6e(dxz,yz) to $7e(\Pi^*_{NO})$ MLCT transition (23). The Raman scattered light was passed through a Spex double monochrometer, detected by using a C31034 photomultiplier and recorded with an electrometer and strip chart recorder. An aqueous solution (~1M) of $K_2[Fe(CN)_5NO]$ was pumped through a needle to produce a roughly 200 um diameter jet stream at the laser focus. Each laser pulse irradiated a fresh 10^{-11}L volume of solution.

The Raman spectra taken at three different pulse energies are shown in figure 4. The lowest trace was taken at the lowest pulse energy and is the spectrum of the ground state molecule. The upper two traces show both the ground state and the excited state Raman peaks. Four new peaks are observed which grow in intensity as the laser pulse energy increases. The intensities of all of the new peaks show a non-linear dependence on the laser pulse energy. Plots of the log of the intensities versus the log of the laser pulse energy are linear with slopes of 1.5 ± 0.2 indicating that the peaks arise from a two photon process with some relaxation of the excited state within the duration of the laser pulse. The energies of the excited state bands are given in table 1.

Table 1. Excited state Raman frequencies in $[Fe(CN)_5NO]^{2+}$ and correlations with ground state normal modes.[a,b]

Observed Excited State Frequency (cm^{-1})	Correlation with ground state stretches	Correlation with ground state bends and stretches
501	400 Fe-C (eq)	462 ν Fe-C (ax)
548	462 Fe-C (ax)	652 ν FeN
716	652 Fe-N	665 δ FeNO
1835	1940 NO	1940 ν NO

a) The excited-state Raman spectra were taken by using the 406-nm excitation from an excimer-pumped dye laser. All values are accurate to 5 cm^{-1}.

b) The same ground-state Raman frequencies were obtained by using pulsed 406 nm and CW 514 nm excitation with the exception of the 400 cm^{-1} mode which was obscured by the Rayleigh scattering and ASE with 406 nm excitation.

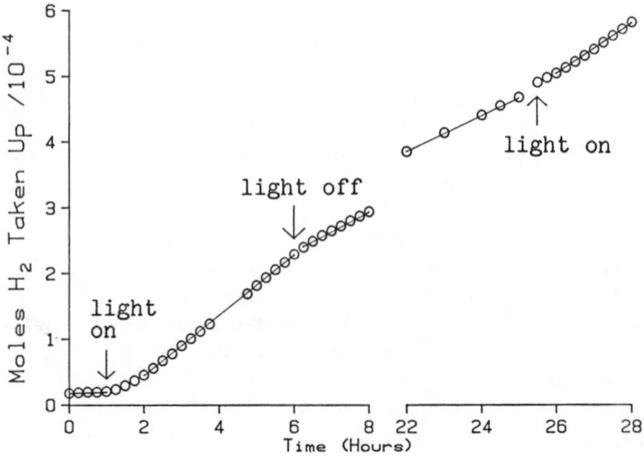

Figure 3. Photohydrogenation catalysis by $Rh(PPh_3)_3NO$.

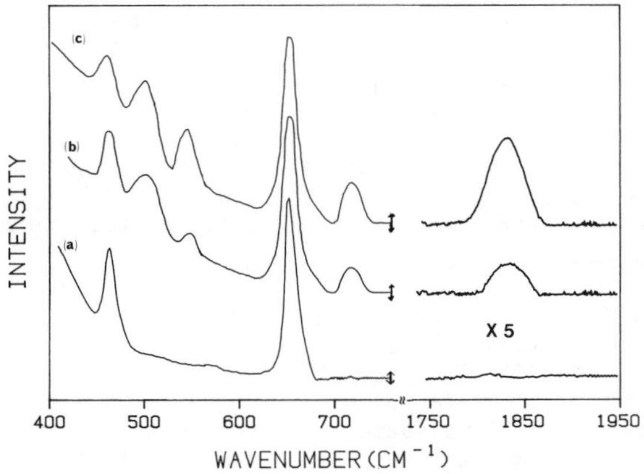

Figure 4. Raman spectra of aqueous solutions of $K_2[Fe(CN)_5NO]$. All spectra are normalized with respect to the 652 cm^{-1} peak. The 1850 cm^{-1} region is shown magnified five times. The magnitude of the background noise, smoothed during digitization, is shown by the arrows. The broad band widths are caused by the large slit widths which were required. The weak ground state NO stretch at 1940 cm^{-1} is observed at a scale larger than that shown. The laser pulse energies were a) 1.1 mJ/pulse; b) 3.0 mJ/pulse; c) 4.3 mJ/pulse.

The new peak which grows in at 1835 cm^{-1} is assigned to the NO stretch in the excited state molecule. Its energy is reduced by 105 cm^{-1} from that of the NO stretch in the electronic ground state. The lower frequency is expected for the bent nitrosyl where the formal bond order is reduced from three to two. The decrease is similar in magnitude to that observed in $RuCl(NO)_2(PPh_3)_2$ where both a linear and a bent nitrosyl is observed (24).

Two correlations between the observed low frequency excited state modes and the corresponding ground state modes are given in the table (25). The first is a one-to-one correlation with the totally symmetric metal-ligand stretches. In this interpretation, all of the metal-ligand stretching frequencies increase in the excited state. The magnitudes of the increases are larger than those expected for a one-electron oxidation. For example, in a series of $Fe(CN)_5X^{n-}$ complexes, the changes in the M-C stretching frequencies are less than 30 cm^{-1} when the iron is oxidized from Fe(II) to Fe(III) (26). The large observed excited state frequencies are consistent with the large increase in metal formal charge expected in the excited state. The second correlation associates the 716 cm^{-1} excited state band with an Fe-NO bending mode and the 548 cm^{-1} band with the Fe-N stretching mode as given on the right of the table. This correlation is consistent with the trends observed for these modes in ground state cobalt complexes containing linear and bent NO geometries (27). The 501 cm^{-1} mode is correlated with an Fe-C stretch which is increased in frequency by about 40 cm^{-1}. These changes, together with the decrease in the NO stretching frequency, are consistent with a linear to bent FeNO geometry change and a concomitant large increase in the positive charge on the metal.

Literature Cited

1. Yersin, H.; Otto, H.; Zink, J.I.; Gliemann, G. J. Am. Chem. Soc. 1980, 102, 951, and references therein.
2. Tutt, L.; Tannor, D.; Heller, E.J.; Zink, J.I. Inorg. Chem. 1982, 21, 3859.
3. Wrighton, M.S.; Hammond, G.S..; Gray, H.B.; J. Am. Chem. Soc. 1971, 93, 4336.
4. Dahlgren, R.M.; Zink, J.I. Inorg. Chem. 1977, 16, 3154.
5. Heller, E.J. Acc. Chem. Res. 1981, 14, 368.
6. Heller, E.J.; Sundberg, R.L.; Tannor, D. J. Phys. Chem., 1982, 86, 1822.
7. Wrighton, M.S.; Abrahamson, H.B.; Morse, D.L. J. Am. Chem. Soc. 1976, 98, 4105.
8. Schreiner, A.F.; Amer, S.; Duncan, W.M.; Ober, G.; Dahlgren, R.M.; Zink, J.I. J. Am. Chem. Soc. 1980, 102, 6871.
9. Incorvia, M.J.; Zink, J.I. Inorg. Chem., 1974, 13, 2489.
10. Zink, J.I. Inorg. Chem. 1973, 12, 1018.
11. Zink, J.I. J. Am. Chem. Soc., 1972, 94, 8039.
12. Wrighton, M.; Gray, H.B.; Hammond, G.J. Mol. Photochem. 1973, 5, 165.

13. Van Quickenbourne, H.G.; Ceulemans, A.J. J. Am. Chem. Soc. 1977, 99, 2208.
14. Zink, J.I. J. Am. Chem. Soc. 1974, 96, 4464.
15. Enemark, J.H.; Feltham, R.D. Coord. Chem. Rev. 1974, 13, 339.
16. Evans, W.; Zink, J.I. J. Am. Chem. Soc. 1981, 103, 2635.
17. Zink, J.I. Laser Chem. 1983, ; Evans, W.E., Ph.D. thesis, UCLA, 1980.
18. Dallinger, R.F.; Woodruff, W.H. J. Am. Chem. Soc. 1979, 101, 4391-3.
19. Dallinger, R.F.; Miskowski, V.M.; Gray, H.B.; Woodruff, W.H. J. Am. Chem. Soc. 1981, 103, 1595-6.
20. Smothers, W.K.; Wrighton, M.S. J. Am. Chem. Soc., 1983, 105, 1067-9.
21. Foster, M.; Hester, R.E. Chem. Phys. Lett., 1981, 81, 42.
22. Schindler, J.W.; Zink, J.I. J. Am. Chem. Soc., 1981, 103, 5968-9.
23. Manoharan, P.T.; Gray, H.B. Inorg. Chem. 1966, 5, 823-838.
24. Pierpont, C.G.; Van Derveer, D.G.; Durland, W.; Eisenberg, R. J. Am. Chem. Soc., 1970, 92, 4760-2. Linear: 1845 cm^{-1}, bent: 1687 cm^{-1}.
25. Work is in progress to differentiate between these correlations by using ^{15}NO isotopic substitution.
26. Brown, D.B. Inorg. Chimica Acta 1971, 5, 314-6.
27. Quinby-Hunt, M.; Feltham, R.D. Inorg. Chem. 1978, 17, 2515-2520.

RECEIVED November 8, 1985

5

Investigation of One-Dimensional Species and of Electrochemically Generated Species

Use of Resonance Raman Spectroscopy

Robin J. H. Clark

Christopher Ingold Laboratories, University College London, 20 Gordon Street, London WC1H 0AJ, United Kingdom

Resonance Raman spectroscopy, well known to be a sensitive probe of the nature of charge-transfer excited states, is now established to be a sensitive probe of intervalence states. In particular, one-dimensional systems prove to be very amenable to study, and the results on a variety of linear-chain platinum complexes of both the Wolffram's red sort as well as the pop sort (pop = $H_2P_2O_5^{2-}$) are outlined. Brief mention is made of the application of resonance Raman spectroscopy to the study of electrochemically generated species.

In any discussion of one-dimensional materials much mention is, justifiably, placed on linear-chain complexes of the KCP type (KCP = $K_2Pt(CN)_4.Br_{0.30}.3H_2O$). The striking optical properties and conductivities of this type of complex, which are associated with their very short Pt-Pt distances, have been studied in detail by synthetic chemists, crystallographers, theoretical physicists and materials scientists alike, particularly over the past 15 years (1,2). However, they are not the only type of one-dimensional system to command attention. More recently it has been realised that halogen-bridged chain-complexes give rise to interesting electronic, Raman and resonance Raman spectra. In particular, mixed-valence chain complexes such as Wolffram's red, $[Pt^{II}(C_2H_5NH_2)_4]$-$[Pt^{IV}(C_2H_5NH_2)_4Cl_2]Cl_4.4H_2O$ give rise to very intense resonance Raman spectra from which information on the nature of the intervalence state may be obtained; this is a matter of considerable contemporary interest (3). It is the purpose of this article to summarise the key results obtained by way of resonance Raman studies on different chain complexes and to draw attention to the likely implications of these results vis à vis this particular type of excited state, the intervalence state (4). The relevance of these results to the study of electrochemically generated species is also outlined.

0097–6156/86/0307–0057$06.00/0

Wolffram's Red Type Salts

Chain complexes of the Wolffram's red sort have structures of the general type

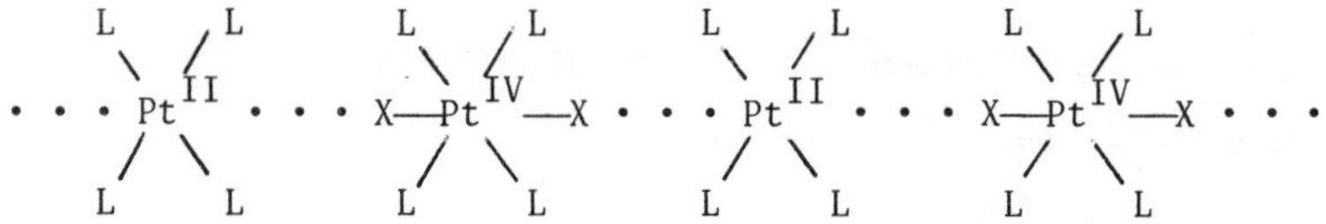

where the neutral equatorial ligand, L, is an amine such as NH_3, CH_3NH_2 or $C_2H_5NH_2$ and X = Cl, Br or I. Bidentate ligands $\widehat{LL}$ can also form complexes of this sort, where $\widehat{LL}$ = 1,2-diaminoethane, 1,2-diaminopropane, 1,3-diaminopropane, 1,2-diaminobutane, 1,2-diaminocyclopentane, 1,2-diaminocyclohexane, etc., as can certain terdentate amines $L\widehat{L}L$ such as diethylenetriamine and N-methyldiethylenetriamine (5). This chain structure persists through the five different possible charge types for this series of complexes, viz. $[Pt^{II}L_4][Pt^{IV}L_4X_2]Z_4$, $[Pt^{II}L_3Y][Pt^{IV}L_3X_2Y]Z_2$, $[Pt^{II}L_2Y_2][Pt^{IV}L_2X_2Y_2]$, $A_2[Pt^{II}LY_3][Pt^{IV}LX_2Y_3]$, $A_4[Pt^{II}Y_4][Pt^{IV}Y_6]$, where X,Y = Cl, Br or I, Z = singly charged anion such as Cl^-, Br^-, I^-, HSO_4^-, ClO_4^-, BF_4^-, NO_3^-, and A is a singly charged cation such as K^+.

The $Pt^{II} \rightarrow Pt^{IV}$ intervalence transitions of such chain complexes occur in the regions 25,000-18,200 cm^{-1}, 23,600-14,300 cm^{-1} and 20,600-7,500 cm^{-1} for chloro-, bromo-, and iodo-bridged complexes, respectively, the trend Cl > Br > I being the reverse of that of the conductivity of the complexes. The transition wavenumbers may be determined either by Kramers-Kronig analysis of specular reflectance measurements or from plots of the excitation profiles of Raman bands enhanced at or near resonance with the Pt^{II}-Pt^{IV} intervalence band. The maxima have been found to be related to the Pt^{II}---Pt^{IV} chain distance, the smaller the latter the less being the intervalence transition energy (3).

The Raman spectra of such halogen-bridged mixed-valence complexes of platinum obtained at resonance with $Pt^{II} \rightarrow Pt^{IV}$ intervalence band are characterised by an enormous enhancement to the Raman band attributed to the symmetric X-Pt^{IV}-X chain-stretching mode (ν_1), together with the development of long and intense overtone progressions $v_1\nu_1$ (v_1 = vibrational quantum number of ν_1). ν_1 occurs over the ranges 309.1-297.8, 175.7-172.0, and 122.3-114.2 cm^{-1} for chlorine-, bromine-, and iodine-bridged complexes, respectively (2). Progressions reaching as far as $17\nu_1$ have been observed in the resonance Raman spectra of some chain complexes, these long progressions enabling values for the harmonic wavenumbers (ω_1) and anharmonicity constants (x_{11}) to be obtained under the usual approximations (6). The considerable length of the ν_1 progression at resonance with the intervalence band implies a substantial (~ 0.1 Å) change in the position of the bridging atom on excitation to this state.

Recent synthetic and spectroscopic work has concentrated on the study of structurally related linear-chain palladium complexes (7,8), on the possibility of synthesising related nickel complexes, and on the study of mixed-metal mixed-valence complexes of the sort

$[Ni^{II}(en)_2][Pt^{IV}(en)_2Cl_2][ClO_4]_4$ and $[Pd^{II}(en)_2][Pt^{IV}(en)_2Cl_2][ClO_4]_4$ (9). The intervalence band maxima are found to vary in the order Cl > Br > I and Pd^{II}/Pt^{IV} > Ni^{II}/Pt^{IV} > Pt^{II}/Pt^{IV} > Pd^{II}/Pd^{IV}, implying that the valence electrons are most delocalized for the Pd^{II}/Pd^{IV} complexes; this implication is consistent with the relatively high chain conductivity of such complexes (10).

The immense amount of spectroscopic work carried out on these complexes, particularly on their electronic, infrared, Raman and resonance Raman spectra (2), has led to Wolffram's red being regarded as the archetypal class II or localized valence complex in which the two metal atoms differ in their oxidation states by two, cf. Prussian blue is so regarded for mixed-valence complexes in which the two metal atoms differ in their oxidation states by one.

The principal points of current interest concerning Wolffram's-red type complexes and on which research is being concentrated are

(a) the nature of the dependence of the $X\text{-}M^{IV}\text{-}X$ symmetric chain-stretching mode and of the absorption edge of the intervalence band on pressure (11).

(b) the understanding of the origin of the luminescence emitted by these complexes and of its pressure dependence.

(c) the chlorine isotopic structure of the ν_1 band of Pt^{II}/Pt^{IV} complexes, which differs from 9 : 6 : 1 for the fundamental but not for the overtones, and the relationship between the degree of valence delocalization along the chain, the wavenumber of the band gap relative to that of the exciting line, and the quality of the resolution of this structure to ν_1 (which is not isotopic in origin in the case of Pd^{II}/Pd^{IV} complexes).

(d) the definitive characterisation of Ni^{II}/Ni^{IV} linear-chain complexes (10).

Pop Salts of Linear-Chain Complexes

As part of a search for other ligands capable of adopting a square-planar configuration about a metal atom and thus potentially able to form stacked units our attention was drawn to the ligand $H_2P_2O_5^{2-}$ (diphosphonate), usually abbreviated pop. Platinum complexes of this ligand - in particular $[Pt_2(pop)_4]^{4-}$ - have already been subject to interesting studies of their luminescence, electronic, Raman and infrared spectra (12-16). Our initial objectives were to try to incorporate $[Pt^{IV}(en)_2X_2]^{2+}$ (en = 1,2-diaminoethane; X = Br or I) and $[Pt_2^{II}(pop)_4]^{4-}$ alternately into stacked chains. However, the reaction was found to generate a different type of chain complex from that envisaged, viz.

$$2[Pt^{IV}(en)_2X_2]^{2+} + [Pt_2^{II}(pop)_4]^{4-} \rightarrow [Pt^{II}(en)_2][Pt^{IV}(en)_2X_2][Pt_2^{III}(pop)_4X_2].$$

In effect, one $[Pt^{IV}(en)_2X_2]^{2+}$ unit oxidizes $[Pt_2^{II}(pop)_4]^{4-}$ to $[Pt_2^{III}(pop)_4X_2]^{4-}$, itself being reduced to $[Pt^{II}(en)_2]^{2+}$; the last then cocrystallizes with the second $[Pt^{IV}(en)_2X_2]^{2+}$ unit to form a Wolffram's-red type chain complex. The spectroscopic (electronic, infrared, Raman, resonance Raman) evidence on which these conclusions are based is as follows (17).

(a) The absorption spectra are characterised by three bands.

The one of lowest wavenumber occurring at or near the wavenumber known to be characteristic of linear-chain bromine- or iodine-bridged Pt^{II}/Pt^{IV} complexes (17,000 cm^{-1} and 11,000 cm^{-1} for X = Br or I, respectively).

(b) The resonance Raman spectra of the complexes are characterised by a progression in the X-Pt^{IV}-X symmetric stretching mode (ν_1') of the chain at resonance with the cation-chain intervalence band mentioned above, but by a progression in the Pt^{III}-Pt^{III} stretching mode (ν_1) of the dimeric anion $[Pt_2^{III}(pop)_4X_2]^{4-}$ at resonance with second electronic band in each case (assigned to the $\sigma \rightarrow \sigma^*$ transition of the anion). A summary of the key results on these chain complexes is given in Table I.

Table I. Summary of Spectroscopic Data on the Complexes $[Pt^{II}(en)_2][Pt^{IV}(en)_2X_2][Pt^{III}(pop)_4X_2]$[a]

	X = Br	X = I
Crystals	copper needles	gold-green needles
Powder	black	black
$\tilde{\nu}(Pt^{II} \rightarrow Pt^{IV})/cm^{-1}$	17000	11000
$\tilde{\nu}(d_\sigma \rightarrow d_{\sigma*})/cm^{-1}$	23300	18000
$\omega_1'(X\text{-}Pt^{IV}\text{-}X)/cm^{-1}$	175.7 ± 0.3	∿ 116
x_{11}'/cm^{-1}	-0.39 ± 0.03	-
$\omega_1(Pt^{III}\text{-}Pt^{III})/cm^{-1}$	138.8 ± 0.3	121.6 ± 0.3
x_{11}/cm^{-1}	-0.04 ± 0.03	-0.04 ± 0.03
$\tilde{\nu}_2(Pt^{III}\text{-}X)/cm^{-1}$	228.6	199.6
x_{12}/cm^{-1}	-0.1 ± 0.05	+0.35 ± 0.05
$v_1\nu_1$ progression	v_1 = 9(12)	v_1 = 9
$v_1\nu_1 + \nu_2$	v_1 = 7(9)	v_1 = 7
$v_1\nu_1 + 2\nu_2$	v_1 = 5(8)	v_1 = 2
$v_2\nu_2$	v_2 = 2(2)	v_2 = 2
$v_1'\nu_1'$	v_1' = 10(11)	v_1' = 5

[a] Raman data relate to measurements taken on samples at ca. 80 K. Those in parentheses relate to measurements on single crystals at ca. 15 K.

The complexes are unique in that they involve (a) platinum in three different formal oxidation states and (b) both class I and class II mixed-valence interactions within a single complex.

Pop Complexes $K_4[Pt_2(pop)_4X].nH_2O$

Recent synthetic work (14-19) on $K_4[Pt_2(pop)_4].2H_2O$ has shown that it can be partially oxidized to a new type of semi-conductor $K_4[Pt_2(pop)_4X].nH_2O$, X = Cl, Br or I, in which the average oxidation state of the platinum atoms is 2.5. These complexes form as golden metallic-looking crystals with chain conductivities ($\sigma_{\|}$ = 10^{-4}-10^{-3} Ω^{-1} cm^{-1}) which are many orders of magnitude greater than those found for Wolffram's-red type complexes (10^{-12}-10^{-8} Ω^{-1} cm^{-1}). X-ray crystallographic work shows that both the chloride (18,19) and the bromide (14,16) consist of linear chains of -X-Pt-Pt-X-, the Pt atoms being linked, as is usual for this type of complex, pairwise by four pop bridges; clearly, the chains provide an effective pathway for electrical conductivity.

The electronic spectra of these complexes are in each case dominated by a very intense, broad band centred at 19,500, 15,500 and 11,400 cm^{-1} for X = Cl, Br, and I, respectively. In all its characteristics (wavenumber, intensity, breadth) this band behaves like an intervalence band of a halogen-bridged species. The Raman spectrum of $K_4[Pt_2(pop)_4Cl].3H_2O$ at resonance with the intervalence band is dominated by a band at 291 cm^{-1} which is attributed to the symmetric PtCl stretching mode, somewhat lowered (on account of bridging) from its value (305 cm^{-1}) for the discrete anion $[Pt_2(pop)_4Cl_2]^{4-}$. Six-membered progressions in the 291 cm^{-1} band are observed in the Raman spectrum under resonance conditions. The important implication of these results is that the chlorine atom cannot be centrally bridging between the $Pt_2(pop)_4$ units for, in that case, the symmetric PtCl stretching mode would be Raman inactive. The results also indicate that the principal structural change on excitation to the intervalence state is along the Pt-Cl coordinate.

There are two possible structures for such a chain, Figure 1. The former structure consists of stacked polar dimers, which would seem to be unlikely since the bridging pop ligands are themselves symmetric (polar dimers are, however, well established where the bridging ligands are unsymmetric). The latter structure is the Wolffram's red analogue and, in view of the large number of such complexes known (vide supra), was considered to be the more probable. These possibilities may in principle be distinguished by consideration of infrared and Raman band activities, but in practice this proves to be very difficult. Strong evidence that the complex consists of a stacked polar dimer has, however, been obtained by X-ray crystallography, the Pt-Pt, Pt^{III}-Cl and $Pt^{II}\cdots$Cl distances being 2.813(1), 2.367(7) and 2.966(8) Å, respectively (19). Thus $K_4[Pt_2(pop)_4Cl].3H_2O$ is a localized-valence (Pt^{II}/Pt^{III}) complex on account of the fact that the Pt-Cl distances differ very substantially (by 0.60 Å). This difference is entirely comparable with that found for Wolffram's-red type salts, viz. the chain Pt^{IV}-Cl distances for seven such structures cover the range 2.30 ± 0.04 Å while the chain $Pt^{II}\cdots$Cl distances cover the range 3.06 ± 0.23 Å.

Both crystallographic (14,16,19) as well as spectroscopic results (18) indicate that the analogous bromide and iodide complexes are much nearer to being delocalized-valence species (as are the bromide and iodide versions of Wolffram's-red type salts), consistent

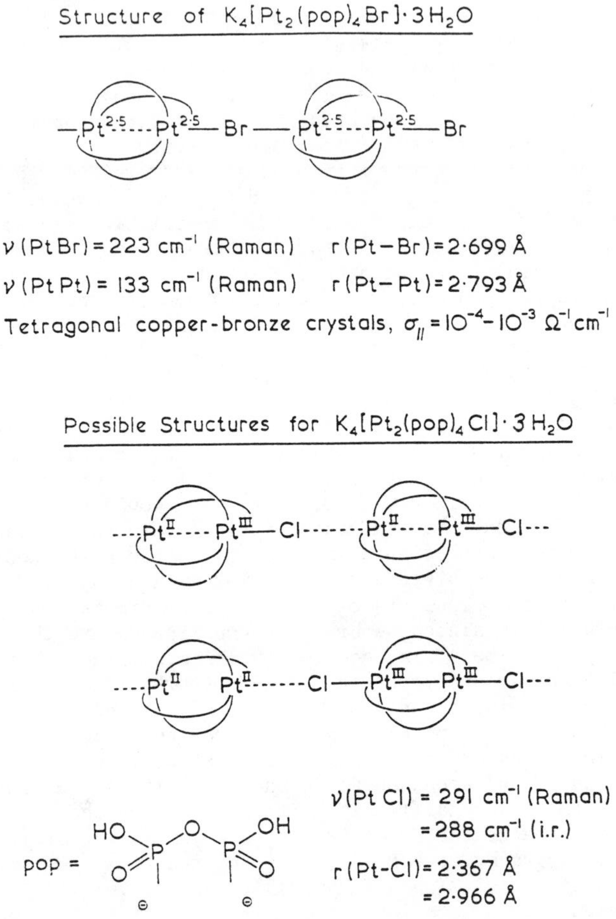

Figure 1. Structure of $K_4[Pt_2(pop)_4Br].3H_2O$ and possible structures for $K_4[Pt_2(pop)_4Cl].3H_2O$.

with their chain conductivities being higher than that of the chloride. Although crystallographic work (16) has been interpreted to imply exact central bridging by bromide for $K_4[Pt_2(pop)_4Br].3H_2O$, the spectroscopic results indicate that there must be a distortion, albeit slight, from this geometry, consistent with the expected Peierl's distortion of a symmetric linear chain. The bromide and iodide are thus closely similar in both structures and properties to the dithioacetate complexes of platinum and nickel, $M_2(CH_3CS_2)_4I$, which are likewise semiconductors with nearly but not exactly equal chain M-I bond lengths (2.975 and 2.981 Å for Pt, 2.928 and 2.940 Å for Ni) (20,21).

Spectroelectrochemically-generated Species

The chemistry of cluster complexes, e.g. of the sort $[Fe_4S_4(SR)_4]^{2-}$, is of particular interest since such complexes are known to be close representations or synthetic analogues of the redox centres present in various iron-sulphur proteins. It is important to know whether the valence electrons are localized or delocalized in such complexes - in fact several studies by e.s.r., n.m.r., and, more recently, resonance Raman spectroscopy have shown that such clusters are delocalized rather than trapped-valence species. This result is linked with the most important biophysical property of iron-sulphur proteins, viz. that of electron transfer. Rapid electron transfer is possible if any consequential geometric rearrangements around the metal atom sites are small, as implied by many resonance Raman results on such cluster complexes (cf. the small-displacement approximation, which provides a basis for enhancement to fundamental but not to overtone bands) (22). Initial studies of $[MS_4]^{2-}$ ions (M = Mo or W) (23,24) have since been supplemented by studies of dinuclear species e.g. $[(PhS)_2FeS_2MS_2]^{2-}$ (25) and cluster species (26) such as the copper(I) tetrathiomolybdate(VI) anions $[MS_4(CuL)_n]^{2-}$ (M = Mo or W; L = CN, SC_6H_5, $SC_6H_4CH_3$, Cl, or Br; n = 1 - 4).

The question of whether electrons added to a complex ion become localized or delocalized is important, not only for the type of complex mentioned above, but also for much wider ranges of complexes. For such studies the use of an OTTLE (optically transparent thin layer electrochemical) cell is most appropriate (27). Such cells can be adopted not only to electronic and infrared studies, but also to Raman and in particular, resonance Raman studies. The time scale of OTTLE measurements, being rapid by comparison with that of conventional bulk electrolysis, permits rapid spectral sampling of the product. Moreover, by establishing the electronic band maxima of the electrochemically generated species, it is then possible to resonance enhance Raman bands of this species using an exciting line which is off-resonance for the reactant(s); thus the effects of interfering reactant bands are removed from the required spectrum. The optical cell used consisted of a platinum minigrid (52 mesh, 0.1 mm diameter wire) with a transparency of 60%; a similar cell has also been used successfully for both Raman (Spex 14018, R6) and infrared (Bruker 113 V interferometer) studies.

Many inorganic systems are currently under study by these means, viz. the tris-dithiolates of Cr, Mo and W and V in all of which the electrons are found to be added to, or removed from, extensively delocalized orbitals on the complex ion (a conclusion again based on the small-displacement approximation), and the confacial bioctahedral complexes $[L_3RuCl_3L_3]^{2+}$ and $[L_{3-x}Cl_xRuCl_3RuCl_yY_{3-y}]^{n+}$, where L = PEt_2Ph, $As(tol)_3$ or PPh_3 (28). Such complexes display at least one and usually two stepwise, reversible one-electron transfer reactions without there being any gross structural change. Such work has already established that the extent of valence-electron delocalization depends on the degree of asymmetry (y - x) of the complex and on the basicity of the terminal ligands. Moreover it is established that, where y - x is zero, e.g. for $[(PEt_2Ph)_3RuCl_3Ru(PEt_2Ph)_3]^{2+}$, the intervalence band occurs at low wavenumber (4350 cm^{-1}), implying that the valence electrons are highly delocalized. However, as y - x increases, the intervalence band loses intensity and moves well into the visible region, e.g. for $[(PEt_2Ph)_3RuCl_3RuCl_2(PEt_2Ph)]$ it occurs at 13,500 cm^{-1}, clearly a localized valence species. Extensive Raman and resonance Raman studies of these systems are currently in progress in order to establish (a) which mode(s) change wavenumber and/or intensity on oxidation or reduction, and (b) by implication, therefore, where in the molecule the odd electron enters on reduction or is removed from on oxidation. The nature of the HOMO and LUMO may then be able to be established.

Thin layer electrochemistry thus offers a very convenient way of controlling the oxidation state of a very thin ($\leqslant$ 1 mm) layer of an electrochemically generated species.

Conclusion

The combination of several spectroscopic techniques, but particularly with involvement of resonance Raman spectroscopy, offers an effective way of studying the nature of the intervalence state in a wide variety of complexes ranging from linear-chain complexes of the Wolffram's red and ---$Pt_2(pop)_4$--- sorts to electrochemically generated species.

Literature Cited

1. "Extended Linear Chain Compounds", Miller, J.S., Ed., Plenum: New York. Vol.1, 1982.
2. Clark, R.J.H. Chem. Soc. Rev. 1984, 13, 219.
3. Clark, R.J.H. "Advances in Infrared and Raman Spectroscopy", Clark, R.J.H.; Hester, R.E., Eds., Wiley-Heyden: Chichester. Vol.11, 1984, p.95.
4. Allen, S.D.; Clark, R.J.H.; Croud, V.B.; Kurmoo, M. Phil. Trans. Roy. Soc. Lond. A 1985, 314, 131.
5. Fanizzi, F.P.; Natile, G.; Lanfranchi, M.; Tiripicchio, A.; Clark, R.J.H.; Kurmoo, M. J. Chem. Soc. (Dalton Trans.) in press.
6. Clark, R.J.H.; Stewart, B. Structure and Bonding 1979, 36, 1.
7. Clark, R.J.H.; Croud, V.B.; Kurmoo, M. Inorg. Chem. 1984, 23, 2499.
8. Clark, R.J.H.; Croud, V.B. J. Chem. Soc. (Dalton Trans.) 1985, 815.

9. Clark, R.J.H.; Croud, V.B. Inorg. Chem. 1985, 24, 588.
10. Yamashita, M.; Ito, T. Inorg. Chim. Acta 1984, 87, L5.
11. Tanino, H.; Koshizuka, N.; Kobayashi, K.; Yamashita, M.; Hoh, K. J. Phys. Soc. Japan 1985, 54, 483. Clark, R.J.H.; Croud, V.B. unpublished results.
12. Sperline, R.P.; Dickson, M.K.; Roundhill, D.M. J. Chem. Soc. (Chem. Comm.) 1977, 62.
13. Filomena Das Remedios Pinto, M.A.; Sadler, P.J.; Neidle, S.; Sanderson, M.R.; Subbiah, A. J. Chem. Soc. (Chem. Comm.) 1980, 13.
14. Che, C.-M.; Schaeffer, W.P.; Gray, H.B.; Dickson, M.K.; Stein, P.; Roundhill, D.M. J. Am. Chem. Soc. 1982, 104, 4253.
15. Stein, P.; Dickson, M.K.; Roundhill, D.M. J. Am. Chem. Soc. 1983, 105, 3489.
16. Che, C.-M.; Herbstein, F.H.; Schaefer, W.M.; Marsh, R.E.; Gray, H.B. J. Am. Chem. Soc. 1983, 105, 4604.
17. Clark, R.J.H.; Kurmoo, M. J. Chem. Soc. (Dalton Trans.) 1985, 579.
18. Kurmoo, M.; Clark, R.J.H. Inorg. Chem. in press.
19. Clark, R.J.H.; Kurmoo, M.; Dawes, H.M.; Hursthouse, M.B. Inorg. Chem. in press.
20. Bellito, C.; Flamino, A.; Gastaldi, L.; Scaramuzza, L. Inorg. Chem. 1983, 22, 444.
21. Bellito, C.; Dessy, G.; Fares, V. Mol. Cryst. Liq. Cryst. 1985, 120, 381.
22. Clark, R.J.H.; Dines, T.J. Mol. Phys. 1981, 42, 193.
23. Clark, R.J.H.; Dines, T.J.; Wolf, M.L. J. Chem. Soc. (Faraday Trans.) 1982, 78, 679.
24. Clark, R.J.H.; Dines, T.J.; Proud, G.P. J. Chem. Soc. (Dalton Trans.) 1983, 2019.
25. Clark, R.J.H.; Dines, T.J.; Proud, G.P. J. Chem. Soc. (Dalton Trans.) 1983, 2229.
26. Clark, R.J.H.; Joss, S.; Zvagulis, M.; Garner, C.D.; Nicholson, J.R. J. Chem. Soc. (Dalton Trans.) in press.
27. Heineman, W.R. J. Chem. Ed. 1983, 60, 305.
28. Heath, G.A.; Lindsay, A.J.; Stevenson, T.A.; Vattis, D.K. J. Organometal. Chem. 1982, 233, 353.

RECEIVED November 7, 1985

6

Metal-Ligand Charge Transfer Photochemistry

Metal-Metal Bonded Complexes

D. J. Stufkens, A. Oskam, and M. W. Kokkes

Anorganisch Chemisch Laboratorium, University of Amsterdam, Nieuwe Achtergracht 166, 1018 WV Amsterdam, The Netherlands

The complexes $(CO)_5MM'(CO)_3(\alpha\text{-diimine})$ (M,M'=Mn,Re) show a different photochemistry in 2-Me-THF at $T > 200K$ and at $T < 200K$ upon irradiation into the low-energy M' to α-diimine charge tranfer band. At higher temperatures homolysis of the metal-metal bond takes place, at lower temperatures breaking of a metal-nitrogen bond in the case of $(CO)_5MnRe(CO)_3(\alpha\text{-di-imine})$ and photosubstitution of CO by 2-Me-THF for $(CO)_5MMn(CO)_3(\alpha\text{-diimine})$ followed by a disproportionation into $[M(CO)_5]^-$ and $[Mn(CO)_3(\alpha\text{-diimine})(2\text{-Me-THF})]^+$ upon raising the temperature. For these reactions an energy vs distortion diagram with two close-lying excited states $^3\sigma_b\pi^*$ and $^3d_\pi\pi^*$ is proposed, based on the electronic absorption , resonance Raman and UV-photoelectron spectra of these complexes. Irradiation of several of these complexes in THF in the presence of $P(n\text{-}Bu)_3$ leads to a photocatalytic disproportionation reaction for which a highly reducing 18 electron radical $Mn(CO)_3(\alpha\text{-diimine})(P(n\text{-}Bu)_3)$ is proposed. Photolysis of $(CO)_5MnRe(CO)_3(i\text{-Pr-DAB})$ in a CH_4-matrix at 10K leads to a change of coordination of the i-Pr-DAB ligand from σ,σ in the parent compound into $\sigma,\sigma,\eta'^2,\eta'^2$ in the photoproduct $(CO)_3Mn(i\text{-Pr-DAB})Re(CO)_3$. A similar reaction is observed for $(CO)_5MnMn(CO)_3(i\text{-Pr-DAB})$ in a PVC film.

Low-valence transition metal complexes of α-diimine ligands are highly colored because of the presence of low-energy metal to α-diimine charge transfer (MLCT) transitions. For a series of d^6-$M(CO)_4$ (α-diimine) (M=Cr,Mo,W) and d^8- $M'(CO)_3$ (α-diimine) (M'=Fe, Ru) complexes, we have studied the spectroscopic and photochemical properties (1-10). The α-diimine ligands used are 1,4-diaza-1,3-butadiene (R-DAB), pyridine-2-carbaldehyde-imine (PyCa), 2,2'-bipyridine (bipy) or 1,10-phenanthroline (phen) molecules. A close relationship was deduced between the photochemical behavior of these complexes and their resonance Raman (rR) spectra, obtained by excitation into the low-energy MLCT band.

Recently, this work has been extended to the binuclear complexes $(CO)_5MM'(CO)_3(\alpha\text{-diimine})$ (M,M'=Mn, Re) (Figure 1) in order to study the influence of the MLCT photochemistry on the metal-metal bond. Wrighton

0097-6156/86/0307-0066$06.00/0

et al. studied the photochemistry of several of these complexes and observed a homolytic splitting of the metal-metal bond (11). On the basis of this result they assigned the lowest energy absorption band of these complexes to an electronic transition from the metal-metal bonding orbital (σ_b) to the lowest π^*-level of the α-diimine ligand. In that case the metal-metal bond will be weakened in the excited state which explains the photochemistry. However, if this assignment were correct we would expect a strong rR effect for ν_s (M-M'), the symmetrical metal-metal stretching mode, upon excitation into this absorption band. Preliminary results showed that this is not the case, which prompted us to investigate in more detail the spectroscopy and photochemistry of these complexes. In this paper a survey is given of the results of this study and of the conclusions derived from them (12).

Spectroscopic properties

Information about the relative energies of the m.o.'s of molecules can be derived from their UV-photoelectron spectra. For the complexes $(CO)_5MM'(CO)_3$(i-Pr-DAB) (i-Pr= iso-propyl) these spectra have been measured by He(I) and He(II) excitation (16). Figure 2. shows the spectra of $(CO)_5MnRe(CO)_3$(i-Pr-DAB). Bands A,B and C at the low ionization energy side of the spectra, are assigned to ionizations from the metal-d orbitals because of their relatively high cross section in the He(II) spectra. Band A belongs to the ionization from σ_b (M-M'), the metal-metal bonding orbital, which has a much lower I.P. (7.15 eV) than the corresponding orbital of $(CO)_5MnRe(CO)_5$ (8.08eV) (17). Bands B and C belong to ionizations from the metal-d_π orbitals not involved in the metal-metal bond, which are responsible for the π-backbonding from the metals to the i-Pr-DAB and CO ligands. The l.u.m.o. of these complexes is the lowest π^* level of the α-diimine ligand and transitions to this orbital are allowed from σ_b (M-M') as well as from the d_π(M') orbitals of the $M'(CO)_3(\alpha$-diimine) moiety. However, the σ_b and π^* orbitals will hardly overlap and the transition between them can not be responsible for the high intensity (ε_{max} = 7-12,10^3 $dm^3mol^{-1}cm^{-1}$) of the low-energy absorption band of these complexes (Figure 3). The intensity, energy and solvatochromism of this band closely resembles that of the corresponding mononuclear $M(CO)_4(\alpha$-di-imine)(M=Cr,Mo,W) complexes, which has been assigned to $d_\pi(M)\rightarrow\pi^*(\alpha$-di-imine) transitions (1-2, 5-6).

Support for a similar assignment in the case of these binuclear complexes is presented by the rR spectra of these complexes. The rR spectra of two representative complexes, measured in a N_2 matrix at 10K, are shown in Figure 4. These spectra strongly differ in character. The spectrum of $(CO)_5MnRe(CO)_3$(i-Pr-DAB) shows a strong rR effect for ν_s (CN) and ν(CC) which means that the CN and CC bonds of the i-Pr-DAB ligand have changed character by the charge transfer from the metal to the π^* level of this ligand. In the spectrum of $(CO)_5ReRe(CO)_3$(i-Pr-DAB) on the other hand, both these Raman bands have nearly disappeared whereas two deformation modes of the i-Pr-DAB ligand (800-1000 cm^{-1}) and the metal-ligand stretching modes have increased in intensity. This means that the electronic transition involved has lost its charge transfer character. This change of character of the MLCT transition has been observed before for the mononuclear d^6-complexes $MX(CO)_3(\alpha$-diimine)(M=Mn,Re; X=Cl,Br) and $M(CO)_4(\alpha$-diimine) (M=Cr,Mo,W) and ascribed to a change of interaction (π-backbonding) between the d_π(M) and π^*-orbitals involved in this transition (3). When the π-backbonding increases, the transition loses its

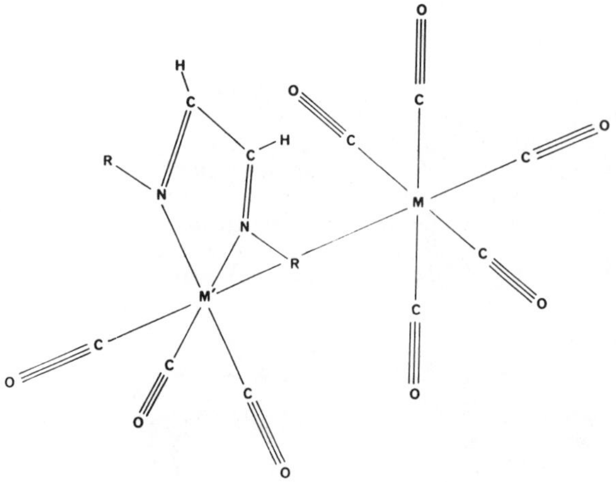

Figure 1. Structure of $(CO)_5MM'(CO)_3$(R-DAB)(M,M'=Mn, Re).

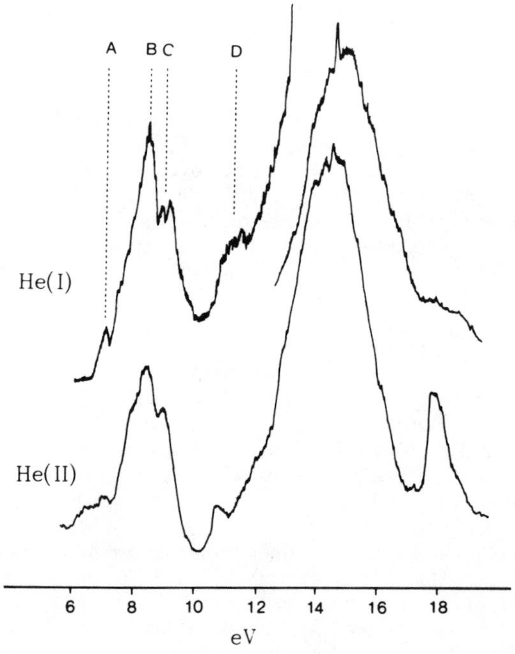

Figure 2. He(I) and He(II) photoelectron spectra of $(CO)_5MnRe(CO)_3$-(i-Pr-DAB). (Reproduced with permission from Ref. 16. Copyright 1985, Elsevier Sequoia S.A.).

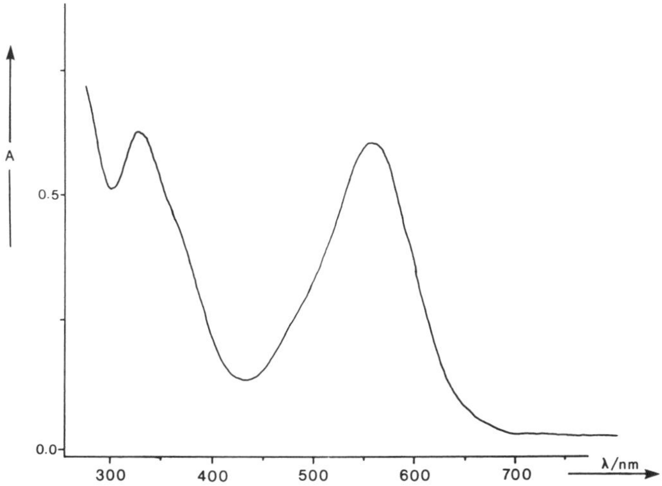

Figure 3. Electronic absorption spectrum of $(CO)_5MnMn(CO)_3$-(i-Pr-DAB) in n-pentane.(Reproduced from Ref. 12. Copyright 1985, American Chemical Society).

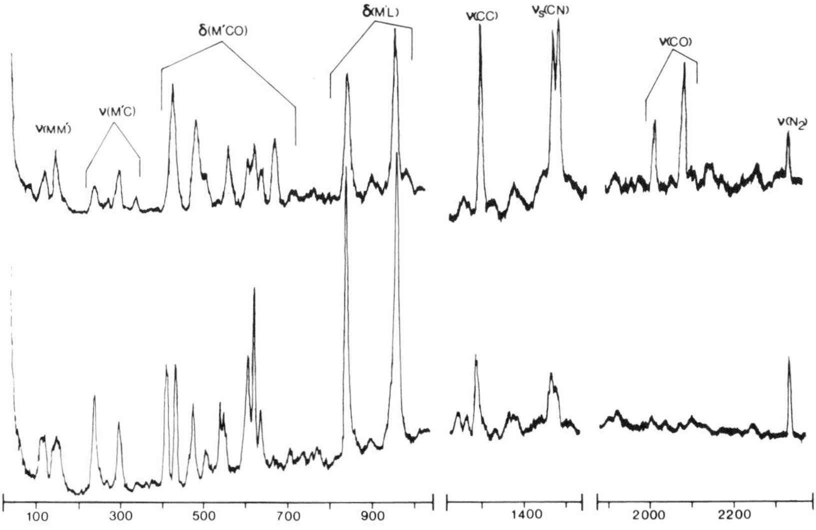

Figure 4. rR spectra of $(CO)_5MnRe(CO)_3$(i-Pr-DAB)(top) and $(CO)_5Re$-$Re(CO)_3$(i-Pr-DAB)(bottom) in a N_2 matrix at 10 K ($\lambda_{exc.}$ = 514.5 nm).

MLCT character and obtains more metal-ligand bonding to anti-bonding character. At the same time the i-Pr-DAB ligand becomes distorted in the excited state as can be seen from the high rR intensity of the ligand deformation modes. Because of this analogy with the mononuclear α-diimine complexes, the absorption band is assigned to an electronic transition from a $d_{\pi}(M')$ orbital of the $M'(CO)_3$(α-diimine) moiety, not involved in the metal-metal bond, to the lowest π^* orbital of the α-diimine ligand. This assignment is confirmed by the observation that ν(M-M') at about 150 cm^{-1} is far too weak in the rR spectra for excitation into the $\sigma_b(M-M')\rightarrow\pi^*$ transition. On the basis of these results the following energy vs distortion diagram is proposed (Figure 5). Absorption takes place to the $^1d_{\pi}\pi^*$ state and not to $^1\sigma_b\pi^*$. From $^1d_{\pi}\pi^*$, however, intersystem crossing may occur to both $^3d_{\pi}\pi^*$ and $^3\sigma_b\pi^*$, thus affording two thexi states from which reactions may occur. The reaction from $^3\sigma_b\pi^*$ will lead to splitting of the metal-metal bond. This splitting may be homolytic or heterolytic depending on the difference in electronegativity between both metal fragments. The reaction from $^3d_{\pi}\pi^*$ will be similar to that of the corresponding mononuclear α-diimine complexes, viz. release of CO (6,8,10). We have studied the photochemistry of these complexes in order to find out whether both kinds of reactions do in fact occur (12-14)

Photolysis in 2-Me-THF (133K $\leq T < 298K$)

The complexes were photolyzed in 2-Me-THF, a solvent in which many organometallic compounds readily dissolve and which can be used over a large temperature range ($T > 133K$). The green line (λ = 514.5 nm) of an argon-ion laser was used as irradiation source. Figure 6 shows the IR bands of the photoproducts in the CO-stretching region upon photolysis of $(CO)_5MnMn(CO)_3$(i-Pr-DAB) at different temperatures. At 230K $Mn_2(CO)_{10}$ is formed with CO-stretching modes at 2045, 2009 and 1977 cm^{-1}. The changes in the electronic absorption spectrum accompanying the reaction at 230K are shown in Figure 7. The $\sigma_b\rightarrow\sigma^*$ transition shifts from 340 to 350 nm due to the formation of $Mn_2(CO)_{10}$ while the MLCT band shifts from 550 to 745 nm. The formation of $Mn_2(CO)_{10}$ is the result of a homolytic splitting of the metal-metal bond (scheme 1).

Apparently the $Mn(CO)_5$ radicals formed react to $Mn_2(CO)_{10}$. At the same time the $Mn(CO)_3$(i-Pr-DAB) radicals react to $Mn_2(CO)_6$(i-Pr-DAB)$_2$, a binuclear metal-metal bonded complex with an i-Pr-DAB ligand at each metal fragment. Such complexes have been identified before as thermally unstable side-products of the reaction between $[Mn(CO)_5]^-$ and $Mn(CO)_3X$(R-DAB) (X=Cl, Br or I) (18). The corresponding complex $Mn_2(CO)_6$ (bipy')$_2$ (bipy'=4,4'-dimethyl-2,2'-bipyridine) has been described by Morse and Wrighton (11). The MLCT band at 745 nm and the three CO-vibrations at 1946, 1898 and 1888 cm^{-1} are assigned to the binuclear complex. The shift of the MLCT band to lower energy agrees with the difference in electron withdrawing power between the $Mn(CO)_5$ group of the parent compound and the $Mn(CO)_3$(i-Pr-DAB) fragment of the photoproduct $Mn_2(CO)_6$(i-Pr-DAB)$_2$. This photoproduct is partly split into its radicals which have been identified with ESR (12). Furthermore, this complex is thermally unstable. Raising the temperature to 293 K causes the disappearance of the 745 nm band and a shift of the CO-stretching modes. The same decomposition product, presumably $Mn_2(CO)_6$(i-Pr-DAB)-(2-Me-THF)$_2$, is found as a photoproduct when the photolysis takes place at 293K.

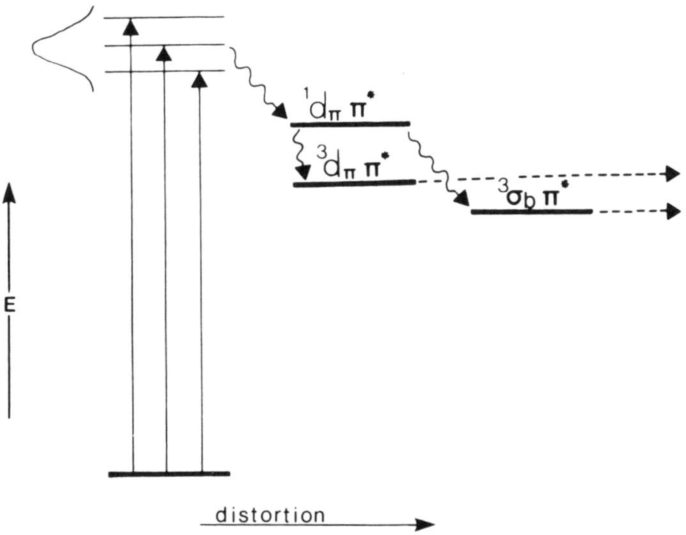

Figure 5. Energy vs distortion diagram. (Reproduced from Ref. 12. Copyright 1985, American Chemical Society).

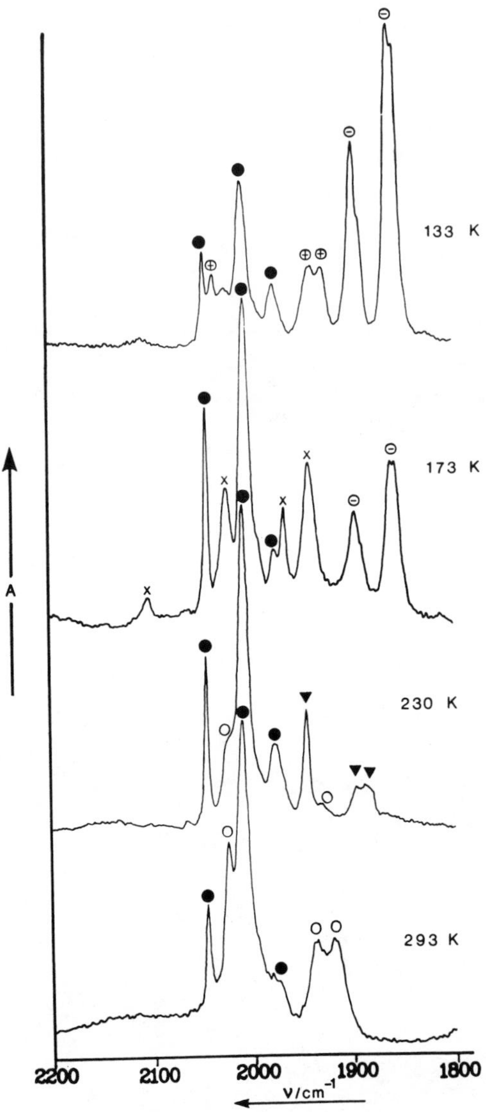

Figure 6. CO stretching modes (IR) of the products formed upon photolysis of $(CO)_5MnMn(CO)_3$(i-Pr-DAB) in 2-Me-THF at different temperatures. ●=$Mn_2(CO)_{10}$; ▼=$Mn_2(CO)_6$ $(i\text{-}Pr\text{-}DAB)_2$; o=$Mn_2(CO)_6$(i-Pr-DAB)$(2\text{-}Me\text{-}THF)_2$; X=$Mn_2(CO)_9$(2-Me-THF); ⊖= $[Mn(CO)_5]^-$; ⊕=$[Mn(CO)_3$(i-Pr-DAB)(2-Me-THF)$]^+$. (Reproduced from Ref. 12. Copyright 1985, American Chemical Society).

$$(CO)_5Mn\text{-}M(CO)_3(\alpha\text{-diimine}) \xrightarrow{h\nu} Mn(CO)_5 + M(CO)_3(\alpha\text{-diimine})$$

$$2\ Mn(CO)_5 \longrightarrow Mn_2(CO)_{10}$$

$$2\ M(CO)_3(\alpha\text{-diimine}) \rightleftharpoons M_2(CO)_6(\alpha\text{-diimine})_2$$

$$M_2(CO)_6(\alpha\text{-diimine})_2 \xrightarrow[\text{2-Me-THF}]{\Delta} M_2(CO)_6(\alpha\text{-diimine})(\text{2-Me-THF})_2 + (\alpha\text{-diimine})$$

Scheme I

This homolytic splitting of the metal-metal bond is observed for all complexes $(CO)_5MnM(CO)_3(\alpha\text{-diimine})$ (M=Mn,Re) at $T > 200K$. When however the photolysis of $(CO)_5MnMn(CO)_3$(i-Pr-DAB) in 2-Me-THF takes place below 200K but above the temperature at which the solvent solidifies to a glass ($T \simeq 130K$), new bands show up in the CO-stretching region at the expense of those belonging to $Mn_2(CO)_{10}$ and $Mn_2(CO)_6$(i-Pr-DAB)$_2$. This effect is strongest when the photolysis takes place at 133K. From the new bands, those at 1884, 1861 and 1857 cm^{-1} ($\ominus$ in Figure 6) belong to $[Mn(CO)_5]^-$ (19). The bands at 2043, 1939 and 1928 cm^{-1} are assigned to the solvated cation $[Mn(CO)_3(\text{i-Pr-DAB})(\text{2-Me-THF})]^+$ and indicated by $\oplus$ in the IR spectrum. This assignment is based on the close similarity between these bands and those of $[Mn(CO)_3(\text{i-Pr-DAB})(THF)]^+[OTF]^-$ ($OTF=CF_3SO_3$), which compound has been prepared separately. In the IR spectrum of the photoproducts obtained at 173K extra bands (indicated with X) are observed, which do not show up upon photolysis at 230K and which are very weak in the 133K spectrum. These bands apparently belong to a complex which is only formed when $Mn_2(CO)_{10}$ is a major photoproduct and which is unstable at higher temperatures. Indeed, these bands disappear when the solution, after photolysis at 173K, is raised in temperature. These bands are therefore assigned to the complex $Mn_2(CO)_9$(2-Me-THF) and this assignment is supported by the close agreement between these frequencies and those of other $Mn_2(CO)_9L$ complexes (20). At higher temperatures 2-Me-THF is substituted by CO from the solution.

The ions formed by photolysis at $T < 200K$ are not stable at higher temperatures. Raising the temperature causes them to react back to the parent compound with loss of 2-Me-THF. When, however, $P(n\text{-}Bu)_3$ is added to a solution of the ions, $[Mn(CO)_3(\text{i-Pr-DAB})(P(n\text{-}Bu)_3]^+$ $[Mn(CO)_5]^-$ is formed, which is a stable compound at room temperature.

From these results one might conclude that photolysis leads to homolytic or heterolytic splitting of the metal-metal bond, depending on the temperature of the solution. If instead of $(CO)_5MnMn(CO)_3$(i-Pr-DAB), $(CO)_5MnMn(CO)_3$(phen) or $(CO)_5MnMn(CO)_3$(bipy) are photolyzed at 133K, it becomes clear that heterolytic splitting of the metal-metal bond is not a primary photoprocess. In that case neither $Mn_2(CO)_{10}$ nor $[Mn(CO)_5]^-$ is formed but instead a photosubstitution product. Figure 8. shows the IR spectral changes upon photolysis of $(CO)_5MnMn(CO)_3$(phen) at 133K. Only a small amount of $[Mn(CO)_5]^-$ is formed. Instead, free CO shows up with $\nu = 2132\ cm^{-1}$. Furthermore, all CO-vibrations shift to lower frequencies, especially those of the metal fragment with the α-diimine ligand. At the same time the MLCT band at 520 nm dissappears and a broad band shows up between 580 and 700 nm. These data point to a photosubstitution of CO of the $Mn(CO)_3$(phen) moiety by 2-Me-THF. The photoproduct $(CO)_5MnMn$-$(CO)_2$(phen)(2-Me-THF) cannot be isolated since it disproportionates upon raising the temperature. A similar disproportionation reaction was observed

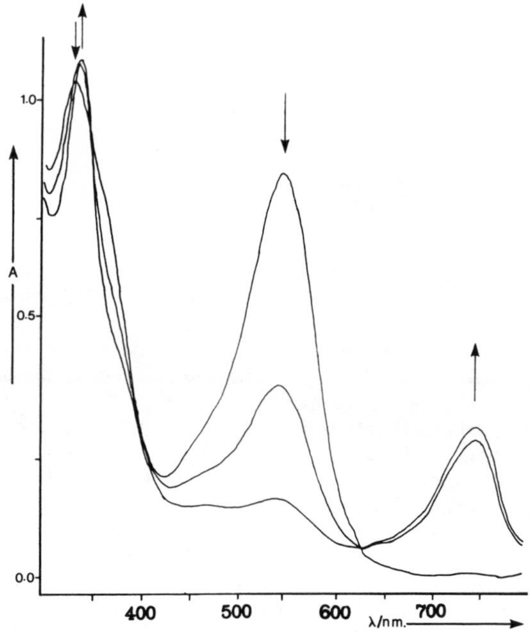

Figure 7. Changes in the electronic absorption spectrum upon photolysis of $(CO)_5MnMn(CO)_3$(i-Pr-DAB) in 2-Me-THF at 230K. (Reproduced from Ref. 12. Copyright 1985, American Chemical Society).

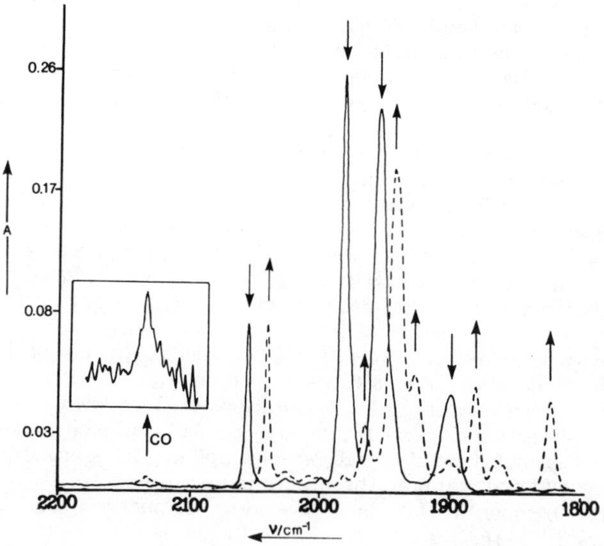

Figure 8. IR spectral changes upon photolysis of $(CO)_5MnMn(CO)_3$-(phen) in 2-Me-THF at 133K. (Reproduced from Ref. 12. Copyright 1985, American Chemical Society).

for the phosphine substituted product formed by the reaction of a phosphine ligand with the 2-Me-THF substituted complex. The corresponding photoproduct $(CO)_5ReMn(CO)_2$(i-Pr-DAB)(PPh_3) did however not disproportionate and could be isolated and identified. As mentioned above, the photosubstitution products of $(CO)_5MnMn(CO)_3$(α-diimine) disproportionate upon raising the temperature. Up to now only the cations $[Mn(CO)_3(\alpha\text{-diimine})(2\text{-Me-THF})]^+$ could de identified, which means that the cation $[Mn(CO)_2(\alpha\text{-diimine})(2\text{-Me-THF})_2]^+$, formed by heterolytic splitting of the metal-metal bond, readily reacts with CO from the solution.

On the basis of these results we propose the mechanism shown in Figure 9 for the photolysis of these complexes at T < 200K. This reaction is observed for all complexes $(CO)_5MMn(CO)_3$(α-diimine). Irradiation into the MLCT band causes photosubstitution of CO of the $Mn(CO)_3$(α-diimine) moiety by 2-Me-THF. The difference in electronegativity between both metal fragments is then so large that raising the temperature causes a heterolytic splitting of the metal-metal bond. The cation formed reacts with CO from the solution. When the temperature is raised to room temperature the ions recombine to the parent compound because 2-Me-THF is then released.

A different photolysis behavior at 133 K is found for the complexes $(CO)_5MnRe(CO)_3$(α-diimine). Irradiation of these complexes causes the dissappearance of the MLCT band. When the temperature is raised to 203 K the MLCT band returns. A similar reversible behavior is observed in the IR spectra. These spectra do not show free CO and the CO- vibrations of the main photoproduct do not agree with those of any photoproduct found so far. Apart from this photoproduct a small amount of ions is observed. The disappearance of the MLCT band cannot be the result of complete loss of the α-diimine ligand since different CO frequencies are found for the photoproducts of the R-DAB and phen complexes. For this reaction we propose the breaking of a metal-nitrogen bond by which a complex $(CO)_5MnRe(CO)_3$(σ-α-diimine)(2-Me-THF) is formed in which the α-diimine ligand is σ-monodentately coordinated to Re. Such σ-monodentately bonded α-diimine ligands occur e.g. in $Cr(CO)_5$(R-DAB)(5), $M(CO)_5$(bipy') (M=Cr,Mo or W) (21) and $MCl_2(PPh_3)$(t-Bu-DAB) (M=Pd or Pt; t-Bu=tertiary-butyl) (22). After breaking of the metal-nitrogen bond, 2-Me-THF coordinates to Re at the open site. The same photoproduct is formed by photolysis of $(CO)_5MnRe(CO)_3$(i-Pr-DAB) in a PVC film at 198K. Since this film is cast from THF, the solvent molecules, still present in the film, will stabilize the photoproduct. When these complexes are irradiated with light of higher energy (λ=350nm), disproportionation takes place just as for the $(CO)_5MnMn(CO)_3$(α-diimine) complexes. These photolysis reactions are shown schematically for $(CO)_5MnRe(CO)_3$(R-DAB) in Figure 10.

The photochemical mechanism

It has been argued that the low-energy absorption band of these $(CO)_5MM'(CO)_3$(α-diimine) complexes has to be assigned to a MLCT transition from a d_π(M') orbital not involved in the metal-metal bond to the lowest π* level of the α-diimine ligand. The electronic transition will therefore be directed to $^1d_\pi\pi^*$ and from this state intersystem crossing may occur to both $^3d_\pi\pi^*$ and $^3\sigma_b\pi^*$. From both states a reaction may occur. The reaction from $^3\sigma_\pi\pi^*$ leads to splitting of the metal-metal bond and our results show that this splitting is homolytic and that it is the main reaction in 2-Me-THF at T > 200K. Just as in the case of the corresponding d^6-complexes $M(CO)_4$(α-diimine) (M=Cr,Mo or W) (6, 10) and d^8-complexes

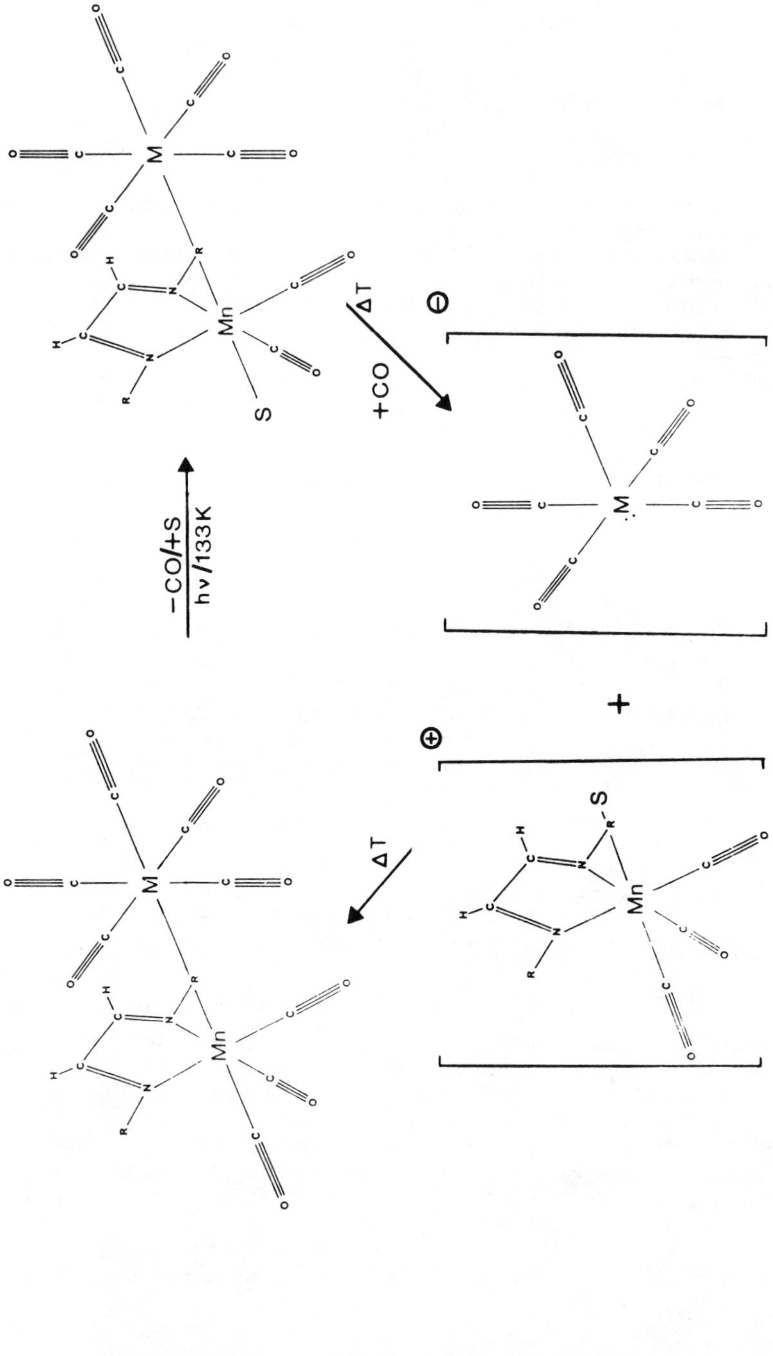

Figure 9. Photolysis of $(CO)_5MMn(CO)_3(R\text{-}DAB)$ in 2-Me-THF(S) at $T < 200K$.

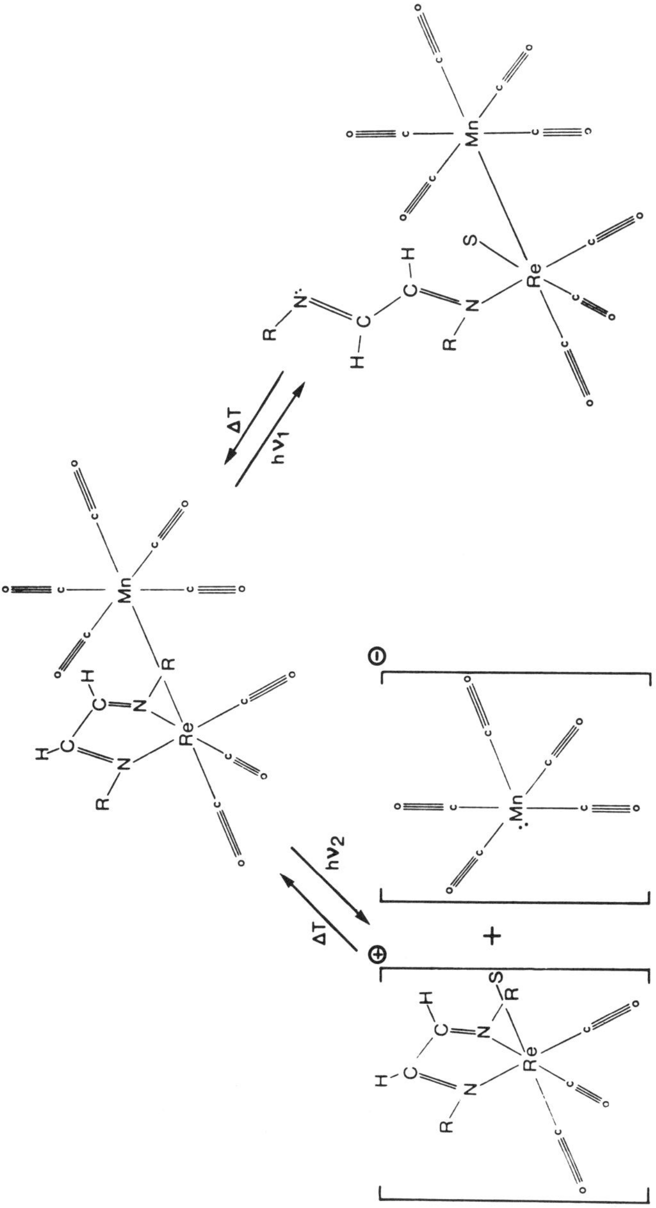

Figure 10. Photolysis of $(CO)_5MnRe(CO)_3(R\text{-}DAB)$ in 2-Me-THF (S) at $T<173K$. $\nu_2>\nu_1$ (Reproduced from Ref. 12. Copyright 1985, American Chemical Society).

$Fe(CO)_3$(α-diimine) (8), a photochemical reaction from $^3d_{\pi}\pi^*$ will lead to release of CO. This is in fact observed for several of these complexes in 2-Me-THF at 133K. It is unlikely that one of these reactions occurs from a ligand field state since the corresponding spin- allowed transitions are found below 400 nm and irradiation takes place at 514.5 nm.

We therefore conclude that the high temperature reaction occurs from the $^3\sigma_b\pi^*$ state and the low temperature one from $^3d_{\pi}\pi^*$. It is not yet clear whether this change of reaction is a mere temperature effect or that the increase of viscosity of the solvent plays an important role here. In the first case we deal with a $^3\sigma_b\pi^*$ state higher in energy than $^3d_{\pi}\pi^*$ and only occupied at higher temperatures. This situation is then similar to that of the complexes $[Ru(NH_3)_5(4R\text{-}Py)]^{3+}$ (23) and $M(CO)_5L$ (L=4R-pyridine, pyridazine; M=Cr or W)(24), in which a ^{3}LF state is close in energy to a ^{3}MLCT state. It is however quite possible that the viscosity of 2-Me-THF is of importance here. This viscosity increases drastically upon cooling. The radicals formed by the homolytic splitting of the metal-metal bond can then not diffuse through the solution and will recombine to the parent compound. As a result the quantum yield for the homolysis decreases and the much slower reaction from the $^3d_{\pi}\pi^*$ state (release of CO) can be observed.

The breaking of a metal-nitrogen bond instead of CO release for the complexes $(CO)_5MnRe(CO)_3$(α-diimine) agrees with the mechanism proposed for the photosubstitution of CO in $Fe(CO)_3$(α-diimine) (8). The primary photoprocess of this reaction was proposed to be breaking of a metal-nitrogen bond with formation of an intermediate in which the α-diimine ligand is σ-monodentately bonded to Fe. A nucleophilic ligand then attacks the open site, CO is released and the σ,σ-coordination of the α-diimine ligand is restored. The same mechanism is proposed for the photosubstitution of CO by 2-Me-THF in the complexes $(CO)_5MnMn(CO)_3$(α-diimine). Apparently, this reaction stops after the primary photoprocess in the case of the $(CO)_5MnRe(CO)_3$(α-diimine) complexes because the Re-CO bond is too strong.

Photochemical reactions with PR_3

Photolysis in THF or 2-Me-THF at 293K in the absence of PR_3 leads to the formation of $Mn_2(CO)_{10}$ for all complexes $(CO)_5MnM(CO)_3$(α-diimine). This means that the primary photoprocess is a homolytic splitting of the metal-metal bond. In the presence of PR_3 different reactions are observed depending on the complex and on the PR_3 ligand. Thus, upon photolysis of $(CO)_5MnMn(CO)_3$(α-diimine) in the presence of PPh_3, homolytic splitting of the metal-metal bond occurs and $Mn_2(CO)_8(PPh_3)_2$ is formed, provided the complex is irradiated by visible light with a low photonflux (e.g. medium pressure Hg-lamp, 500 nm filter). If instead an argon ion laser is used with a much higher photonflux (e.g. λ = 514.5 nm, p=20 mW) both $Mn_2(CO)_8(PPh_3)_2$ and $Mn_2(CO)_{10}$ are formed. Such a photonflux dependence has e.g. been observed by Stiegman and Tyler (25, 26). Upon irradiation with low intensity light the concentration of $Mn(CO)_5$ radicals formed is small compared with that of PPh_3. The reaction with PPh_3 is then favored with respect to the formation of $Mn_2(CO)_{10}$. At a higher photonflux, the concentration of $Mn(CO)_5$ radicals is much larger and this promotes the formation of $Mn_2(CO)_{10}$. When the photolysis of $(CO)_5MnMn(CO)_3$(α-diimine) takes place in the presence of $P(n\text{-}Bu)_3$, neither $Mn_2(CO)_{10}$ nor $Mn_2(CO)_8(P(n\text{-}Bu)_3)_2$ is formed. Instead, the ions $[Mn(CO)_3(\alpha\text{-diimine})(P(n\text{-}Bu)_3]^+$ and $[Mn(CO)_5]^-$ are formed. This cation has been identified with ^{1}Hand ^{31}P-NMR and with IR after performing this photochemical reaction on a preparative scale. Moreover, the CO-frequencies of this cation closely resemble those of the cation in $[Mn(CO)_3(bipy')(THF)]^+[OTF]^-$ (OTF=-CF_3SO_3), which compound was prepared by reaction of $Mn(CO)_3(bipy')Br$ with Ag(OTF).

A remarkable property of this reaction is its very high quantum yield ($\Phi \simeq 10$). Such a high quantum yield points to a chain reaction. In analogy with the mechanisms of the photodisproportionation of $Mn_2(CO)_{10}$ in N-donor solvents (27) and of $(\eta^5\text{-}CH_3C_5H_4)_2Mo_2(CO)_6$ in the presence of phosphine (29), the following chain mechanism is proposed for the disproportionation of $(CO)_5MnMn(CO)_3$(α-diimine) in the presence of $P(n\text{-}Bu)_3$:

Reaction

$$(CO)_5MnMn(CO)_3L + P \xrightarrow[\Phi \simeq 10]{h\nu} [Mn(CO)_3LP]^+ + [Mn(CO)_5]^-$$

Initiation

$$(CO)_5MnMn(CO)_3L \xrightarrow{h\nu} Mn(CO)_5 + Mn(CO)_3L$$

Propagation

$$Mn(CO)_3L + P \longrightarrow Mn(CO)_3LP$$
$$Mn(CO)_3LP + (CO)_5MnMn(CO)_3L \longrightarrow [Mn(CO)_3LP]^+ + [(CO)_5MnMn(CO)_3L]^-$$
$$[(CO)_5MnMn(CO)_3L]^- \longrightarrow [Mn(CO)_5]^- + Mn(CO)_3L$$

Termination

$$Mn(CO)_3LP + Mn(CO)_5 \longrightarrow [Mn(CO)_3LP]^+ + [Mn(CO)_5]^-$$

Scheme 2

The initation reaction is homolysis of the metal-metal bond just as in the absence of $P(n\text{-}Bu)_3$. The propagation steps start from the radical $Mn(CO)_3$(α-diimine). The ESR spectra show that the unpaired electron is localized at the α-diimine ligand, which makes the radical a 16-electron species, contrary to the 17-electron radical $Mn(CO)_5$. This intermediate takes up $P(n\text{-}Bu)_3$ forming the 18-electron species $Mn(CO)_3$(α-diimine)-($P(n\text{-}Bu)_3$), which is the key factor in the disproportionation reaction. Such intermediates are assumed to play an important role in the photodisproportionation of $Mn_2(CO)_{10}$ (27) and $(\eta^5\text{-}CH_3\text{-}C_5H_4)_2Mo_2(CO)_6$ (28,29). Thus, Stiegman and Tyler proposed that the photodisproportionation of $Mn_2(CO)_{10}$ in N-donor solvents takes place via electron transfer from the 19-electron intermediate $Mn(CO)_3N_3$ to $Mn_2(CO)_{10}$ (27). Electron transfer from $Mn(CO)_3$(α-diimine)($P(n\text{-}Bu)_3$) can take place to both $Mn(CO)_5$ and to the parent compound. The first reaction is a terminating step, the latter one leads to the formation of the unstable anion $[(CO)_5MnMn(CO)_3$(α-diimine$]^-$ which decomposes into $[Mn(CO)_5]^-$ and $Mn(CO)_3$(α-diimine). The latter radical is responsible for the chain reaction. The ions $[Mn(CO)_3$(α-diimine)($P(n\text{-}Bu)_3$)$]^+$ and $[Mn(CO)_5]^-$ are the only products observed and therefore no other terminating reactions are assumed to occur. The factors influencing the quantum yield of this reaction and the chemical properties of the highly reducing radical $Mn(CO)_3$(α-diimine)($P(n\text{-}Bu)_3$) are subject to further study.

Photolysis in rigid media

If the photolysis takes place in an inert gas matrix, both the homolytic splitting of the metal-metal bond and the breaking of a metal-nitrogen bond will be followed by a fast backreaction to the parent compound. The radicals formed by homolysis of the metal-metal bond can not diffuse from the matrix site and will recombine to the parent compound. Moreover, the photoproduct obtained by breaking of a metal nitrogen bond, will not be stabilized by a coordinating solvent molecule and therefore react back to the parent compound. Because of this the photochemistry of some of these complexes has also been studied in a CH_4-matrix at 10K and for comparison in a PVC film, which is a less rigid medium than the matrix especially at room temperature.

When the complex $(CO)_5MnRe(CO)_3$(i-Pr-DAB) is photolyzed in a CH_4-matrix five new CO bands show up with simultaneous loss of CO. The new bands do not belong to any of the photoproducts observed so far. The IR spectral changes, shown in Figure 11, exhibit several isosbestic points indicating a well-defined clean reaction. The MLCT band disappears and no new band shows up in the visible region. Annealing a CO-matrix after the photolysis did not cause a reaction which means that the photoproduct is coordinatively satured. Drastic changes also occur in the 1200-1500 cm^{-1} region of the IR spectrum. The band at 1479 cm^{-1} belonging to ν_s (CN) of the coordinated i-Pr-DAB ligand and the 1294 cm^{-1} band belonging to ν(CC) disappear while two new bands show up at 1389 and 1304 cm^{-1}, respectively. These bands are assigned to ν_s (CN) and ν(CC), respectively, of the i-Pr-DAB ligand in the photoproduct. This low frequency of ν_s (CN) points to a large involvement of the π^*level of the i-Pr-DAB ligand in the bonding since this orbital is anti-bonding between C and N. A similar low frequency for ν_s(CN) has been found for the photoproduct of $Fe(CO)_3$ (R-DAB), in which the R-DAB ligand is η^2,η^2 coordinated to Fe(8)

On the basis of these results we propose that the complex $(CO)_3Mn$(i-Pr-DAB)$Re(CO)_3$ (30) is formed (Figure 12) in which the i-Pr-DAB ligand is σ,σ-coordinated to Re and η^2,η^2 to Mn. This proposal is strongly supported by the observations of Adams (31) and Keijsper (32). Adams accidentally synthesized the complex $(CO)_3Mn$(Me-DAB(CH_3,CH_3)) $Mn(CO)_3$ and established the structure by X-ray diffraction. Keijsper synthesized in low yield an analogous complex $(CO)_3Mn$(t-Bu-DAB)$Mn(CO)_3$ by treating $Mn(CO)_3$(t-Bu-DAB)Br with $[CpFe(CO)_2]^-$. The CO-stretching frequencies of these complexes agree very well with those of our photoproduct.

Photolysis of the other $(CO)_5MM'(CO)_3$(i-Pr-DAB) (M,M'=Mn,Re) complexes in a CH_4-matrix did not lead to the formation of $(CO)_3M$(i-Pr-DAB)-$M'(CO)_3$. If these complexes are however photolyzed in a PVC film, cast from THF, three of the four complexes $(CO)_5MM'(CO)_3$(i-Pr-DAB)-(M,M'=Mn,Re except M=M'=Re) can be converted into $(CO)_3M$(i-Pr-DAB)$M'(CO)_3$. Figure 13 shows the IR spectral changes accompanying the photolysis of $(CO)_5MnMn(CO)_3$(i-Pr-DAB) in a PVC film at 293K. At this temperature no free CO is observed due to the broadness of the IR band. Now, when this complex is photolyzed in the film at 193K, five extra CO bands show up which disappear upon raising the temperature with formation of the parent compound. Apparently, a THF substituted complex is formed. This substitution does not take place at the $Mn(CO)_3$(i-Pr-DAB) moiety since the CO-vibrations do not agree with those of the 2-Me-THF substituted complex formed in 2-Me-THF at 133K. The bands are assigned to (THF) $(CO)_4MnMn(CO)_3$ (i-Pr-DAB). Raising the temperature will lead to backsubstitution of THF by CO, still present in the film.

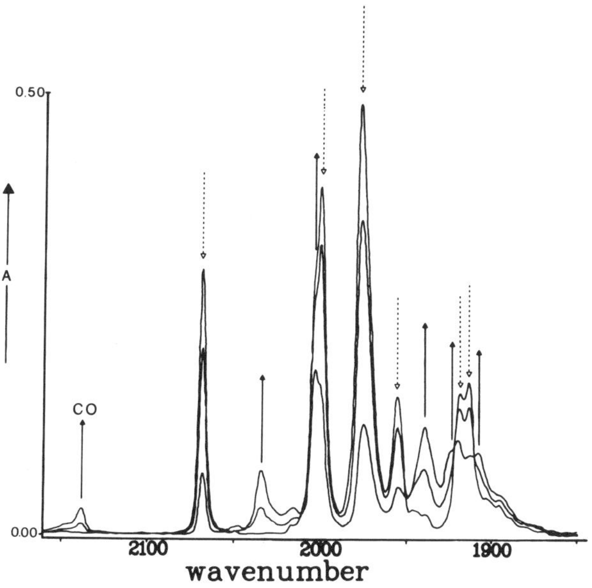

Figure 11. IR spectral changes in the CO-stretching region upon photolysis of $(CO)_5MnRe(CO)_3$(i-Pr-DAB) in a CH_4-matrix at 10K.

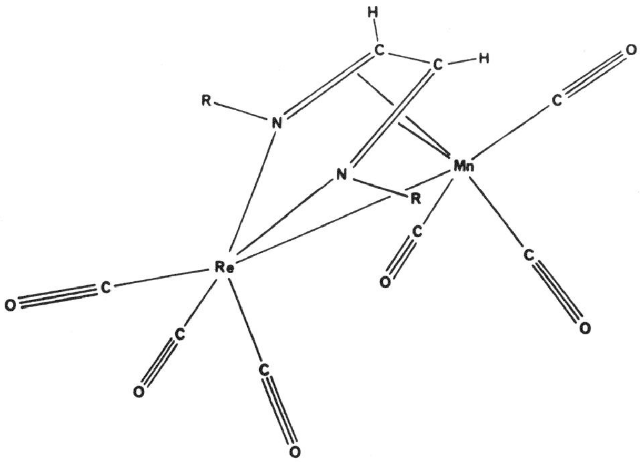

Figure 12. Structure proposed for the photoproduct of $(CO)_5$Mn-$Re(CO)_3$(R-DAB) in a CH_4- matrix at 10K.

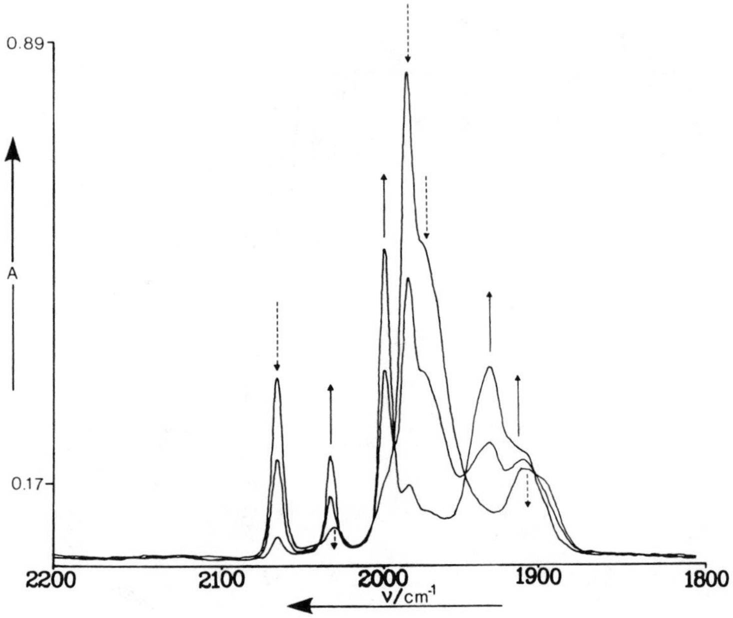

Figure 13. IR spectral changes in the CO-stretching region upon photolysis of $(CO)_5MnMn(CO)_3$(i-Pr-DAB) in a PVC film at 293K. (Reproduced from Ref. 13. Copyright 1985, American Chemical Society).

The formation of these two photoproducts can be explained with a homolytic splitting of the metal-metal bond. Most radicals will react back to the parent compound. Some of the $M(CO)_5$ radicals will however react, presumably associatively (33,34), with the R-DAB ligand of the $M'(CO)_3$ (R-DAB) radical or with THF in the PVC film with formation of these photoproducts and release of CO.

In 2-Me-THF we observed, apart from the homolytic splitting of the metal-metal bond, breaking of a metal-nitrogen bond for $(CO)_5MnRe$-$(CO)_3$(α-diimine), followed by release of CO in the case of $(CO)_5MMn$-$(CO)_3$(α-diimine) (M=Mn,Re), at 133K. This reaction is not observed for any of these complexes in the matrix. However, contrary to its behavior in the matrix, $(CO)_5MnRe(CO)_3$(i-Pr-DAB) shows this reaction in the film at 193K upon irradiation into the MLCT band (upon irradiation with u.v. light $(CO)_3$Mn(i-Pr-DAB)$Re(CO)_3$ is formed). The MLCT band disappears and no CO is released. Raising the temperature causes a backreaction to the parent compound just as in 2-Me-THF. The different behavior of this complex in the PVC film and the CH_4-matrix is due to the higher rigidity of the matrix and to the presence of stabilizing THF in the film.

The intriguing photochemistry of these complexes in relationship to their excited state properties certainly deserves more attention.Subject to further study are also the stability and electron distribution of the radicals $M'(CO)_3$(α-diimine) and the identification and chemical properties of the highly reducing species $M'(CO)_3$(α-diimine)(PR_3).

Literature Cited

1. R.W. Balk, D.J. Stufkens and A. Oskam, Inorg. Chim. Acta 1978, 28, 133.
2. R.W. Balk, D.J. Stufkens and A. Oskam, Inorg. Chim. Acta 1979, 34, 267.
3. R.W. Balk, D.J. Stufkens and A. Oskam, J. Chem. Soc., Dalton Trans. 1982, 275.
4. R.W. Balk, D.J. Stufkens and A. Oskam, J. Chem. Soc., Dalton Trans. 1981, 1124.
5. L.H. Staal, D.J. Stufkens and A. Oskam, Inorg. Chim. Acta 1978, 26, 255.
6. R.W. Balk, D.J. Stufkens and A. Oskam, Inorg. Chem., 1980, 19, 3015.
7. M.W. Kokkes, D.J. Stufkens and A. Oskam, J. Chem. Soc., Dalton Trans. 1983, 439.
8. M.W. Kokkes, D.J. Stufkens and A. Oskam, J. Chem. Soc., Dalton Trans. 1984, 1005.
9. P.C. Servaas, H.K. van Dijk, T.L. Snoeck, D.J. Stufkens and A. Oskam, Inorg. Chem., in press.
10. H.K. van Dijk, P.C. Servaas, D.J. Stufkens and A. Oskam, Inorg. Chim. Acta, in press.
11. D.L. Morse and M.S. Wrighton, J. Am. Chem. Soc. 1976, 98, 3931.
12. M.W. Kokkes, D.J. Stufkens and A. Oskam, Inorg. Chem., in press.
13. M.W. Kokkes, D.J. Stufkens and A. Oskam, Inorg. Chem., in press.
14. M.W. Kokkes, W.G.J. de Lange, D.J. Stufkens and A. Oskam, J. Organomet. Chem., in press.
15. M.W. Kokkes, T.L. Snoeck, D.J. Stufkens, A. Oskam, M. Cristophersen and C.H. Stam, J. Mol Struct., in press.

16. R.R. Andréa, D.J. Stufkens and A. Oskam, J. Organomet. Chem. 1985,290,63.
17. R.R. Andréa, A. Terpstra, D.J. Stufkens and A. Oskam, Inorg. Chim. Acta 1985, 96, L57.
18. L.H. Staal, G. van Koten and K. Vrieze, J. Organomet. Chem. 1979, 175, 73.
19. N. Flitcroft, D.K. Huggins and H.D. Kaesz, Inorg. Chem. 1964, 3, 1123.
20. M.L. Ziegler, H. Haas and R.K. Sheline, Chem. Ber. 1965, 98, 2454.
21. R.J. Kazlauskas and M.S. Wrighton, J. Am. Chem. Soc. 1982, 104, 5748.
22. H. van der Poel, G. van Koten and G.C. van Stein, J. Chem. Soc., Dalton Trans. 1981, 2164.
23. P.C. Ford in "Progress in Inorganic Chemistry"; Lippard, S.J., Ed.; John Wiley: New York, 1983; vol. 30, pp. 213-271.
24. A.J. Lees and A.W. Adamson, J. Am. Chem. Soc. 1982, 104, 3804.
25. D.R. Tyler, J. Photochem. 1982, 20, 101.
26. A.E. Stiegman and D.R. Tyler,J. Photochem. 1984, 24, 311.
27. A.E. Stiegman and D.R. Tyler, Inorg. Chem. 1984, 23, 527.
28. A.S. Goldmann and D.R. Tyler, J. Am. Chem. Soc. 1984, 106, 4067.
29. A.E. Stiegman, M. Stieglitz and D.R. Tyler, J. Am. Chem. Soc. 1983, 105, 6032.
30. This notation is used to indicate that the i-Pr-DAB ligand is σ-N, σ-N coordinated to Re an η^2-CN, η^2-CN to Mn.
31. R.D. Adams, J. Am. Chem. Soc. 1980, 102, 7476.
32. J. Keijsper, G. van Koten, K. Vrieze, M. Zoutberg and C.H. Stam, Organometallics 1985, 4, 1306.
33. H. Yeasaka, T. Kobayashi, H. Yasufuko and S. Nagakuru, J. Am. Chem. Soc. 1983, 105, 6249.
34. Q.Z. Shi, T. Richmond, W.C. Trogler and F. Basolo, J. Am. Chem. Soc. 1984, 106, 71.

RECEIVED November 8, 1985

7

Manipulation of Doublet Excited State Lifetimes in Chromium(III) Complexes

John F. Endicott, Ronald B. Lessard, Yabin Lei, Chong Kul Ryu, and R. Tamilarasan

Department of Chemistry, Wayne State University, Detroit, MI 48202

The relaxation rates observed for the lowest energy doublet excited states (^2E) of chromium(III) complexes can be represented, $k_{re} = k^o_{re} + k_{re}(T)$. The limiting low temperature excited state lifetime, $\tau = (k^o_{re})^{-1}$, is a molecular property which is nearly independent of temperature and the condensed phase environment, but τ^o does decrease with such molecular properties as the number of N-H vibrational modes available to function as acceptors for the electronic excitation energy. The thermally activated decay, $k_{re}(T)$, is a function of the solvent and the coordinated ligands. When $k_{re}(T)$ is fitted to an Arrhenius function, values of ln A vary more than do the apparent activation energies, and variations in $k_{re}(298)$ are more often determined by the pre-exponential factor than by E_a. The transition between τ^o and thermally activated relaxation occurs at a temperature, T_{tr}, which is a strong function of the medium and of the coordinated ligands. The observed room temperature ^2E lifetimes are more strongly correlated with T_{tr} than with the apparent Arrhenius activation energy. It is suggested that the thermally activated relaxation channel(s) involves a strongly coupled, but spin forbidden surface crossing to the potential energy surface of a reaction intermediate. However, some preliminary observations suggest that there may be more than one possible decay channel.

Vibrationally equilibrated electronic excited states in molecules are unique chemical species. These are metastable species with unusual electronic configurations, and on this basis alone they might be expected to exhibit unusual patterns of reactivity. However, it is probably a more striking feature

0097–6156/86/0307–0085$06.00/0

of these electronic excited states that they have stored a considerable amount of energy in a predominately electronic form, and that if this electronic energy were converted to other energy forms, it would exceed the energy requirements for many simple chemical processes. The conversion of electronic excitation energy into a more useful energy form (e.g., vibrational, electrical, etc.) can be very selective. The principles governing this selectivity are not always well understood.

The lowest energy excited electronic state of the chromium(III) complexes considered here is designated the ^2E state (for convenience, even in low symmetry complexes). This excited state differs from the ground state, $^4A_{2g}$ (in O_h symmetry, see Figure 1), in spin multiplicity, but not in orbital population. As a consequence the ^2E and $^4A_2$ states have nearly identical molecular geometries, and the difference in energy content of these states is entirely electronic (a small entropy correction must be made when considering reaction driving forces). The typical excited state-ground state energy difference is about 14×10^3 cm^{-1} and it exceeds the energy requirements for simple substitution, isomerization, etc., reactions of the ground state. Thus it is not surprising that (^2E)Cr(III) species are often very unstable with respect to ligand replacement or isomerization reactions (**1-3**). However, describing such chemical processes is not simple since the ^2E and $^4A_2$ electronic state potential energy surfaces must be very similar in shape (**1-4**), at least near their equilibrium nuclear configurations. Furthermore, many of these reactions have been found to be stereoselective and to occur in competition with phosphorescence emission and non-radiative relaxation of the excited state (**1-4**).

In this report we describe some of the studies which have been initiated to investigate the factors contributing to the behavior of the lowest doublet excited state in chromium(III). The principle goal of our work has been to explore those steric constraints, introduced by the ligands, which greatly alter the photophysical properties of the ^2E excited state. In pursuit of this goal, we have re-investigated some features of well known amine and polypyridyl complexes in order to obtain internally consistent reference systems.

In principle one must determine or take into account a variety of factors. Among these are:

1. the energy differences, ΔE^*, between the lowest energy electronically excited quartet and doublet states;
2. the energy and role, if any, of higher energy doublet excited states (especially components of the $^2T_1$, state);
3. distortions of the ^2E excited state potential energy surface;

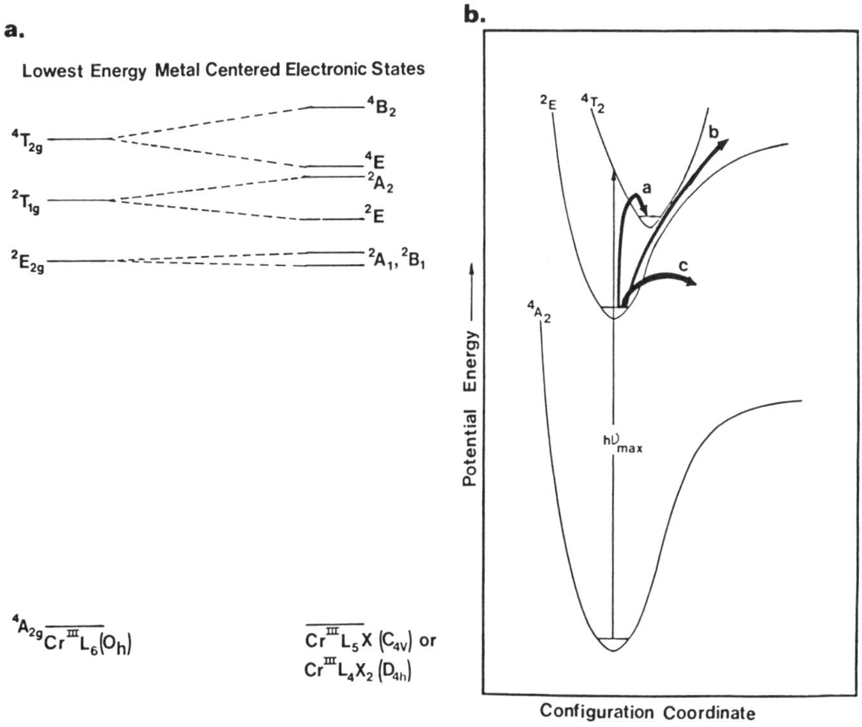

Figure 1. State energies (a) and qualitative potential energy surfaces (b) for $Cr(NH_3)_6^{3+}$. Alternative mechanistic proposals for $(^2E)Cr(III)$ decay are illustrated in 1(b): **a**, back intersystem crossing; **b**, direct reaction to yield electronically correlated products; **c**, surface crossing to some ground state intermediate potential energy surface.

4. the significance of variations in the chemical and electronic properties of Cr(III) induced by variations in the ligands;
5. the nature and chemical behavior of any photogenerated reaction intermediates.

In practice, and despite much study of Cr(III) systems, there is only a little definitive information about any of these factors. In this report we discuss some current studies of chromium complexes with sterically constrained ligands. These studies are providing information mostly about the fourth factor listed above.

Temperature Dependence of (^2E)Cr(III) Lifetimes

The lifetimes $\tau(^2E)$, of the ^2E electronic excited state of Cr(III) complexes is a strong function of temperature. The detailed temperature dependence differs from complex to complex and from solvent to solvent. In general, $\tau(^2E)$ is strongly temperature dependent at high temperatures and temperature independent at low temperatures (**3**, **5**, **6**). Thus,

$$[\tau(^2E)]^{-1} = k_{re} = k^o_{re} + k_{re}(T) \qquad (1)$$

where $k^o_{re} = (\tau^o)^{-1}$ is the limiting low temperature rate constant for excited state relaxation and the rate of radiative relaxation is assumed to be small. In many instances the temperature dependent term, $k_{re}(T)$, can be represented as a simple Arrhenius temperature dependence (**6**) but more often a more complex function is required to fully accommodate the curvature (**3**, **6**, **7**, **8**).

(^2E)Cr(III) Lifetimes in the Low Temperature Regime. As the temperature decreases, k_{re} approaches a limiting value, k^o_{re} which is usually independent of the condensed phase medium. The low temperature lifetimes of $Cr^{III}N_6$ complexes and cyano-amine-Cr(III) complexes are usually found to be nearly temperature independent over an appreciable temperature range. Values of k^o_{re} can therefore be regarded as functions of structural features of the chromium complexes. Since the ^2E and $^4A_2$ potential energy surfaces are nested, excited state relaxation in this regime is attributed to limiting weak coupling (**9**) between the excited and ground electronic states. In such a limit, the non-radiative transition between surfaces depends on nuclear and electronic tunneling. The more important molecular parameters which are expected (**9**, **10**) to contribute to relaxation rates in this regime are: (1) the energy difference, ΔE^o, between the excited and ground electronic states; (2) the number of high frequency vibrational modes available to dissipate this energy; (3) the magnitude of the difference between the excited state and the ground state nuclear coordinates for these modes; (4) the frequencies of the nuclear promoting modes; and, (5) spin-orbit coupling. Of these factors, the contributions of the high frequency acceptor modes have been best documented, (**3**, **11-13**); e.g., in $Cr(NH_3)_6^{3+}$ and $Cr^{III}(NH_3)_5X$ complexes, perdeuteration increases τ^o by nearly two orders of magnitude, more or less consistent with expectation. No simple dependence on ΔE^o has emerged, but the compounds investigated span a range of only, at most 2000 cm^{-1}, in $E(^2E)$, and the energies of their promoting modes probably vary over a comparable range. Selected observations are summarized in Table I.

Thermally Activated Relaxation Rates

Solvent Mediation of Excited State Lifetimes: Examples from the Behavior of Polypyridyl Complexes. Most Cr(III) complexes exhibit a strong solvent dependence of their ^2E excited state lifetimes. The most extreme examples of this behavior are probably found among the polypyridyl complexes. For example, $\tau(^2E)$ is about 50 ns for $Cr(tpy)_2^{3+}$ in water at 25°C (**14**), but increases to 20 μs in the solid state (perchlorate salt; 25°C) (**15**). This kind of behavior has been extensively investigated, especially by Kemp and co-workers (**6**), for $Cr(phen)_3^{3+}$. The effect of solvent on $\tau(^2E)$ for this complex appears to be manifested by compensation between temperature dependent ($\Delta H^\ddagger$) and temperature independent components, ($\Delta S^\ddagger$): Values of $\tau(^2E)$ have usually been found to be solvent dependent in the thermally activated regime, and to varying degrees these solvent dependencies are manifested by variations in both the Arrhenius activation energies and pre-exponential factors (**3**, **6**). Thus, in the thermally activated regime the lifetime of the lowest doublet excited state cannot be rigorously regarded as an intrinsic molecular property. Rather, values of $\tau(^2E)$ are determined by some interaction between the molecular excited state and its environment.

Photophysical Properties of Simple Ammine (Amine) Complexes. Many systems have been studied; we will only consider a representative few. A more extensive recent review of the literature can be found elsewhere (**3**).

In many ways the photophysical behavior of $(^2E_g)Cr(NH_3)_6^{3+}$ is paradigmatic of ammine and amine chromium(III) systems. Near ambient conditions in fluid solution, this electronically excited complex has the following properties: (1) a highly structured emission (Figure 2) exhibiting an intense 0-0 line and resolved vibronic components; this is strong evidence for similar equilibrium nuclear configurations of the excited and ground electronic states (**3**, **16**); (2) a strongly temperature dependent excited state lifetime (Figure 3); $\tau(^2E_g) \sim 2$ μs at

Table I. Spectroscopic Parameters of Some Chromium(III) Complexes

Complex	$E_{oo}(^2E)^b$ $(cm^{-1}/10^3)$	$E_{max}(^4T$ or $^4E)$ $(cm^{-1}/10^3)$	$\tau^o(N\text{-}H)^{c,d}$ (μs)	$\tau^o(N\text{-}D)^{c,e}$ (μs)
$Cr(NH_3)_6^{3+}$	15.2 (**16**)	21.64 (**11**)	75 ± 5 (**15, 23, 24**)	3500 (**15, 25**)
$Cr(NH_3)_5Cl^{2+}$	14.8 (**26**)	19.25 (**27**)	41 (**28, 29**)	
$Cr(NH_3)_5CN^{2+}$	14.7 (**30, 31**)	22.2 (**31**)	100 (**32**)	9662 (**32**)
$Cr(en)_3^{3+}$	15.0 (**33**)	21.88 (**34, 35**)	108 ± 12 (**15, 23, 24**)	2560 (**15**)
$Cr(sen)^{3+}$	14.8 (**17**)	22.2 (**36**)	120 (**17**)	
$Cr(phen)_3^{3+}$	13.7 (**37**)	~23.9 (**34, 35**)	5000 (**37**)	
$Cr(tpy)_2^{3+}$	13.0 (**38**)	~21.1 (**38**)	540 (**39**)	
$Cr(phen)_2(NH_3)_2^{3+}$	14.2 (**40**)	23.6 (**41**)	196 (**40**)	
trans-$Cr(L_1)(CN)_2^+$	14.0 (**32**)	24.16 (**42**)	370 ± 20 (**8, 20**)	3660 (**32**)
trans-$Cr(L_1)(NH_3)_2^+$	~15[f] (**18**)	22.5 (**18**)	175 ± 5 (**18**)	3690 ± 30 (**18**)
cis-$Cr(L_1)(NH_3)_2^+$	~15[f] (**18**)	21.4 (**18**)	116 (**18**)	1580 (**18**)
trans-$Cr(L_1)Cl_2^+$	14.4 (**17**)	17.6 (**43**)	88 (**17**)	
cis-$Cr(L_1)Cl_2^+$	14.2 (**17**)	18.9 (**43**)	52 (**17**)	
trans-$Cr(L_1)(NCS)_2^+$	14.1 (**17**)	20.7 (**44**)	>80 (**17**)	
cis-$Cr(L_1)en^{3+}$	14.8 (**11**)	21.5 (**11**)	136 (**11**)	
trans-$Cr(L_2)(CN)_2^+$	14.3 (**32**)	23.5 (**32**)	379 ± 50 (**32**)	5600 (**32**)

trans-$Cr(L_2)Cl_2^+$	14.4 (**17**)	17.4 (**45**)	94 (**17**)	
trans-$Cr(L_2)(NCS)_2^+$	14.1 (**17**)	20.0 (**45**)	>108 (**17**)	
cis-$Cr(L_3)(CN)_2^+$	13.8 (**32**)	21.6 (**32**)	208 (**32**)	1854 (**32**)
cis-$Cr(L_3)Cl_2^+$		16.7 (**45**)	94 (**17**)	
cis-$Cr(L_3)(NCS)_2^+$	13.8 (**17**)	19.0 (**45**)	56 (**17**)	
cis-$Cr(L_3)en^{3+}$	14.7 (**40**)	20.1 (**40**)	91 (**40**)	
$Cr(L_4)_2^{3+}$	14.7 (**12**)	22.8 (**12**)	370 ± 30 (**12**)	3200 (**12**)
$Cr(L_5)$	14.1 (**40**)	19.5 (**21**)	430 (**40**)	
$Cr(L_6)^{3+}$	13.8 (**40**)	20.8 (**46**)	50 (**40**)	

NOTES FOR TABLE I:

Abbreviations: en = ethylenediamine; sen = 4,4',4"-Ethylidynetris(3-azabutan-1-amine)
phen = 1,10-phenanthraline; tpy = terpyridine;
L_1 = 1,4,8,11-Tetraazacyclotetradecane(cyclam);
L_2 = 5,12-*meso*-5,7,7,12,14,14-hexamethyl-1,4,8,11-tetraazacyclotetradecane (teta);
L_3 = 5,12-rac-5,7,7,12,14,14-hexamethyl-1,4,8,11-tetraazacyclotetradecane (tetb);
L_4 = 1,4,7-triazacyclononane;
L_5 = 1,4,7-tris(acetato)-1,4,7-triazacyclononane (TCTA);
L_6 = 1,2-bis(1,4,7-triazacyclononane)-ethane (BCNE)

[a] Data from (**3**, **6**, **8**, **11**, **12**, **15–18**).
[b] Electronic origin of the lowest energy doublet state, except as indicated.
[c] Low temperature limiting lifetime (in a rigid glass matrix).
[d] Coordinated amines (ammines) in the proteo form.
[e] Coordinated amines (ammines) perdeuterated.
[f] Unresolved emission spectrum (**20**).

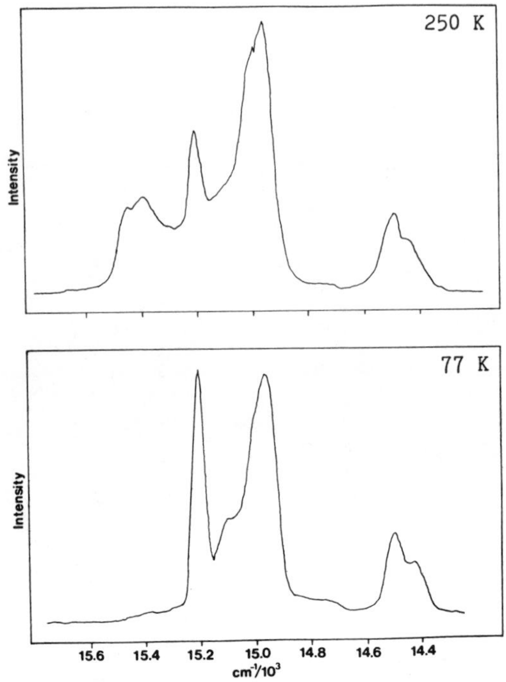

Figure 2. Emission spectra for $Cr(NH_3)_6(ClO_4)_3$ at 250 K and 77K.

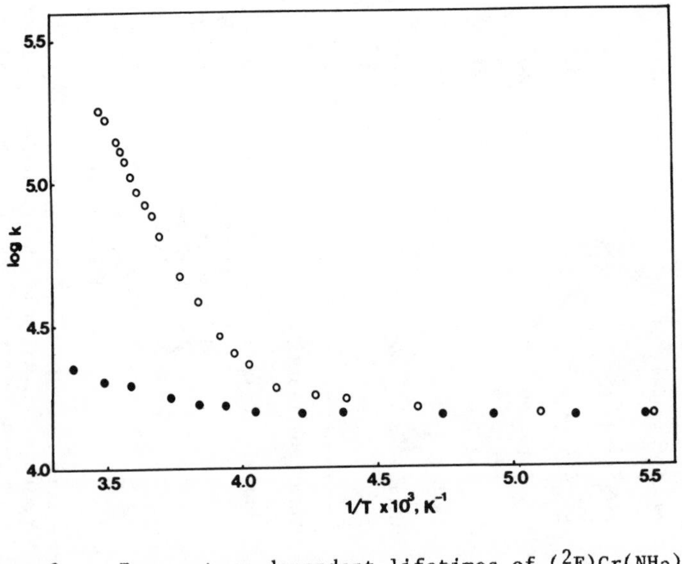

Figure 3. Temperature-dependent lifetimes of $(^2E)Cr(NH_3)_6^{3+}$ in $DMF:CHCl_3$ solutions (upper curve) and in $Ru(NH_3)_6^{3+}$ crystals (Ru:Cr = 30:1); from (**8**).

25°C and $E_a \sim 3.8 \times 10^3$ cm^{-1} in DMF (**8**); (3) a large probability for substitution, $\eta_D \sim 0.5$, and the apparent rate constant for substitution is at least 10^{11} times larger for the ^2E_g excited state than for the ground state (**2**); (4) $\tau(^2E_g)$ becomes nearly temperature independent for T < 250 K and approaches a matrix independent low temperature limiting value, $\tau^o(^2E_g) = 75 \pm 5$ μs; (5) the excited state lifetime is increased by perdeuteration even in the strongly temperature dependent regime; (6) $\tau(^2E_g)$ is dependent on the condensed phase environment; this is most dramatically illustrated by the contrast between $(^2E)Cr(NH_3)_6^{3+}$ in DMF/$CHCl_3$ solution and doped into the $Ru(NH_3)_6Cl_3$ solid (Figure 3).

The thermally activated behavior of $(^2E_g)Cr(NH_3)_6^{3+}$, (**15**), $(^2E)Cr(NH_3)_5CN^{2+}$ (**17**), and $Cr(NH_3)_5Cl^{2+}$ (**17**) provide some instructive contrasts. The 77 K lifetime of the hexammine is bracketed by those of the pentammines (75, 100 and 42 μs, respectively). The room temperature lifetimes are in the same order, but span many orders of magnitude (in H_2O *ca*: 2, 14 and $<10^{-3}$ μs, respectively); the apparent Arrhenius activation energies are slightly smaller for the hexammine than for these pentammines ($[3.8, 4.5$ and $4.4] \times 10^3$ cm^{-1}, respectively). The variation in room temperature lifetimes can be correlated with a systematic variation in the temperatures, T_{tr}, for the transition between the temperature independent and the temperature dependent lifetime behavior (*ca* 250, 287, and 170 K, respectively).

The relatively large values of $k_{re}(T)$ and E_a found for $(^2E)Cr(NH_3)_5Cl^{2+}$ and trans-$Cr([14]aneN_4)Cl_2^+$ in a DMSO-H_2O glass result in exceptionally large pre-exponential factors (*ca* 10^{21} s^{-1}). There are scattered reports of other large pre-exponential factors found in Arrhenius treatments of $k_{re}(T)$ for certain other $(^2E)Cr(III)$ systems (**3**, **6**, **18**), but the extra large A values for these chloro complexes could be exaggerated by the glassy medium. Arrhenius activation parameters which require very large pre-exponential factors must be viewed with caution, but they could have a physical meaning in a semiclassical formalism describing an excited state solvolysis process. This view is developed further below.

The room temperature values of $\tau(^2E)$ for most hexaamine chromium(III) complexes fall into a fairly narrow range in fluid solutions (*ca* $10^{-6\pm1}$ s^{-1}). This small range, and the variety of solvent media used by different investigators tends to make any correlations difficult. If we take $Cr(en)_3^{3+}$ and $Cr(sen)^{3+}$ to be representative of an "average" and a short-lived hexaamine complex, then in a single medium the variations in $\tau(^2E)$ (*ca* 2 and 0.05 μs, respectively, at 298 K) again parallel the variations in T_{tr} (*ca* 240 and 190K, respectively).

Observations such as those outlined above raise a number of mechanistic issues and questions:

1. thermal activation suggests some sort of surface crossing;
2. an isotope effect in the thermally activated regime suggests that the activation barrier involves nuclear rearrangements and that these nuclear motions have large quanta compared to k_BT;
3. are the relaxation and reaction channels independent or coupled? (curvature in the Arrhenius plots could indicate separate channels);
4. how do solvent species contribute to the relaxation process?
5. what models are useful in describing the thermally activated relaxation process?

The most apparent alternatives for the surface crossing process involve either: (a) back intersystem crossing to populate some very reactive higher energy excited state (usually assumed to be the lowest energy quartet excited state) (**2**); or, (b) an electronically forbidden crossing to a lower energy surface (e.g., of a reaction intermediate) which has the ground state electronic configuration (**3, 19**). Crossing from (^{2}E)Cr(III) into the potential energy surface of species with the ground state electronic configurations cannot be efficient between species with the equilibrium nuclear coordinates. A crossing between displaced potential energy surfaces is plausible if the immediate "product" species has nuclear coordinates so different from those of (^{2}E)Cr(III) that this "product" must be considered a new chemical species, be it a reaction "intermediate" or an "excited state" (see Figure 1b). Very recent work, in our laboratory and elsewhere, has used constrained ligand systems to restrict the possible nuclear motions. This work is providing a useful perspective on some of the issues raised above.

The Macrocyclic Ligand Effect. A few very recent studies have demonstrated that simple *trans*-$Cr^{III}(N_4)X_2$ complexes (N_4 = a macrocyclic amine ligand, X = NH_3 or CN^-) have exceptionally long lifetimes in fluid solution at room temperature (**8, 18, 20**). These long lifetimes turn out to be a consequence of values of T_{tr} which are near to or greater than 300 K; i.e., the room temperature fluid solution lifetimes of these compounds are nearly identical with the 77 K, rigid matrix lifetimes (**8, 17, 18**).

In point of fact, values of T_{tr} are not well defined for the *trans*-$Cr(N_4)(CN)_2^+$ complexes, since their doublet excited states are strongly quenched by OH^-, even in moderately acidic solutions (Figure 4). It is clear that T_{tr} for the OH^--independent pathway is well above room temperature; e.g. $T_{tr} \geq$ 360 K for *trans*-$Cr(N_4)(CN)_2^+$ complexes. The N-deuterated analogs have somewhat smaller values of T_{tr} in D_2O (CH_3CO_2D), but even so their room temperature, fluid solution lifetimes are extraordinarily long: *ca* 1.4 ms (**8, 17, 18**).

In very striking contrast, the corresponding *cis*-Cr^{III}-$(N_4)X_2$ complexes are quenched by OH^- only at higher pH, have much shorter lifetimes under ambient conditions, and they have correspondingly smaller values of T_{tr}. Much larger values of T_{tr} for *trans*- than for *cis*-$Cr^{III}(N_4)X_2$ complexes seem to be a characteristic feature of $(^2E)Cr(III)$ behavior when N_4 is in a macrocyclic tetraamine ligand. However, the values of T_{tr} also depend very strongly on the ligands X, for both the *trans*- and *cis*-complexes, with T_{tr} decreasing in the order, $CN^- > NH_3 > NCS^- > Cl^-$. The much shorter ambient solution lifetimes found for the chloro complexes raises the possibility of the intervention of second relaxation channel; e.g., such a channel could involve direct participation of the lowest energy quartet excited states, which are probably reasonably close in energy to the ^2E state in these chloro complexes. Any additional relaxation channel would require modification of equation 1, so that

$$k_{re}(T) = k^a_{re}(T) + k^b_{re}(T) \tag{2}$$

If the two channels were truly distinct, then domination of $k_{re}(T)$ by the chloride mediated relaxation channel (designated "b") should lead to a repression of (and ultimately eliminate) characteristic features of relaxation channel "a". If the longer lifetime for the *trans*- than for the *cis*-geometries be taken as a characteristic feature of $k^a_{re}(T)$, then this feature does appear to be repressed when the coordination of chloride increases $k_{re}(T)$ by more than three orders of magnitude. This argues that the chloride mediated relaxation pathway is distinct. A possible means for accounting for these observations is that the ^2E relaxation involves crossing to the potential energy surfaces of reaction intermediates (i.e., into some, not necessarily the lowest energy, reaction channel or channels of the electronic ground state) which have quartet spin multiplicity. Such electronically forbidden crossings would be facilitated by spin orbit coupling, and ^2E-$^4T_2$ spin orbit coupling increases as $\Delta E^* = E(^4T_2) - E(^2E)$ decreases.

<u>Ligands with a Tendency Towards Trigonal Distortions</u> The observations summarized in the preceding section seem to indicate that the nuclear distortions which effect relaxation of the $(^2E)Cr(III)$ excited state are more easily accomplished from a *cis*-$Cr^{III}(N_4)X_2$ complex than from a *trans*-$Cr^{III}(N_4)X_2$ complex. This suggests that ligands which facilitate certain types of nuclear motions should reduce the excited state lifetimes. With this hypothesis in mind, we have been investigating the potential for highly strained (in the electronic ground state) ligands to induce excited state distortions by

employing a series of substituted 1,4,7-triazacyclononane complexes. These N-substituted ligands tend to promote a trigonal prismatic geometry, a potential distortion which might open a reaction channel not available in the tetragonal complexes discussed above.

The parent, $Cr([9]aneN_3)_2^{3+}$, is relatively long lived under ambient conditions (**12**). The tri-acetato derivative, $(^2E)Cr(TCTA)$, and $(^2E)Cr([9]aneN_3)_2^{3+}$ have comparable values of k^o_{re}, but $(^2E)Cr(TCTA)$ has an exceptionally short lifetime under ambient conditions. These variations in $k_{re}(T)$ are not manifested in E_a, but in very different values of T_{tr} (or A). The TCTA ligand has a tendency to favor a trigonal prismatic geometry, but Cr(TCTA) is only slightly distorted from an antiprismatic geometry (**21**). If a trigonal twisting mechanism promoted relaxation, then one would expect a smaller value of E_a for $(^2E)Cr(TCTA)$ than for $(^2E)Cr([9]aneN_3)_2^{3+}$. This is not our observation. Once again there are large differences in the lowest quartet excited state energies, and the arguments developed in the preceding section, for a chloride mediated pathway, may also be applicable here. The behavior of the $Cr(BCNE)^{3+}$ complex is more in line with expectation based on the tendency of a strained ligand to facilitate entry into a relaxation channel. The Arrhenius activation energy is exceptionally small and a relatively broad doublet emission is observed at 77 K (fwhh *ca* 340 cm^{-1}). Thus, it would appear that some of the ground state strain energy is relaxed through excited state distortions, and that these distortions reduce the barrier for the relaxation process.

The contrast in behavior of the $Cr(en)_3^{3+}$ and $Cr(sen)^{3+}$ complexes might be attributed to the effects of a trigonally strained ground state. Thus, $E(^4T_2)$ is larger, but τ(298 K) is much smaller in the sen complex. These complexes differ only in the capping of one trigonal face of the sen complex. The capping $CH_3-C(CH_2-)_3$ moiety is appreciably strained in the ground state. However, the enthalpic component of this strain does not make a clear contribution to the difference in value of k_{re}(298 K). Once again the effect appears to be "entropic" (manifested in T_{tr} or A) and the emission band widths are comparable. This does not rule out the possibility that the difference in ligand strain dominates variations in τ(298 K) for these two complexes, but the strain contribution is apparently not manifested in an excited state distortion in $Cr(sen)^{3+}$.

An Attempt to Elucidate the Role of the $^2T_1$, State. We have some preliminary observations which may bear on the role of the $^2T_1$ excited state in the relaxation process. This state can often be found 200 to 1000 cm^{-1} above the (^2E) state in simple amine and ammine complexes (**22**). As the symmetry of the complexes decreases, the $^2T_1$ state splits into two or three components, and the energy of the lowest of these can approach $E(^2E)$. For example $E(^2E) - E(^2T_1) \sim 600\ cm^{-1}$ in $Cr(phen)_3^{3+}$, but in $Cr(phen)_2(NH_3)_2^{3+}$ the lowest component of $^2T_1$ origin appears at about 150 cm^{-1} higher energy (Figure 5a). At 77 K both of electronic origins are resolved, but the vibronic structure is weak enough that only the structure associated with the ^2E emission is easily detected (Figure 5a). As the temperature is increased the higher energy electronic origin is populated sufficiently that the vibronic structure associated with it is superimposed on the structured (^2E) emission (222 K spectrum) resulting in a net, broadened emission spectrum. A similar effect appears in the more complex spectrum of $Cr(teta)(CN)_2^+$ (Figure 5b). In the 222 K emission spectrum of this complex the superposition of the components of emissions from the different electronic origins gives a net spectrum which appears to be Stokes shifted at ambient temperatures.

Since $\tau(^2E)$ is essentially independent of temperature for $Cr(teta)(CN)_2^+$, it is clear that thermal population of low lying electronic components of the higher energy doublet state do not provide an efficient relaxation pathway in this system.

Some Inferences About the Mechanism for Solvent Mediated, Thermally Activated, Non-Radiative Relaxation of $(^2E)Cr(III)$. The $(^2E)Cr(III)$ relaxation rate has been formulated, in equation 1, as a composite of the contributions of mechanistically distinct pathways. The temperature independent contribution, k^o_{re} seems well behaved, in accord with expectation for nested excited and ground state potential energy surfaces. The temperature dependent component, $k^o_{re}(T)$, exhibits a number of important general features:

1. $k^o_{re}(T)$ tends to vary with the solvent medium, and these variations appear in both E_a and A for the thermally activated relaxation rates;
2. ^2E is quenched by base for many of the amine compounds, and this sensitivity is most strikingly manifested in variations in T_{tr};
3. Arrhenius pre-exponential factors (A) vary over a considerable range;
4. the overwhelming majority of compounds studied in a single solvent have very similar Arrhenius activation energies (E_a = (3.6 ± 0.5) x 10^3 cm^{-1} in DMSO-H_2O, but many of the Arrhenius plots are curved (**3**, **6**, **8**);
5. there is no correlation of E_a with ΔE^*;
6. ligands in which the nuclear motions are either sterically inhibited or promoted have profound effects on the high temperature lifetimes, manifested mostly as variations in T_{tr} (or A);
7. deuteration of coordinated amines (ammines) results in a decrease in $k_{re}(T)$ in several systems.

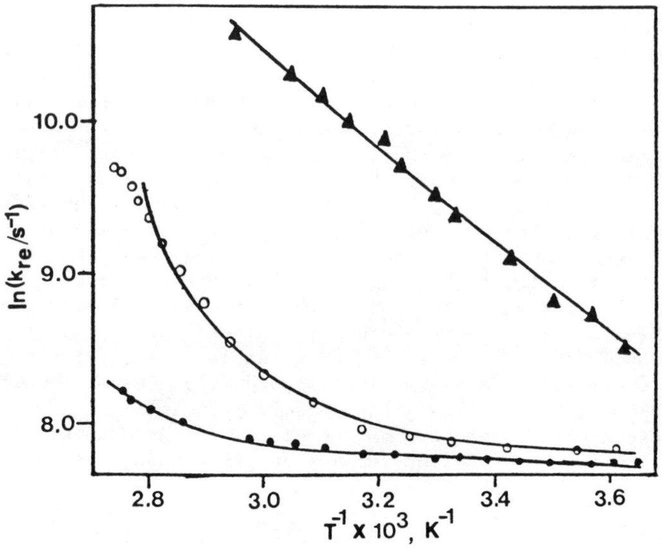

Figure 4. Acid dependence of the (^2E)Cr(teta)$(CN)_2^+$ lifetime. in water with acid or base added: 3 x 10^{-6}M NaOH, triangles; 3 x 10^{-6}M HCl, open circles; 2.5 x 10^{-6}M HCl, closed circles.

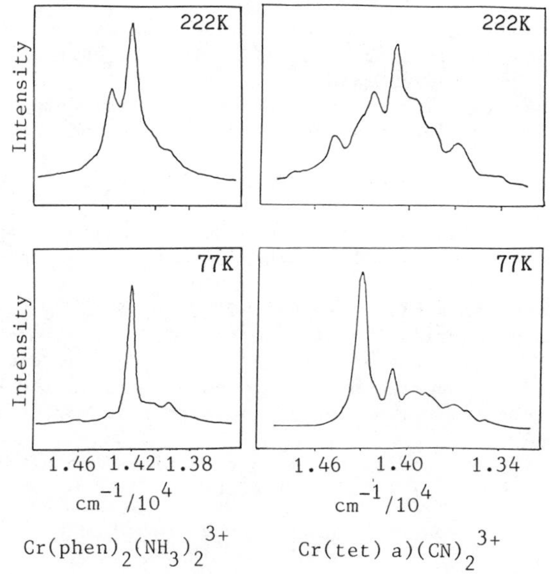

Figure 5. Doublet emission spectra of Cr(PHeu)$_2$(NH$_3$)$_2^{3+}$ and Cr(teta)$(CN)_2^+$ at 222 K and 77 K.

These observations force us to conclude that large nuclear displacements make a major contribution to $k_{re}(T)$. That the effects of solvent are manifested in both the temperature dependent and the temperature independent components of $k_{re}(T)$ is reminiscent of the behavior of classical solvolysis reactions or conformational rearrangements. Thus, the relaxation dynamics might better be discussed in terms of free energy contributions than in terms of potential energies. In such a view the variation in the Arrhenius pre-exponential term translates into a $\pm$ 24 J K^{-1} mol^{-1} (2σ value) variation around a mean $\Delta S^{\ddagger} = 1$ J K^{-1} mol^{-1}; only for complexes with constrained ligands are the $\Delta S^{\ddagger}$ values for entries in Table II significantly negative; i.e., for *trans*-$Cr(N_4)(CN)_2^+$ and for $Cr([9]aneN_3)_2^{3+}$. These are also the complexes with the highest energy quartet excited states.

In summary, our photophysical studies indicate that the thermally activated relaxation pathways of $(^2E)Cr(III)$ very likely involve ^2E-to-^{4}(intermediate) surface crossing. These ^{4}(intermediates) can be associated with some, not necessarily the lowest energy, transition state (or transition states) for ground state substitution. The Arrhenius activation barriers for thermally activated relaxation are remarkably similar from complex to complex, but they can be altered in systems with highly strained ligands. Some of this work indicates that the steric and electronic perturbations of the ligands dictate the choice among possible relaxation channels.

ACKNOWLEDGMENTS

We are grateful to Professor Karl Wieghardt for providing samples of the substituted $[9]aneN_3$ ligands. Professor N. A. P. Kane-Maguire kindly provided us with a number of useful details about the cyclam complexes prior to publication. The research described in this paper has been generously supported by the National Science Foundation.

Table II. Photophysical Parameters for Some Chromium(III) Complexes

Complex[a]	δE[b] ($cm/10^3 cm^{-1}$)	$\tau(298K)$[c] (μs)	T_{tr}[d] (K)	E_a[c] ($cm^{-1}/10^3$)	A[e] ($s^{-1}/10^{13}$)	Ref
$Cr(NH_3)_6^{3+}$	4	2.2	244	3.2	0.1	(8)
$Cr(NH_3)_5Cl^{2+}$	2.0	$(2x10^{-6})$[g]	170	4.5	$(2x10)^8$[h]	(17)
$Cr(NH_3)_5CN^{2+}$	5.0	14	287	4.3	88	(32)
$Cr(en)_3^{3+}$	4.4	1.2	234	3.4	1.2	(15)
$Cr(sen)^{3+}$	5.0	(0.02)[g]	202	3.5	90	(17)
$Cr(phen)_3^{3+}$	7.5	126	235	3.2	$6x10^{-3}$	(39)
$Cr(tpy)_2^{3+}$	5.8	$(1x10^{-3})$[g]	178	4.0	$2x10^4$	(39)
$Cr(phen)_2(NH_3)_2^{3+}$	7.0	3.6	233	2.9	0.03	(39)
trans-$Cr(L_1)(CN)_2^+$	7.7	361	314	i		(32)
trans-$Cr(L_1)(NH_3)_2^{3+}$	~5	136[j]	~308	5.6	$4x10^2$	(18)
cis-$Cr(L_1)(NH_3)_2^{3+}$	~4	1[j]	~235	3.6	5	(18)
trans-$Cr(L_1)Cl_2^+$	0.7	$(8x10^{-4})$[g]	168	4.1	$(2x10)^8$ [h]	(17)
cis-$Cr(L_1)Cl_2^+$	2.3	$(1x10^{-2})$[g]	156	2.1	0.4	(17)
trans-$Cr(L_1)(NCS)_2^+$	4.2	80				(17)
cis-$Cr(L_1)en^{3+}$	4.3		243	3.7	2	(32)
trans-$Cr(L_2)(CN)_2^+$	6.7	380	331	i		(32)

cis-$Cr(L_3^+)(CN)_2$	5.4	2.1	237	3.8	4.2	(**32**)
cis-$Cr(L_3)(NCS)_2^+$			181	2.3	0.1	(**17**)
cis-$Cr(L_3)en^{3+}$	5.2	$(2x10^{-3})^g$	196	3.5	150	(**24**)
$Cr(L_4)_2^{3+}$	5.7	40	260	2.8	$1.1x10^{-3}$	(**12**)
$Cr(L_5)$	3.0	$(5x10^{-3})^g$	140	2.5	38	(**40**)
$Cr(L_6)^{3+}$	7	$(6x10^{-3})^g$	134	1.5	0.025	(**40**)

NOTES FOR TABLE II:

a See note a Table I for abbreviations.

b Relative values of $\Delta E^* = E(^2E) - E(^4T$ or $^4E)$ obtained relative to δE set equal to 4×10^3 for $Cr(NH_3)_6{}^{3+}$ and assuming band widths do not change.

c In DMSO-water (or H_2O) except as noted

d Temperature of the transition between temperature dependent and temperature independent regimes of τ (in DMSO-H_2O except as noted)

e Arrhenius activation parameters

f DMF-$CHCl_3$

g Value extrapolated from thermally activated behavior at very low temperatures

h Data collected in a rigid glass.

i Temperature dependence seems correlated with OH^- quenching. See figure 4.

j 293 K

LITERATURE CITED

1. Zinato, E.; Riccieri, P. In Concepts of Inorganic Photochemistry, Adamson, A. W.; Fleischauer, P. D., Eds.; Wiley: New York, 1975, p. 203.
2. Kirk, A. D. Coord. Chem. Rev., 1981, 39, 225.
3. Endicott, J. F.; Ramasami, T.; Tamilarasan, R.; Lessard, R. B.; Ryu, C. K.; Brubaker, G. R. Coord. Chem. Rev., submitted.
4. Fleischauer, P. D.; Adamson, A. W.; Sartori, G. Progr. Inorg. Chem., 1972, 17, 1.
5. Pfeil, A. J. Am. Chem. Soc., 1971, 93, 5395.
6. Allsopp, S. R.; Cox, A.; Kemp, T. J.; Reed, W. J. J. Chem. Soc. Faraday I, 1980, 76, 162.
7. Targos, W.; Forster, L. S. J. Chem. Phys., 1966, 44, 4342.
8. Endicott, J. F.; Tamilarasan, R.; Lessard, R. B. Chem. Phys Lett., 1984, 112, 381.
9. Englman, R.; Jortner, J. Mol. Phys., 1970, 18, 145.
10. Freed, K. F.; Jortner, J. J. Chem. Phys., 1970, 52, 6272.
11. Kuhn, K.; Wasgestian, F.; Kupka, H. J. Phys. Chem., 1981, 85, 665.
12. Ditze, A.; Wasgestian, F. J. Phys. Chem., 1985, 89, 426.
13. Robbins, D. J.; Thomson, A. J. Mol. Phys., 1973, 25, 1103.
14. Brunschwig, B. S.; Sutin, N. J. Am. Chem. Soc., 1978, 100, 7568.
15. Tamilarasan, R.; Endicott, J. F. unpublished work.
16. Flint, C. D.; Greenough, P. J. Chem. Soc. Faraday II, 1972, 68, 897.
17. Lessard, R. B.; Endicott, J. F., unpublished work.
18. Kane-Maguire, N. A. P.; Wallace, K. C.; Miller, D. B. Inorg. Chem., 1985, 24, 597.
19. Endicott, J. F.; Ferraudi, G. J. J. Phys. Chem., 1976, 80, 949.
20. Miller, P. K.; Crippen, W. S.; Kane-Maguire, N. A. P. Inorg. Chem., 1983, 22, 696.
21. Wieghardt, K.; Bossek, U.; Chandhuri, P.; Herrmann, W.; Menke, B. C.; Weiss, J. J. Am. Chem. Soc., 1982, 104, 4308.
22. Lever, A. B. P. Inorganic Electronic Spectroscopy; Elsevier: New York, 1984, 2nd. edn.
23. Forster, L. S.; Rund, J. V.; Castelli, F.; Adams, P. J. Phys. Chem., 1982, 86, 2395.
24. Walters, R. J.; Adamson, A. W. Acta Chem. Scand., 1973, A33, 53. 25. Mønsted, L.; Mønsted, O. Acta Chem. Scand., 1974, A28, 569.
26. Flint, C. D.; Mathews, A. P. J. Chem. Soc. Faraday II, 1972, 68, 419.
27. Tyano, E.; Kamada, M.; Tanaka, N. Bull. Chem. Soc. Japan, 1967, 40, 1848.
28. Forster, L. S.; Rund, J. V.; Fucalaro, A. F. J. Phys. Chem., 1984, 88, 5012.
29. Fucaralo, A. F.; Forster, L. S.; Rund, J. V.; Lim, S. H. J. Phys. Chem., 1983, 87, 1796.
30. Zinato, E.; Riccieri, P. Inorg. Chem., 1973, 12, 1451.
31. Riccieri, P.; Zinato, E. Inorg. Chem., 1980, 19, 3279.
32. Lessard, R. B., M.S. Thesis, Wayne State University, Detroit, 1985.

33. Flint, C. D.; Mathews, A. P. J. Chem. Soc. Faraday II, 1976, 72, 379.
34. Forster, L. S. Transition Metal Chem., 1969, 5, 1.
35. Kane-Maguire, N. A. P.; Langford, C. H. Chem. Commun., 1971, 1895.
36. Endicott, J. F.; Schwarz, C. L., unpublished work.
37. Jamieson, M. A.; Serpone, N.; Hoffman, M. Z. Coord. Chem. Rev., 1981, 39, 121.
38. Serpone, N.; Jamieson, M. A.; Henry, M. S.; Hoffman, M. Z.; Bolletta, F.; Maestri, M. J. Am. Chem. Soc., 1979, 101, 2907.
39. Endicott, J. F.; Lei, Y., unpublished work.
40. Endicott, J. F.; Ryu, C. K., unpublished work.
41. Josephsen, J.; Schäffer, C. E. Acta Chem. Scand., 1977, A31, 813.
42. Kane-Maguire, N. A. P.; Bennet, J. A.; Miller, P. K. Inorg. Chim. Acta, 1983, 76, L123.
43. Ferguson, J.; Tobe, M. L. Inorg. Chim. Acta, 1970, 109, 4.
44. Poon, C. K.; Pun, K. C. Inorg. Chem., 1980, 19, 568.
45. House, D. A.; Hay, R. W.; Ali, M. A. Inorg. Chim. Acta, 1983, 72, 239.
46. Wieghardt, K.; Tolksdorf, I.; Herrmann, W. Inorg. Chem., 1985, 24, 1230.

RECEIVED December 23, 1985

8

Static Spectral Sensitization of Photocatalytic Systems

Formation of Mixed-Valence Compounds as a Possible Route

H. Hennig and D. Rehorek

Sektion Chemie, Karl-Marx-Universität Leipzig, DDR-7010 Leipzig, Talstr. 35
German Democratic Republic

Models for the static spectral sensitization of coordination compounds are discussed. Among them, ion pairs with oppositely charged metal complexes, where intervalence charge-transfer interactions occur, appear to be promising examples. The spectra and the electron transfer kinetics of mixed-valence ion pairs of the type $M^{n+}/[Mo(CN)_8]^{4-}$, where $M^{n+} = Fe^{3+}$, Cu^{2+}, UO_2^{2+}, VO^{2+}, are discussed in detail.

Very recently (1), in an attempt to overcome the confusion surrounding the definition of photocatalysis, considering the contributions by Salomon (2), Moggi (3), Wrighton (4), Carassiti (5), Mirbach (6) and others (7-9), we have suggested the following terms be used for the description of the different light-induced reactions leading to the catalysed conversion of diverse substrates: (a) photoinduced catalytic reactions, (b) photoassisted reactions, (c) sensitized photoreactions, and (d) catalysed photoreactions. The various types of photocatalytic reactions are illustrated in Figure 1.

The term photoinduced catalytic reaction implies the photochemical generation of a catalyst C from a thermally stable and catalytically inactive precursor ML_n. Unlike the formation of the catalyst C, which is a photochemical reaction, the subsequent transformation of the substrate A into the product B is an exclusively thermal process catalysed by C. If the catalyst is not consumed in the course of the reaction, the quantum yield, i.e. the number of product molecules B formed per photons absorbed by ML_n, may exceed unity.

On the other hand, in a photoassisted reaction a catalytically inactive precursor ML_n is transformed photochemically into a photoassistor (here: ML_{n-1} or L) which may bring about the transformation of the substrate A into the product B, either thermally or by absorption of a second photon while the photoassistor itself is converted back to the precursor ML_n. In this process ML_n is not consumed, however, the overall quantum yield is ≤ 1. In addition, unlike the photoinduced catalytic reaction, which may continue to proceed after the flux of incident light quanta

0097-6156/86/0307-0104$06.00/0

is terminated, a photoassisted reaction requires continous irradiation. So far, the photoassisted reaction resembles the sensitized photoreaction. In the latter, however, the catalytically active species is an electronically excited molecule.

Finally, in a catalysed photoreaction the conversion of the excited substrate molecule $\underline{A}^*$ into the product $\underline{B}$ is catalysed by ML_n in its electronic ground state.

Both photoinduced catalytic and photoassisted reactions are of increasing interest with respect to e.g. storage and conversion of solar energy, homogeneous complex catalysis, uncommon organic syntheses, modelling of light-sensitive metallo enzymes and unconventional photographic processes.

We have extensively studied (10) the photoinduced catalytic system (Figure 2) which is based upon the photochemical generation of cyanide ions from various cyanometallates, especially from octacyanomolybdate(IV) and octacyanotungstate(IV) ions. Upon photolysis both compounds form cyanide ions with relatively high quantum yields. Cyanide ions may act as a catalyst for the dimerization of appropriate heterocyclic carb-2-aldehydes to enediols. Since this photocatalytic system has proved to operate also in solid layers, it may be used for an unconventional photographic process (11,12).

However, due to the spectroscopic properties of the octacyanomolybdate(IV) ion, which restricts the photosensitivity to the UV and blue region of the light, a broader application of this photocatalytic system, e.g. for the color photography, requires its spectral sensitization.

The general problem of spectral sensitization of photocatalytic systems

The aim of spectral sensitization of a coordination compound ML_n is, broadly speaking, the photochemical conversion of the catalytically inactive coordination compound into a product $\underline{C}$ which may act as a catalyst or a photoassistor in a subsequent photoinduced catalytic or photoassisted cycle.

$$ML_n \xrightarrow[S]{h\nu_{sens}} C + \ldots \qquad (1)$$

While irradiation of ML_n with light of the energy $h\nu_{sens}$ in the absence of a sensitizer $\underline{S}$ leads to no or only neglibible photoreaction, photochemical conversion into $\underline{C}$ occurs when $\underline{S}$ is present. Usually, the excitation energy $h\nu_{sens}$ of the sensitized reaction should be less than the energy $h\nu$ which is required for the direct excitation of ML_n.

The general pathways of spectrally sensitized photoreactions of coordination compounds as well as organometallics is best described by the Scheme 1 given below.

Besides the usual deactivation processes of S^*, interactions with ML_n may lead to further deactivation of S^*. Among the quenching processes, energy transfer and electron transfer are particularly interesting because they may regarded as the starting

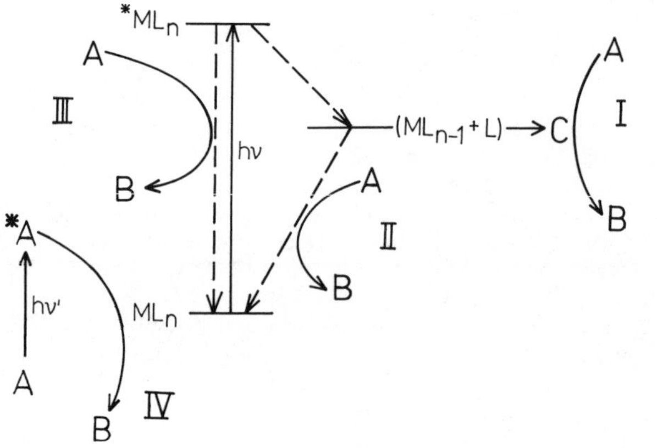

Figure 1. Simplified Jablonski-type diagram of photoinduced catalytic (I), photoassisted (II), and sensitized photoreaction (III) as well as catalysed photolysis (IV). $(ML_{n-1} + L)$ represents coordinatively unsaturated species, free ligands as well as ligands and/or coordination compounds with changed formal oxidation number generated by photo-redox and/or photodissociation reactions.

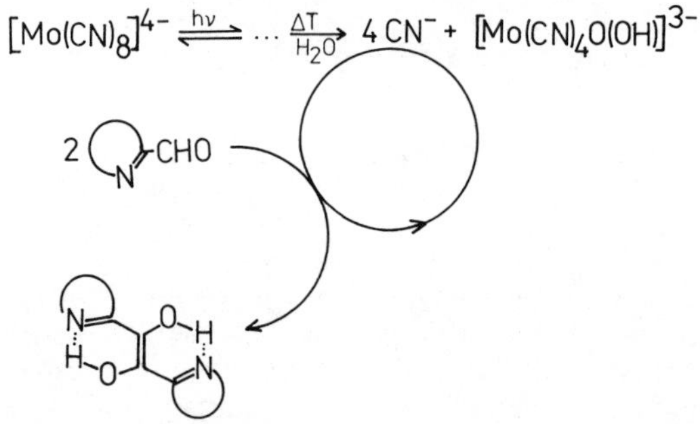

Figure 2. The photoinduced catalytic system $[Mo(CN)_8]^{4-}$/ heterocyclic carb-2-aldehyde.

points of spectrally sensitized photoinduced catalytic or photo-assisted reactions.

Radiationless energy transfer may proceed either through dipole-dipole interactions (13) or through exchange interactions (14), whereas electron transfer may be observed only in the case of exchange interactions. However, energy transfer is of only little importance for the long-wavelength spectral sensitization because of the physical restrictions to be met (15).

If the electron transfer proceeds as a bimolecular reaction, the formation of an encounter complex is required. However, besides the bimolecular dynamic sensitization, we have been able to show some advantages in realizing the concept of static sensitization (15).

Dynamic spectral sensitization (16). Based upon the kinetic scheme given above (Scheme 1), the dynamic sensitization pathway leading to the desired product, e.g. by electron transfer, may be written as shown in Equation 2.

$$S \xrightarrow{h\nu} S^* \xrightarrow[k_d]{ML_n} S^* \ldots ML_n \underset{k_r}{\overset{k_{sens}}{\rightleftharpoons}} \text{products} \qquad (2)$$

$$S^* \xrightarrow{\Sigma k_S} \ldots \qquad S^* \ldots ML_n \xrightarrow{\Sigma k_{SML_n}} \ldots\ldots$$

The efficiency of the product formation, η_{prod}, is diminished by a number of competing reactions in a dynamic sensitization process, Equation 3.

$$\eta_{prod} = \frac{k_d \cdot ML_n}{k_d\, ML_n + \Sigma k_S} \quad \frac{k_{sens}}{k_{sens} + k_r + \Sigma k_{SML_n}} \qquad (3)$$

(k_d = diffusion rate constant, Σk_S = sum of the rate constants of all deactivation processes of S^*, k_{sens} = rate constant for the product formation from the encounter complex k_r = rate constant of the electron back transfer, Σk_{SML_n} = sum of the rate constants of all deactivation processes of the encounter complex between S^* and ML_n)

The most critical barrier with respect to the generation of the products is the diffusion-controlled (k_d) formation of the encounter complex, since the generation of $S^* \ldots ML_n$ may be prevented by an inadequate lifetime of S^*, by a high viscosity of the medium or by an unfavourable ionic strength. Further deactivation processes must also be considered, as illustrated previously (Scheme 1). Despite these disadvantages, there are known some examples where dynamic sensitization leads to the formation of products (1).

Static spectral sensitization. To overcome some of the disadvantages of the dynamic sensitization, the concept of static sensiti-

Scheme 1

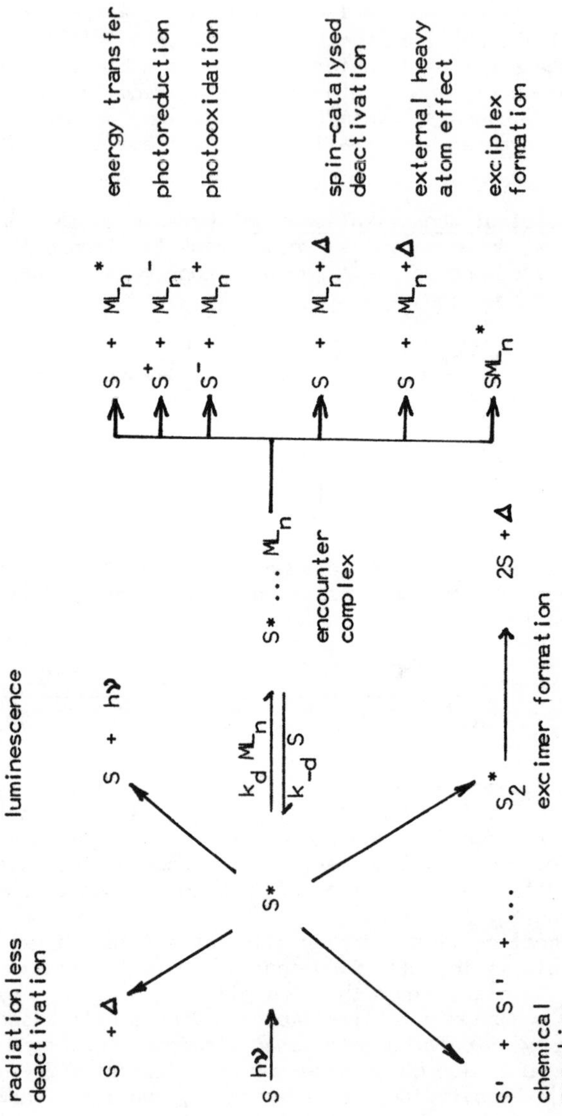
radiationless deactivation
luminescence
chemical reaction
excimer formation
encounter complex
energy transfer
photoreduction
photooxidation
spin-catalysed deactivation
external heavy atom effect
exciplex formation

zation has been proposed (1, 15). The idea behind this concept lies in the linking of a sensitizer S and a metal complex ML_n, which has to be sensitized, in one closed unit $S\text{-}ML_n$ by an ionic, covalent or coordinate bond interaction between both components. Under such circumstances the formation of the encounter complex is no longer a restriction, and the kinetic scheme of sensitization is reduced considerably, Equation 4.

$$S\text{-}ML_n \xrightarrow{h\nu} S\text{-}ML_n^{*} \underset{k_r}{\overset{k_{sens}}{\rightleftharpoons}} \text{products} \qquad (4)$$

$$S\text{-}ML_n^{*} \xrightarrow{k_{S\text{-}ML_n}} \ldots$$

As essential competitive processes, we have only to consider deactivation processes of $S\text{-}ML_n^{*}$ ($k_{S\text{-}ML_n}$) and back electron transfer (k_r). Thus, the efficiency η_{prod} primarily depends on the behavior of the $S\text{-}ML_n$ unit, but it is independent of the solvent properties, provided they do not affect the thermodynamic stability of the sensitizer/complex unit:

$$\eta_{prod} = \frac{k_{sens}}{k_{sens} + k_r + k_{S\text{-}ML_n}} \qquad (5)$$

The following proposals have been made (1, 15) for realizing the concept of static sensitization:

i. Static sensitization by formation of ion pairs with IPCT behavior or with S^{+} being an ionic dye

$$ML_n^{m\pm} + S^{\mp} \rightleftharpoons ML_n^{m\pm} ; S^{\mp}$$

ii. Static sensitization through the formation of mixed-valence compounds with IT (intervalence charge-transfer) behavior

$$L_nM^1X + L_nM^2y \rightleftharpoons L_nM^1\text{-}X\text{-}M^2L_n + y$$

iii. Mixed-ligand complexes with chromophoric ligands (Chr) and/or low-lying CT states

$$ML_n + Chr \rightleftharpoons ML_{n-1}Chr + L$$

Examples for the static sensitization through the formation of mixed-valence cyanometallates are discusses below.

Static sensitization by mixed-valence complexes with IT behavior

Mixed-valence compounds with IT behavior may be described theoretically as proposed by Robin and Day (17) based on the semi-empirical model approach of Hush and Allen (18, 19). Especially interesting concerning the static sensitization process are mixed-valence complexes with weakly interacting metal centers (class II mixed-valence compounds according to the classification by Robin and Day (17)).

The electronic interaction between the two metal centers leads to the appearance of IT bands which can be observed in a region ranging from the ultraviolet to the near infrared part of the spectrum. Optical IT transitions are due to electron transfer from one metal center M^1 to the other M^2. The energy of the IT transition strongly depends on the redox asymmetry of the metal centers as well as the dielectric properties of the solvent. The general behavior of mixed-valence compounds has been reviewed excellently very recently (20).

The synthesis of mixed-valence compounds with optical IT transitions in the visible and near infrared region provides an interesting way to low-energy spectral sensitization. However, experimental results of photochemical reactions initiated by excitation of IT states are very scarce, see e.g. (21, 22).

The IT behavior of mixed-valence cyanometallates (23-26). Aqueous solutions of $[Mo(CN)_8]^{4-}$, $[W(CN)_8]^{4-}$, $[Fe(CN)_6]^{4-}$, and $[Ru(CN)_6]^{4-}$ ions undergo remarkable color changes upon addition of Fe(III), Cu(II), UO_2^{2+}, and VO^{2+} ions as well as some cobalt(III) ammine complexes, whereas the addition of Cr(III), Co(II), Ni(II), Zn(II), Hg(II), and Tl(I) ions leads to absorption spectra which can be described as the sum of the spectra of the single components with no additional bands.

Figure 3 exhibits some typical electronic spectra of mixed-valence complexes of the $[Mo(CN)_8]^{4-}$ ion with various other metal ions. Mixed-valence compounds formed by interaction of Mo(IV), W(IV), Fe(II), and Ru(II) cyanometallates with metal ions such as Fe(III), Cu(II), UO_2^{2+}, and VO^{2+} belong to the class II of the Robin-Day classification:

i. In accordance with the theoretical treatment of Hush, the position of the IT band is strongly correlated to the dielectric properties of the solvent. The energy of the optical IT transition, E_{op}, depends on the inner-sphere reorganization energy, $E_{r,i}$, the outer-sphere reorganization energy, $E_{r,o}$, on the enthalpy changes, ΔE, and on the changes of electrostatic interactions, ΔE_{el}, which are caused by the electron transfer from one metal center to the other, Equation 6 and Figure 6.

$$E_{op} = E_{r,i} + E_{r,o} + \Delta E + \Delta E_{el} \qquad (6)$$

Since the outer-sphere reorganization energy $E_{r,o}$ is a function of the solvent term $(1/D_{op} - 1/D_s)$, as follows from the continuum theory (18,19), μ^{IT}_{max} depends linearly on the dielectric properties of the solvents as shown in Figure 4.

ii. The energy of the IT transition is strongly influenced by the redox asymmetry of the metal centers. Therefore, changing of M^1

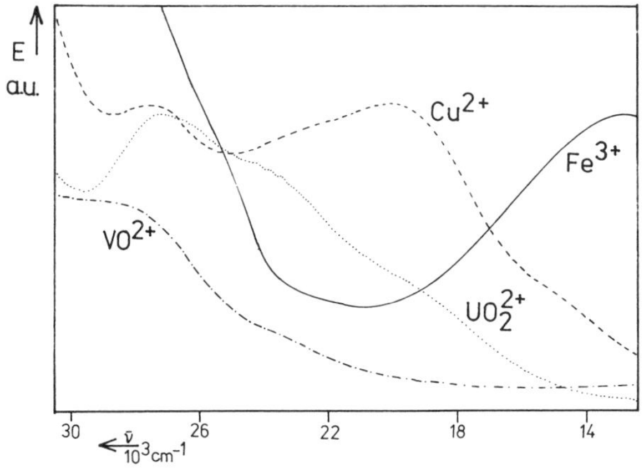

Figure 3. UV/vis spectra of $M^{n+}/[Mo(CN)_8]^{4-}$ mixed-valence compounds.

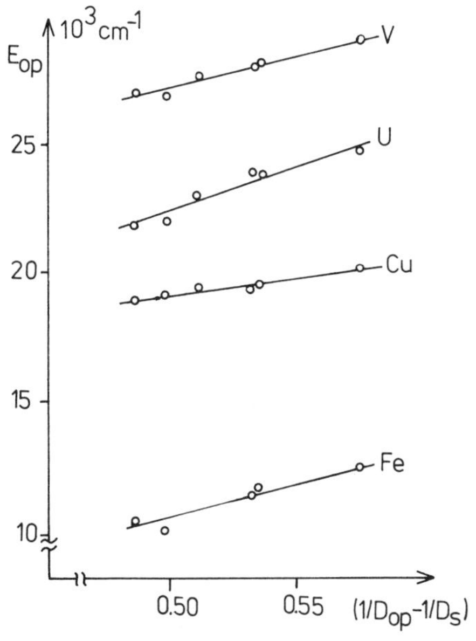

Figure 4. Solvent dependence of the IT transition for $M^{n+}/[Mo(CN)_8]^{4-}$ (M^{n+} = VO^{2+}, UO_2^{2+}, Cu^{2+}, Fe^{3+}).
Dop: Optical dielectric constant of the medium (equal to the square of the refractive index n); D_s: Static dielectric constant of the medium.

and/or M^2 causes a shift of λ_{max}^{IT} as illustrated in Figure 5. The change of the redox asymmetry may also be achieved by variation of the inner coordination sphere of the acceptor site of the mixed-valence cyanometallates. Thus, in the case of Cu(II) it is easily possible to change the redox asymmetry by substitution of the Cu(II) aquo species by either σ-donor or π-acceptor ligands. The variation of the redox asymmetry by changing the first coordination sphere leads to a shift of λ_{max}^{IT} from about 360 nm ($[Cu(en)_2]^{2+}$; $[Mo(CN)_8]^{4-}$) to about 660 nm ($[Cu(dmp)_2]^{2+}$; $[Mo(CN)_8]^{4-}$) as shown in Table I.

However, when $[Cu(dmch)_2]^{2+}$; $[Mo(CN)_8]^{4-}$ ist considered, thermal electron transfer due to the inadaquate redox potential of the Cu(II) unit prevents the formation of a Cu(II); $[Mo(CN)_8]^{4-}$ mixed-valence compound.

Table I. The energies of the IT transitions in CuL_n^{2+};$[Mo(CN)_8]^{4-}$ mixed-valence compounds[a]

CuL_n^{2+}	$\bar{\nu}_{IT}$ (in 10^3 cm^{-1})	E_{Red} (in V)[b]
$Cu(phen)Br_2$	17.8	
$Cu(phen)Cl_2$	17.0	
$Cu(phen)(NO_3)_2$	17.0	
$Cu(phen)_2Br_2$	18.0	
$Cu(phen)_2Cl_2$	18.0	
$Cu(phen)_2(NO_3)_2$	17.5	+ 0.174
$Cu(ach)_2(NO_3)_2$	18.5	
$Cu(bpy)_2(NO_3)_2$	18.0	
$Cu(5\text{-}mp)_2(NO_3)_2$	16.9	+ 0.337
$Cu(dmp)_2(NO_3)_2$	15.2	+ 0.594
$Cu(dmch)_2(NO_3)_2$	c	+ 0.675
$Cu(en)_2(NO_3)_2$	27.8	- 0.38

[a]Data taken from ref.(26); phen - 1,10-phenanthroline bpy - 2,2'-bipyridine, ach - 8-amino-quinoline, 5-mp - 5-methyl-1,10-phenanthroline, dmp - 2,9-dimethyl-1,10-phenanthroline, dmch - 4,4'-dimethyl-3,3'-dimethylene-2,2'-biquinoline, en - ethylenediamine; solvent: methanol; [b] reduction potential of CuL_n^{2+} at 298 K vs. NHE; [c]thermal electron transfer

iii. From the energy of the IT transition, the bandwidth and the intensity, the delocalization parameter α^2 may be calculated, Equation 7, (27-29).

$$\alpha^2 = (4.24 \cdot 10^{-4} \cdot \varepsilon_{max} \cdot \Delta\bar{\nu}_{1/2})/(\bar{\nu}_{IT} \cdot d^2) \qquad (7)$$

(d = distance between donor and acceptor site in the mixed-valence compound)

The α^2 values obtained for the various mixed-valence cyanometallates are ranging between 0.0005 and 0.0121 and, therefore,
provide further support for the classification of these compounds
as Robin-Day class II mixed-valence compounds.
iv. The most interesting advantage of mixed-valence compounds with
respect to their spectroscopic behavior is related to the fact
that simple synthetic variations have a significant influence on
the position of the IT absorption as illustrated in the Scheme 2.

Scheme 2

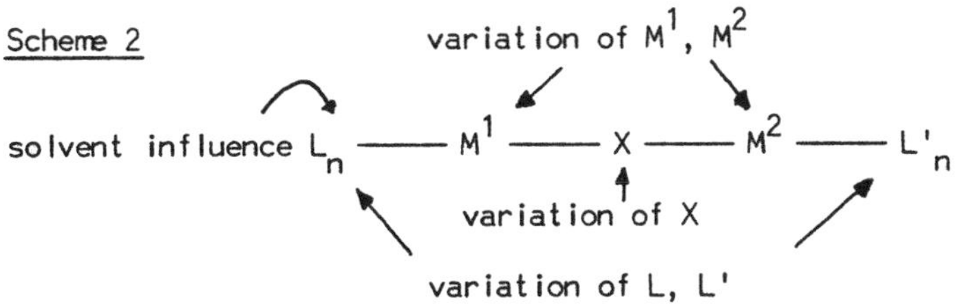

The photochemical behavior of mixed-valence cyanometallates.
The aim of our photochemical investigations of mixed-valence
cyanometallates (23, 26, 30, 31) was to study the spectrally
sensitized formation of free cyanide by excitation of the IT
states of appropriate mixed-valence compounds.
    In this way, cyanide ions may be generated by irradiation
in the low energy region where the photolysis of the pure cyanometallates leads no or neglibible cyanide yields.
    The results obtained with cyanometallates are best illu-
strated by the Scheme 3:

Scheme 3

$$L_n\text{-}M^{n+} \ldots NC\text{-}Mo(CN)_7^{4-} \underset{k_r}{\overset{h\nu_{IT}}{\rightleftarrows}} [Mo(CN)_8]^{3-} + L_nM^{(n-1)+}$$

$$k_1 \downarrow \qquad\qquad \downarrow k_{scav}$$

$$\ldots + CN^- \qquad \ldots$$
$$(\text{thermal catalyst})$$

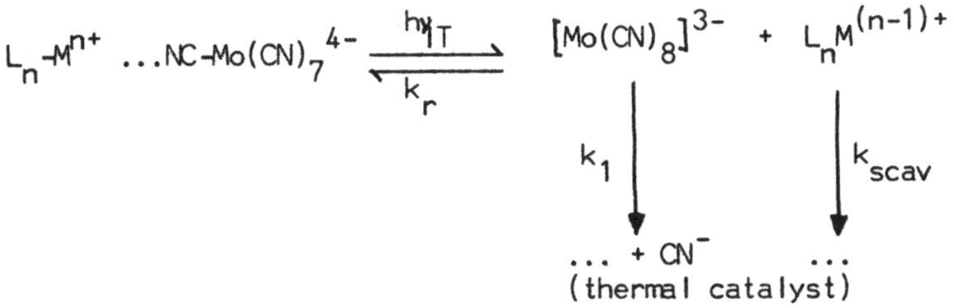

$$k_{scav}, k_1 \gg k_r \;\;!$$

IT excitation of heteronuclear mixed-valence cyanometallates
leads to the formation of a vibronically excited valence-isomeric species (see Figure 6) consisting of octacyanomolybdate
(V) and the corresponding reduced form of the metal center M.
Due to the kinetic lability of $[Mo(CN)_8]^{3-}$ (32-34) fast cyanide
aquation can be expected which competes with the back electron
transfer (k_r).

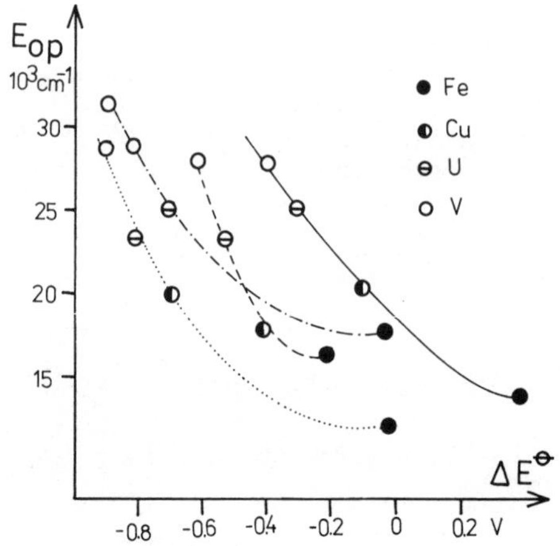

Figure 5. IT transitions of different mixed-valence cyanometallates $M_1{}^{n+}/[M_2(CN)_x]^{4-}$ (______ $[Fe(CN)_6]^{4-}$; ... $[Mo(CN)_8]^{4-}$; _ _ _ $[W(CN)_8]^{4-}$; _._. $[Ru(CN)_6]^{4-}$) vs. redox asymmetry.

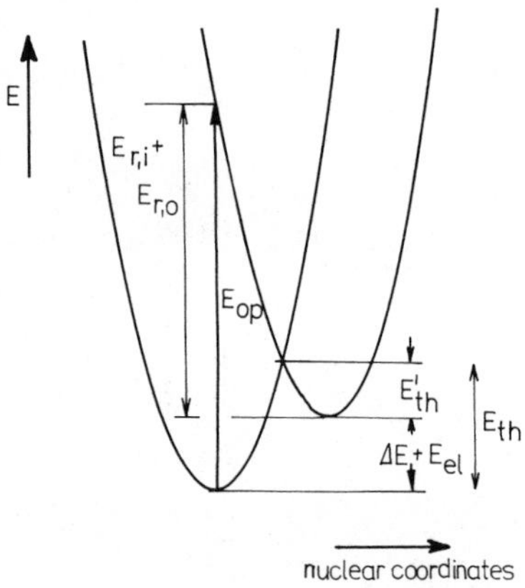

Figure 6. Potential energy diagram for $[M^{n+}/Mo(CN)_8]^{4-}$ mixed-valence compounds. M^{n+}/Mo^{iv}: Precursor complex; $M^{(n-1)}/Mo^{v}$: Successor complex; E_{op}: Energy of the optical IT transition; E_o: Enthalpy difference between precursor and successor compound; E_{th}: Activation energy of thermal electron transfer; E_{th}': Enthalpy of thermal back electron transfer; β: Resonance energy.

Of particular interest are the photochemical investigations of the Cu(II)/$[Mo(CN)_8]^{4-}$ mixed-valence system. Photochemical investigations have been performed by both monochromatic and polychromatic irradiations at selected energy regions. Low concentrations of octacyanomolybdate(IV) and copper(II) have been used by reason of the low solubility of polymeric forms which are formed at higher concentrations. The analytical estimation of free cyanide has been used to monitor the photochemical reactions according to Scheme 3.

Irradiation into the IT region ($\lambda^{irr} \geq 500$ nm) of the Cu(II)/$[Mo(CN)_8]^{4-}$ mixed-valence compound leads to an increased formation of free cyanide as compared with free $[Mo(CN)_8]^{4-}$ ions.

The efficiency of the spectrally sensitized cyanide formation was monitored by the estimation of the photochemical turnover number U instead of the quantum yield for practical reasons.

$$U = n_{CN^-}/(I_o \cdot t) \qquad (8)$$

(n_{CN^-} = moles of cyanide formed; I_o = intensity of incident light; t = irradiation time).

The results summarized in Table II illustrate the increase of the photoinduced formation of cyanide achieved by IT excitation of Cu(II)/$[Mo(CN)_8]^{4-}$ as compared with $K_4[Mo(CN)_8]$. However, despite the increase of cyanide formation the efficiency of the spectral sensitization is rather low. The low efficiency is due to the circumstance that the rate (k_1) of cyanide aquation in the valence isomeric form Cu(I)/$[Mo(CN)_8]^{3-}$ is low compared with the very fast back electron transfer (k_r). In order to make the proper choice of a scavenging reaction (k_{scav}) which may compete successfully with back electron transfer, we have attempted a rough estimate of the rate constant k_r of the back electron transfer following the theoretical treatment proposed by Hush (20).

Table II Results of the photochemical studies of the system $Cu^{2+}_{aq}/[Mo(CN)_8]^{4-}$ [a]

λ^{irr} (in nm)	Turnover U		$U_{IT}/U_{K_4[Mo(CN)_8]}$
	IT system	K_4 Mo(CN)$_8$	
313	1.30	1.95	0.67
436	0.19	0.12	1.58
495	0.048	0.018	2.67
509	0.044	0.009	4.89
546	0.017	0	

[a] Aqueous solutions (0.001 M, 30 ml, d = 2 cm, t^{irr} = 0-5 h); data taken from ref. (26)

Estimation of the back electron transfer rate (k_r). A rough estimate of the rate of the back electron transfer can easily be made using Equation 9 or a modified form as proposed by Grätzel (35):

$$k_r = v_{et} \cdot \exp(-E'_{th}/kT) \tag{9}$$

The value of the barrier of the back electron transfer, E'_{th}, is accessible through the experimental data (E_{op}, E_o and α), while v_{et} may be calculated using Equation 10.

$$v_{et} = 4\,\beta/h \tag{10}$$

The values of k_r (as well as the lifetimes $\tau = 1/k_r$) calculated according to Equations 9 and 10 are shown in Table III.

Table III Calculated parameters (β, E'_{th}, v_{et}) for estimating the rate constant k_r and τ of the mixed-valence systems $M^{n+}/[Mo(CN)_8]^{4-}$

M^{n+}	β (kJ mole^{-1})	E'_{th} (kJ mole^{-1})	v_{et} [a]	k_r [b]	τ (s)
Fe^{3+}	16.44	22.7	1.65	$1.57\ 10^{10}$	$6.37\ 10^{-11}$
Cu^{2+}	15.26	32.4	1.53	$2.78\ 10^{8}$	$3.60\ 10^{-9}$
UO_2^{2+}	6.59	52.0	0.66	$4.09\ 10^{4}$	$2.45\ 10^{-5}$
VO^{2+}	21.10	48.7	2.12	$4.95\ 10^{5}$	$2.02\ 10^{-6}$

[a] (10^{14} s^{-1}), T = 295 K   [b] (s^{-1}), T = 295 K

The results illustrate clearly the short lifetime of the valence isomeric $Cu(I)/[Mo(CN)_8]^{3-}$ mixed-valence complex. The increase of E'_{th} due to the formation of outer-sphere mixed-valence species (25) may account for the higher lifetime of the $UO_2^{+}/[Mo(CN)_8]^{3-}$ complex. Furthermore, the results summarized in Table III illustrate that the lower the energy E_{op} (which is one of the aims of spectral sensitization) the shorter is the lifetime of the valence-isomeric form generated from the precursor mixed-valence complex by IT excitation.

In order to prove the values of k_r and τ experimentally, we have monitored the back electron transfer of $Cu(I)/[Mo(CN)_8]^{3-}$ by low-temperature ESR spectroscopy (30). k_r was found to be $k_r = (2.74+1.2)\ 10^{-4}\ s^{-1}$ which is in a reasonable agreement with the calculated value $k_r = 8.2\ 10^{-4}\ s^{-1}$ for 93 K. Extrapolation to room temperature yields $k_r = 1.3\ 10^{9}\ s^{-1}$ which illustrates

the good approximation following the theoretical approach for
class II mixed-valence compounds proposed by Hush (20).

With respect to the practical application to static spectral
sensitization processes, however, the short lifetime of the
successor complexes generated by IT excitation of the precursor
mixed-valence cyanometallates has to be considered seriously. In
order to overcome back electron transfer processes we have tried
to use scavenging reactions. However, it may be shown by simple
kinetic estimates that high concentrations of scavengers are
required to quench the back electron transfer completely. Thus,
scavenging of $Cu(I)$ from the successor complex $Cu(I)/[Mo(CN)_8]^{3-}$
may be achieved only at scavenger concentrations higher than
about 1 M, even if the scavenging process is diffusion-controlled.

Since some of the potential scavengers, e.g. benzonitrile,
acetonitrile, triaylphosphines, phenol derivatives etc., undergo
thermal redox reactions with $Cu(II)/[Mo(CN)_8]^{4-}$ when they are
present at higher concentrations, it is experimentally impossible
to suppress the back electron transfer by those scavengers. On the
other hand, ionic scavengers such as $[Fe(ox)_3]^{3-}$, $[Co(NH_3)_5N_3]^{2+}$,
and various diazonium compounds give rise to uncontrolled changes
of the association equilibria of the mixed-valence systems. Therefore, it can be assumed that chemical scavenging of back electron
transfer processes can be accomplished preferably by extremely fast
chemical changes taking place within the valence-isomeric species
itself, as it has been demonstrated by Vogler (21, 22).

In addition to chemical scavenging, we have tried to use
photons as physical scavengers to quench the back electron transfer
in excited mixed-valence complexes $M^{(n-1)+}/[Mo(CN)_8]^{3-}$.

Photons as scavengers - unusual secondary photolyses. The photochemistry of the $[Mo(CN)_8]^{3-}$ ion, which is regarded as the primary
product of IT excitation of $M^{n+}/[Mo(CN)_8]^{4-}$ mixed-valence compounds,
is well documented (32). $Mo(CN)_8^{3-}$ is strongly absorbing in the
near UV region where it undergoes efficient photoreduction together
with the oxidation of water (37). Hence, it was suggested that
polychromatic irradiation of $Cu(II)/[Mo(CN)_8]^{4-}$ leads to the mixed-
valence successor complex $Cu(I)/[Mo(CN)_8]^{3-}$ in the primary step
which is then converted to $Cu(I)/[Mo(CN)_8]^{4-}$ by a second photon as
shown in Scheme 4 (31, 36)

Scheme 4

$$Cu(II)/[Mo(CN)_8]^{4-} \underset{k_r}{\overset{h\nu_{IT}}{\rightleftharpoons}} Cu(I)/[Mo(CN)_8]^{3-}$$

$$H_2O \Big\downarrow h\nu_2$$

$$Cu(I)/Mo(CN)_8^{4-} + \cdot OH + H^+$$

(S = 4-nitroso N,N-dimethylaniline,
   spin trap)

$$S \Big\downarrow k_S$$

$$\cdots \qquad S-OH$$

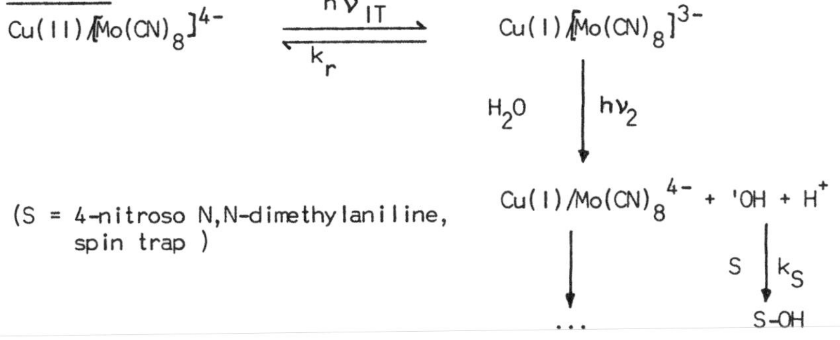

The formation of hydroxyl radicals was monitored by either scavenging with 4-nitroso N,N-dimethylaniline (4-NDMA) or spin trapping (36).

Various experiments confirm the unusual photochemical reaction pathway shown in Scheme 4: Hydroxyl radicals have been detected only during polychromatic irradiation, whereas monochromatic photolysis in the 365 nm or 492 nm region alone did not yield comparable results. Octacyanomolybdate(V) was proved not to form any ·OH radicals thermally. Uncontrolled thermal or photochemical reactions of the scavenger 4-NDMA could also be discarded. Furthermore, direct photochemical generation of $[Mo(CN)_8]^{3-}$ by UV light has been excluded by using cutt-off filters. Finally, thorough calculations of the stationary concentration of the $Cu(I)/[Mo(CN)_8]^{3-}$ mixed-valence intermediate, considering the intensity distribution of the incident light and the absorption by the diverse species present in the solution, do further support the unusual two-photon process given in Scheme 4. Thus, the application of photons as physical scavengers provides a further possibility to overcome fast back electron transfer processes in mixed-valence compounds.

Conclusion. There is an increasing interest in both photoinduced catalytic and photoassisted reactions, particularly in the field of homogeneous complex catalysis and other organic syntheses, in the search of unconventional information recording materials, in the storage and conversion of solar energy, and in modelling lightsensitive metalloenzymes. For a number of applications spectral sensitization of photocatalytic systems is required. It may be achieved by applying the concept of static sensitization. The IT excitation of Robin-Day class II mixed-valence compounds provides an interesting route to static sensitization.

The advantages in applying mixed-valence compounds are due to the fact that the energy of the IT transition may be varied by convenient synthetic procedures. In addition, the photochemical behavior may be predicted using the theoretical treatment proposed by Hush. However, the most serious restriction for the application of mixed-valence compoun ds in the static spectral sensitization arises from the fast back electron transfer. Therefore, very efficient scavenging reactions are required in order to suppress back electron transfer.

Literature Cited

1. Hennig, H.; Rehorek, D.; Archer, R. D. Coord.Chem.Rev. 1985, 61, 1.
2. Salomon, R. G. Tetrahedron 1983, 485.
3. Moggi, L.; Juris, A.; Sandrini, D.; Manfrin, M. F. Rev. Chem.Intermed. 1981, 4, 171.
4. Wrighton, M. S.; Ginley, D. S.; Schroeder, M. A.; Morse, D. L. Pure Appl. Chem. 1975, 4, 671.
5. Carassiti, V. EPA Newsl. 1984, 21, 12.
6. Mirbach, M. J. EPA Newsl. 1984, 20, 16.
7. Wubbels, G. G. Acc.Chem.Res. 1983, 16, 285.

8. Kisch, H.; Hennig, H. EPA Newsl. 1983, 19, 23.
9. Hennig, H.; Thomas, Ph.; Wagener, R.; Rehorek, D.; Jurdeczka, K. Z.Chem. 1977, 17, 241.
10. Hennig, H.; Hoyer, E.; Lippmann, E.; Nagorsnik, E.; Thomas, Ph.; Weissenfels, M. J.Signalaufzeichnungsm. 1978, 6, 31.
11. Hennig, H.; Hoyer, E.; Lippmann, E.; Nagorsnik, E.; Thomas, Ph.; Weissenfels, M.; Epperlein, J.; Dowidat, G. DDR Patent 123 024, 1976.
12. Hennig, H.; Rehorek, D.; Mann, G.; Wilde, H.; Salvetter, J.; Weissenfels, M.; Thomas, Ph.; Epperlein, J.; Rehorek, A. DDR Patent 146 351, 1979.
13. Förster, T. Ann.Phys. 1948, 2, 55.
14. Dexter, D. L. J.Chem.Phys. 1953, 21, 836.
15. Hennig, H.; Thomas, Ph.; Wagener, R.; Ackermann, M.; Benedix, R.; Rehorek, D. J.Signalaufzeichnungsm. 1981, 9, 269.
16. Balzani, V.; Moggi, L.; Manfrin, M. E.; Bolletta, F.; Laurence, G. S. Coord.Chem.Rev. 1975, 15, 321.
17. Robin, M. B.; Day, P. Adv.Inorg.Chem.Radiochem. 1967, 10, 247.
18. Allen, G. C.; Hush, N. S. Prog.Inorg.Chem. 1967, 8, 357.
19. Hush, N. S. Prog.Inorg.Chem. 1967, 8, 291.
20. Brown, D., Ed.; "Mixed-Valence Compunds"; Reidel: Dordrecht, The Netherlands, 1980.
21. Vogler, A.; Kunkely, H. Ber.Bunsenges.Phys.Chem. 1975, 79, 301.
22. Vogler, A.; Kunkely, H. Ber.Bunsenges.Phys.Chem. 1975, 79, 83.
23. Hennig, H.; Rehorek, A.; Rehorek, D.; Thomas, Ph.; Graness, G. Z.Chem. 1982, 22, 388.
24. Hennig, H.; Rehorek, A.; Rehorek, D.; Thomas, Ph. Z.Chem. 1982, 22, 417.
25. Hennig, H.; Rehorek, A.; Ackermann, M.; Rehorek, D.; Thomas, Ph. Z.anorg.allg.Chem. 1983, 496, 186.
26. Hennig, H.; Rehorek, A.; Rehorek, D.; Thomas, Ph. Inorg. Chim.Acta 1984, 86, 41.
27. Marcus, R. A. J.Chem.Phys. 1956, 24, 979.
28. Levich, V. G. In "Physical Chemistry: An Advanced Treatise"; Eyring, H.; Henderson, D.; Jost, W.; Eds.; Academie: New York, 1970; Vol. 9B.
29. Dogonadze, R. R. In "Reactions of Molecules at Electrodes"; Hush, N. S., Ed.; Wiley: New York, 1971.
30. Hennig, H.; Rehorek, A.; Rehorek, D.; Thomas, Ph.; Bäzold, D. Inorg.Chim.Acta 1983, 77, L 11.
31. Hennig, H.; Rehorek, A.; Rehorek, D.; Thomas, Ph. Z.Chem. 1982, 22, 418.
32. Gray, G. W.; Spence, T. J. Inorg.Chem. 1971, 10, 2751.
33. Stasicka, Z.; Bulska, H. Rocz.Chem. 1974, 48, 389.
34. Rehorek, D.; Salvetter, J.; Hantschmann, A.; Hennig, H.; Stasicka, Z.; Chodkowska, A. Inorg.Chim.Acta. 1979, 37, L 471.
35. Kiwi, J.; Kalyanasundaram, K.; Grätzel, M. Structure and Bonding 1982, 49, 37.
36. Rehorek, D.; Rehorek, A.; Thomas, Ph.; Hennig, H. Inorg. Chim.Acta 1982, 64, L 225.
37. Rehorek, D.; Janzen, E. G.; Stronks, H. J. Z.Chem. 1982, 22, 64.

RECEIVED December 2, 1985

9

Electrochemically Generated Transition Metal Complexes

Emissive and Reactive Excited States

A. Vogler, H. Kunkely, and S. Schäffl

Universität Regensburg, Institut für Anorganische Chemie, D-8400 Regensburg, Federal Republic of Germany

A variety of transition metal complexes (A) was subjected to an electrolysis by an alternating current in a simple undivided electrochemical cell. The compounds are reduced and oxidized at the same electrode. If the excitation energy of these compounds is smaller than the potential difference of the reduced (A^-) and oxidized (A^+) forms, back electron transfer may regenerate the complexes in an electronically excited state ($A^+ + A^- \rightarrow A^* + A$). These excited complexes may be emissive ($A^* \rightarrow A + h\nu$) and/or reactive ($A^* \rightarrow B$). Chemical transformations which accompany the ac electrolysis do not only proceed via excited states. As an important alternative the reduced or oxidized compounds can undergo a facile chemical change ($A^- \rightarrow B^-$ or $A^+ \rightarrow B^+$). Back electron transfer merely restores the original charges ($A^+ + B^- \rightarrow A + B$ or $A^- + B^+ \rightarrow A + B$). This mechanism and the ac electrolysis which proceeds via the generation of excited states are not unrelated processes. Hence the photoreaction and the ac electrolysis can lead to the same product irrespective of the intimate mechanism of the electrolysis. However, it is also possible that photolysis and electrolysis generate different products. Examples of ac electrolyses proceeding by these different mechanisms are discussed.

Bimolecular excited state electron transfer reactions have been investigated extensively during the last decade (1-3). Electron transfer is favored thermodynamically when the excitation energy E of an initially excited molecule A^* exceeds the potential difference of the redox couples involved in the electron transfer process.

$$A + h\nu \rightarrow A^*$$

$$A^* + B \rightarrow A^+ + B^- \qquad E(A^*) > E^o(A/A^+) - E(B^-/B)$$

0097-6156/86/0307-0120$06.00/0

Studies of such systems provided a better understanding of the mechanism of electron transfer processes in general. This reaction type is also the basis of almost any type of natural or artificial photosynthesis. Hence it is not surprising that many investigations have been devoted to excited state electron transfer reactions. On the contrary, the reversal of excited state electron transfer has found much less attention although it is certainly not less interesting. In the present paper various aspects of this reaction type are discussed. The products of a redox reaction may be generated in an excited state if to a first approximation the excitation energy is smaller than the potential difference of the associated redox couples.

$$A^{+} + B^{-} \rightarrow A^{*} + B \qquad E(A^{*}) < E^{o}(A/A^{+}) - E^{o}(B^{-}/B)$$

Generally, this energy requirement is only met when a strong oxidant reacts with a strong reductant. The excited state thus produced does not behave differently from that generated by light absorption. It can be deactivated by radiation or chemical transformations. Electron transfer induced emission (chemiluminescence, cl) is such a process. While it is well known for organic systems (4) there are not many observations of cl originating from transition metal complexes (5-12). The reactants can be prepared separately. Upon mixing, electron transfer takes place with concommitant emission of light. While this type of experiment is conceptionally very simple it may be difficult to accomplish due to practical or theoretical limitations. For example, this method cannot be applied when the redox partners A^{+} and B^{-} are not very stable and have only a short lifetime. In this case the redox agents must be prepared in situ. This can be done in two different ways. The redox catalysis represents one possibility. It may apply to highly exoergic redox reactions which do not proceed rapidly due to large activation energies. A suitable redox catalyst may speed up this reaction and finally take up the energy which is released by this redox process.

Redox catalysis leading to cl is illustrated by two examples. The oxidation of oxalate by Pb(IV) does not proceed readily although it is strongly favored thermodynamically. This reaction is catalyzed by $Ru(bpy)_3^{2+}$ with bipy = 2,2'bipyridine according to the following mechanism (13):

$$2Ru(bipy)_3^{2+} + PbO_2 + 4H^{+} \rightarrow 2Ru(bipy)_3^{3+} + Pb^{2+} + 2H_2O$$

$$Ru(bipy)_3^{3+} + C_2O_4^{2-} \rightarrow Ru(bipy)_3^{2+} + CO_2 + CO_2^{-}$$

$$Ru(bipy)_3^{3+} + CO_2^{-} \rightarrow [Ru(bipy)_3^{2+}]^{*} + CO_2$$

$$[Ru(bipy)_3^{2+}]^{*} \rightarrow Ru(bipy)_3^{2+} + h\nu$$

The reaction of $Ru(bipy)_3^{2+}$ with CO_2^{-} is the energy releasing electron transfer step leading to the formation of the electronically excited (*) complex. It cannot be carried out separately. The strong oxidant CO_2^{-} must be prepared in situ since it is a short-lived radical.

The catalyzed decomposition of energy-rich organic peroxides is another typical reaction of this type. It was called "chemically initiated electron-exchange luminescence" (CIEEL) by Schuster, who used organic compounds as redox catalysts (14). However, transition metal complexes work as well. The complex Re(o-phen)$(CO)_3$Cl (R) (o-phen = o-phenanthroline) catalyzes the decomposition of tetraline-hydroperoxide (T) to the ketone α-tetralone (K) and water according to the mechanism (15):

$$R + T \rightarrow R^+ + T^-$$

$$T^- \rightarrow K^- + H_2O$$

$$R^+ + K^- \rightarrow R^* + K$$

$$R^* \rightarrow R + h\nu$$

The reaction of the ketyl radical anion with the oxidized rhenium complex is the energy-releasing electron transfer step. This reaction cannot be carried out separately. While ketyl radical anions are stable species, the oxidized complex is not stable and must be generated as short-lived intermediate.

Electrolysis represents another, very elegant method to prepare suitable redox pairs in situ which are generated by cathodic reduction and anodic oxidation. By application of an alternating current the redox pair is generated at the same electrode. Back electron transfer takes place from the electrogenerated reductant to the oxidant near the electrode surface. At an appropriate potential difference this annihilation reaction leads to the formation of excited products. As a result an emission (electrogenerated chemiluminescence, ecl) may be observed (16). Redox pairs of limited stability can be investigated by ac electrolysis. The frequency of the ac current must be adjusted to the lifetime of the more labile redox partner. Many organic compounds have been shown to undergo ecl (17-19). Much less is known about transition metal complexes. Most of the observations involve $Ru(bipy)_3^{2+}$ and related complexes which possess emissive charge transfer (CT) metal-to-ligand (M→L) excited states (13,20-31). The organometallic compound Re(o-phen)$(CO)_3$Cl is a further example of this category (32). Palladium and platinum porphyrins with emitting intraligand excited states are also ecl active (33). Under suitable conditions ecl was also observed for $Cr(bipy)_3^{3+}$ (27). In this case the emission originates from a ligand field (LF) excited state. Almost all of the ecl active transition metal complexes contain bipy or related ligands. It was therefore of interest to see if ecl could be extended to other types of transition metal compounds which have emitting states of different origin.

Furthermore, excited states generated electrochemically may be not only emissive but also reactive. The possibility of such an "electrophotochemistry" (epc) has been considered before (34). But real examples were discovered only quite recently and will be discussed later (35,36). However, chemical transformations induced by ac electrolysis may not only proceed via excited states. Other mechanisms can be also consistent with these observations. While this extends the range of reaction types of ac electrolysis, it complicates the elucidation of the real mechanism. Examples of the various reaction types are presented in the following sections.

Electrogenerated Chemiluminescence

For our ecl studies a very simple technique was employed. A 1-cm spectrophotometer cell was used as an undivided electrochemical cell. It was equipped with two platinum foil electrodes which were directly connected to a sine wave generator as an ac voltage source. Much more sophisticated methods have been described in the literature (16) but this simple design permitted the observation of ecl which appears at both electrodes.

Recently we observed ecl of the binuclear platinum complex tetrakis(diphosphonato)diplatinate(II) ($Pt_2(pop)_4^{4-}$) (37). This anion has attracted much attention due to its intense green luminescence in room temperature solution (38-40) ($\phi = 0.52$) (41). The excited state of this complex undergoes oxidative (42) and reductive quenching (41). From the quenching experiments the redox potentials were estimated to be $E^o = -1.4$ V vs. SCE for the reduction and $E^o \sim 1$ V for the oxidation of $Pt_2(pop)_4^{4-}$ (41). The potential difference of 2.4 V almost matches the energy of the phosphorescing triplet (~ 2.5 eV) of $Pt_2(pop)_4^{4-}$. Consequently, it should be possible to observe ecl of this complex. However, the reduced ($Pt_2(pop)_4^{5-}$) (43) and oxidized ($Pt_2(pop)_4^{3-}$) (44,45) forms are not stable, but decay rapidly in solution. Hence an ecl of $Pt_2(pop)_4^{4-}$ will only take place if the subsequent generation of both redox partners occurs before they undergo a decay.

The ecl experiment was carried out in a solution of acetonitrile with Bu_4NBF_4 as supporting electrolyte (37). At an ac voltage of 4 V, a frequency of 280 Hz, and a current of 13 mA a green emission appeared at the electrodes. It was identical with the phosphorescence (λ_{max} = 517 nm) of $Pt_2(pop)_4^{4-}$. This observation is consistent with the following reaction sequence:

$$Pt_2(pop)_4^{4-} + e^- \rightarrow Pt_2(pop)_4^{5-} \quad \text{cathodic cycle}$$

$$Pt_2(pop)_4^{4-} - e^- \rightarrow Pt_2(pop)_4^{3-} \quad \text{anodic cycle}$$

$$Pt_2(pop)_4^{5-} + Pt_2(pop)_4^{3-} \rightarrow [Pt_2(pop)_4^{4-}]^* + Pt_2(pop)_4^{4-}$$

$$[Pt_2(pop)_4^{4-}]^* \rightarrow Pt_2(pop)_4^{4-} + h\nu$$

The reduction and oxidation of $Pt_2(pop)_4^{4-}$ takes place at the same electrode. Back electron transfer generates one of the starting ions in the excited triplet state which undergoes phosphorescence. Interestingly, the fluorescence of the complex which appears on photoexcitation at λ_{max} = 407 nm, is not observed in the ecl experiment. This is not surprising since the back electron transfer does not provide enough energy (~ 2.4 V) to populate the emitting singlet (~ 3.3 V).

It should be mentioned here that the processes which are involved in the appearance of an ecl of $Pt_2(pop)_4^{4-}$ are associated with changes in the metal-metal bonding of this binuclear complex (38-40, 42,44,46,47). The Pt-Pt bond order which is zero in the ground state is increased to 0.5 by oxidation as well as by reduction. The annihilation reaction leads to the formation of $Pt_2(pop)_4^{4-}$ as the ground (bond order = 0) and excited state (bond order = 1). A related case which was reported quite recently is the ecl of $Mo_6Cl_{12}^{2-}$. The metal-

metal bonding of the cluster is involved in the redox processes which are associated with the ecl (48).

Electrogeneration of Excited Complexes Undergoing Emission and Reaction

The electrochemical generation of excited states may not only lead to an emission. In addition or as an alternative the excited state can undergo a chemical reaction ("electrophotochemistry", epc) as it would occur upon light absorption (photochemistry). In the ecl experiments the observation of luminescence is by itself a proof for the generation of excited states. But the fact that electrolysis and photolysis both lead to the formation of the same product does not prove the electrochemical generation of an excited state (see below). For this reason it is an advantage to study compounds which are simultaneously photoemissive and photoreactive. A positive correlation between ecl and the electrochemical reaction is a good indication that the chemical transformation is indeed associated with an excited state. In this case the electrochemical reaction is a true epc. Upon ac electrolysis the complex $Ru(bipy)_3^{2+}$ undergoes simultaneously ecl and epc (49).

The well-known photoluminescence of $Ru(bipy)_3^{2+}$ occurs from the lowest excited state which is of the CT (Ru→bipy) type (50,51). The emission appears in aqueous as well as in non-aqueous solutions. While the complex is hardly light-sensitive in water (52) it can undergo an efficient photosubstitution of a bipy ligand in non-aqueous solvents (50,51,53-56). The reactive excited state seems to be a LF state which lies at slightly higher energies but can be populated thermally from the emitting CT state (50-52,55-58). According to these observations the electrochemical generation of excited $Ru(bipy)_3^{2+}$ in non-aqueous solutions should not only be accompanied by the well-known ecl but also by an epc. Moreover, the efficiency of both processes should show a positive correlation. Preliminary experiments indeed provide evidence for a simultaneous occurance of ecl and epc of $Ru(bipy)_3^{2+}$ (49).

An ac electrolysis of $[Ru(bipy)_3]Cl_2$ was carried out in a spectrophotometer cell as an undivided electrochemical cell equipped with platinum foil electrodes. Acetonitrile was used as solvent and Bu_4NBF_4 served as supporting electrolyte. The electrolysis led to the typical ecl of $Ru(bipy)_3^{2+}$ (20,21,23,25). Simultaneously, the complex underwent a chemical change. The spectral variations which accompanied the electrolysis (Figure 1) were very similar to those observed during the photolysis of the same solution (λ_{irr} > 335 nm). The product of electrolysis and photolysis was not yet identified definitely, but according to a preliminary characterization it seems to be $[Ru(bipy)_2(CH_3CN)Cl]^+$. However, it is important to note that all changes of the experimental conditions (e.g. variations of the ac frequency, stirring of the solution) which lead to a change of the ecl intensity also caused a corresponding change of the efficiency of the electrochemical reaction. These observations are good indication that both processes proceed via the generation of excited $Ru(bipy)_3^{2+}$. It is suggested that the ac electrolysis can be described by the following mechanism:

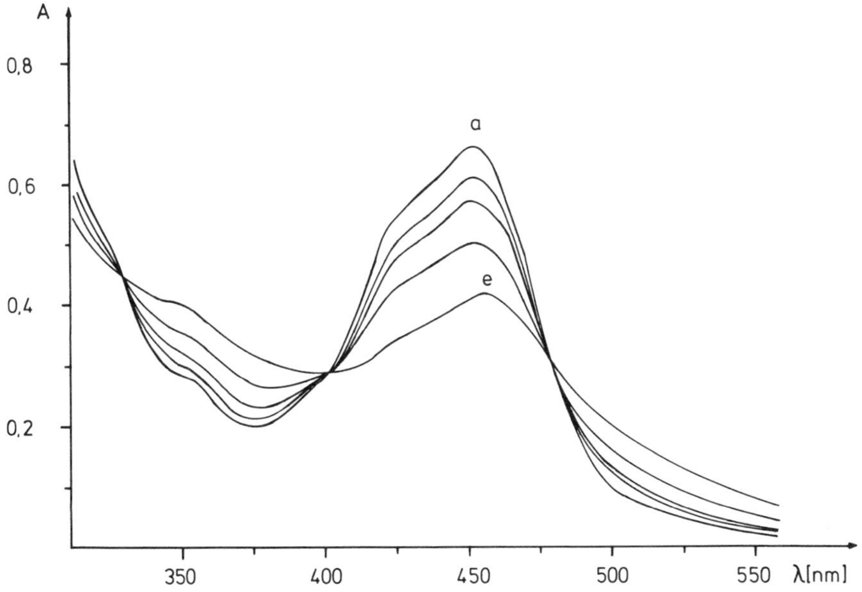

Figure 1. Spectral changes during ac electrolysis of 10^{-4} M $[Ru(bipy)_3]Cl_2$ in acetonitrile/0.1 M Bu_4NBF_4 at (a) 0 and (e) 120 - min electrolysis time at 3 V/20 Hz and 30 mA, 1-cm cell.

$$Ru(bipy)_3^{2+} + e^- \rightarrow Ru(bipy)_3^{+} \quad \text{cathodic cycle}$$

$$Ru(bipy)_3^{2+} - e^- \rightarrow Ru(bipy)_3^{3+} \quad \text{anodic cycle}$$

$$Ru(bipy)_3^{+} + Ru(bipy)_3^{3+} \rightarrow [Ru(bipy)_3^{2+}]^* + Ru(bipy)_3^{2+} \quad \text{annihilation}$$

$$[Ru(bipy)_3^{2+}]^* \rightarrow Ru(bipy)_3^{3+} + h\nu \quad \text{ecl}$$

$$[Ru(bipy)_3^{2+}]^* + Cl^- + CH_3CN \rightarrow Ru(bipy)_2(CH_3CN)Cl^+ + bipy \quad \text{epc}$$

The conclusion that the electrochemical reaction of $Ru(bipy)_3^{2+}$ takes place via an excited state is also supported by other observations. According to electrochemical studies the reduced and oxidized complexes $Ru(bipy)_3^{+}$ and $Ru(bipy)_3^{3+}$ are fairly stable and not expected to undergo rapid chemical transformations (21,23,25,50).

Electrogeneration of Reactive Excited States

Most compounds which undergo a photochemical reaction do not simultaneously show photoluminescence. It is then more difficult to prove that a reaction induced by ac electrolysis proceeds via the intermediate formation of excited states. A different mechanism may be in operation. In this case the chemical transformation occurs in the reduced and/or oxidized form. The back electron transfer merely regenerates the charges of the starting compound:

$$A + e^- \rightarrow A^- \quad \text{cathodic cycle}$$

$$A - e^- \rightarrow A^+ \quad \text{anodic cycle}$$

$$A^- \rightarrow B^- \quad \text{chemical reaction}$$

$$A^+ + B^- \rightarrow A + B \quad \text{annihilation}$$

Nevertheless, the result of the electrolysis may be the same as that of the photolysis, because the origin of the reactivity is similar in both cases. For example, a bond weakening may occur upon reduction or oxidation since an electron is added to an antibonding π^* orbital or removed from a bonding π orbital. The same changes take place upon $\pi\pi^*$ excitation.

A case in question is the ac electrolysis of the complex Re(trans-SP)$_2$(CO)$_3$Cl (SP = 4-styrylpyridine) (59). It was shown before that the coordinated ligand SP undergoes a photochemical trans/cis isomerization (60). The reactive excited state is the lowest $\pi\pi^*$ intraligand (IL) state, which is not luminescent. The ac electrolysis leads also to the trans/cis isomerization of the coordinated ligand (59). Hence it is a reasonable assumption that the electrolysis proceeds via the generation of the $\pi\pi^*$ IL state:

$$Re(trans\text{-}SP)_2(CO)_3Cl + e^- \rightarrow Re(trans\text{-}SP)_2(CO)_3Cl^-$$

$$Re(trans\text{-}SP)_2(CO)_3Cl - e^- \rightarrow Re(trans\text{-}SP)_2(CO)_3Cl^+$$

$$Re(trans\text{-}SP)_2(CO)_3Cl^+ + Re(trans\text{-}SP)_2(CO)_3Cl^-$$

$$\rightarrow Re(trans\text{-}SP)_2(CO)_3Cl^* + Re(trans\text{-}SP)_2(CO)_3Cl$$

$$Re(trans\text{-}SP)_2(CO)_3Cl^* \rightarrow Re(cis\text{-}SP)_2(CO)_3Cl$$

However, as an alternative the isomerization may take place in the reduced and/or oxidized form:

$$Re(trans\text{-}SP)_2(CO)_3Cl^- \rightarrow Re(cis\text{-}SP)_2(CO)_3Cl^-$$

$$Re(trans\text{-}SP)_2(CO)_3Cl^+ \rightarrow Re(cis\text{-}SP)_2(CO)_3Cl^+$$

$$Re(cis\text{-}SP)_2(CO)_3Cl^+ + Re(cis\text{-}SP)_2(CO)_3Cl^-$$

$$\rightarrow 2\ Re(cis\text{-}SP)_2(CO)_3Cl$$

Inspection of some additional data does not lead to a distinction between the two possibilities. The potential difference of the reduced and oxidized complex (2.94 V) exceeds the electronic excitation energy of the neutral complex (~ 2.1 eV) (59). On energetic grounds the electrochemical generation of excited states is certainly possible. The related complex Re(o-phen)$(CO)_3Cl$ is not light sensitive but is photoluminescent and also ecl active (32). By analogy one might assume that the electrolysis of both complexes proceeds by the same mechanism. On the other side, cyclic voltammetry shows that the oxidized form of Re(trans-SP)$_2$(CO)$_3$Cl is fairly stable but the reduced complex decays irreversibly (59). Only at large scan rates (100 Vs^{-1}) the reduction wave shows beginning reversibility. It is then not unreasonable to assume that the ligand isomerization takes place in the reduced complex. The final back electron transfer would merely restore the neutral complex. Of course, in the absence of ecl any direct proof of the electrochemical generation of excited states is difficult to obtain. Nevertheless, indirect but conclusive evidence showed indeed that an excited state mechanism led to the electrochemical isomerization of the complex.

Experiments were carried out to determine if during the ac electrolysis the ligand isomerization requires the formation of the reduced and oxidized form (59). This would indicate an excited state mechanism. If the intermediate formation of the reduced or oxidized complex is sufficient to induce the isomerization, excited states are not required. First support in favor of a true epc was obtained by the results of the ac electrolysis of Re(trans-SP)$_2$(CO)$_3$Cl in the presence of redox buffers. Tetramethyl-p-phenylenediamine (TMPD) was used as reductant and the paraquat cation (PQ^{2+}) served as oxidant. In the presence of an excess of TMPD the complex is still reduced, but TMPD is oxidized during the electrolysis. Since the oxidation potential of TMPD is much lower than that of the complex, the annihilation reaction of the complex anion and $TMPD^+$ does not provide enough energy to generate the complex in the excited state. Quite an analogous situation applies to the electrolysis in the presence of PQ^{2+}. Now the complex is oxidized but PQ^{2+} reduced. Again, the potential difference of the complex cation and PQ^{+} is smaller than the excitation energy of Re(trans-SP)$_2$(CO)$_3$Cl. In both experiments the ligand isomerization was essentially suppressed. Consequently the intermediate formation of the complex cation or anion alone cannot be responsible for the isomerization.

Additional evidence in support of an excited state mechanism was obtained by continuous potential step chronocoulometric experiments (59). When the electrode potential was stepped only over the oxidation potential of the complex at a frequency of 10 Hz a slow net oxidation took place. Potential steps involving only the reduction wave led to rapid net reduction but no ligand isomerization. The isomerization occurred only when the potential steps included both reduction and oxidation of the complex. Since the voltammograms of Re(trans-SP)$_2$(CO)$_3$Cl and Re(cis-SP)$_2$(CO)$_3$Cl are virtually indistinguishable, the ligand isomerization was not accompanied by a potential change. No net Faradaic process was observed.

The conclusion that the ac electrolysis of Re(trans-SP)$_2$(CO)$_3$Cl proceeds via excited states is also supported by the direction of isomerization. In thermal reactions of stilbene derivatives and radicals cis to trans conversions are generally observed (61). Contrary to this behavior the photolysis and ac electrolysis lead to energetically uphill trans to cis isomerization.

AC Electrolysis Without Generation of Excited States

As discussed above, a chemical transformation which occurs during the ac electrolysis does not require the intermediate formation of excited states. The chemical reaction may take place in the reduced and/or oxidized form of a compound. Nevertheless, in this case the electrolysis may still lead to the same products as those of the photolysis due to the obvious relationship between electronic excitation and redox processes. It will be then quite difficult to elucidate the mechanism of electrolysis. This reaction type may apply to the electrochemical substitution of $Cr(CO)_6$ (59).

The ac electrolysis of $Cr(CO)_6$ in CH_3CN was accompanied by the same spectral changes (Figure 2) as those observed in the photolysis of the same solution with 333-nm light. In both cases $Cr(CO)_6$ was converted to $Cr(CO)_5(CH_3CN)$ (59). According to Pickett and Pletcher (62) $Cr(CO)_6$ shows a reversible oxidation wave at 1.52 V vs. SCE; the reduction wave at -2.66 V is irreversible and was attributed to a rapid or even concerted loss of CO from $Cr(CO)_6^-$ to give $Cr(CO)_5^-$. A reverse peak in the cyclic voltammogram at -2.1 V shows the reoxidation of the latter species to the coordinatively unsaturated $Cr(CO)_5$ which can be stabilized by the addition of a solvent molecule as a sixth ligand. Consequently, the ac electrolysis may proceed according to the following reaction scheme without invoking an electronically excited state in the back electron transfer (59):

$Cr(CO)_6 - e^- \rightarrow Cr(CO)_6^+$ anodic cycle

$Cr(CO)_6 + e^- \rightarrow Cr(CO)_6^-$ cathodic cycle

$Cr(CO)_6^- \rightarrow Cr(CO)_5^- + CO$ ligand dissociation

$Cr(CO)_6^+ + Cr(CO)_5^- \rightarrow Cr(CO)_6 + Cr(CO)_5$ annihilation

$Cr(CO)_5 + CH_3CN \rightarrow Cr(CO)_5(CH_3CN)$ ligand addition

This mechanism and the photolysis have in common that the addition of an electron to the antibonding e_g orbitals induces the dissociation of a CO ligand.

As a further possibility the ac electrolysis may lead to other products than those of the photolysis. In this case an excited state mechanism is, of course, excluded. Although there is a certain similarity between the electronic structure of an excited state and the reduced or oxidized form of a molecule, they are not identical. Consequently, it is not surprising when photolysis and electrolysis do not yield the same product. Another reason for such an observation may be the different lifetimes. An excited state can be extremely short-lived. Non-reactive deactivation could then compete successfully with a photoreaction. The compound is not light-sensitive. On the contrary, the reduced and oxidized intermediates generated by ac electrolysis should have comparably long life times which may permit a reaction. The ac electrolysis of Ni(II)(BABA)(MNT) (BABA = biacetyl-bis(anil) and MNT^{2-} = disulfidomaleonitrile) is an example of this reaction type (63).

The complex Ni(BABA)(MNT) (64) is not light sensitive (λ_{irr} > 400 nm) in solutions of acetonitrile but undergoes an ac electrolysis which is accompanied by spectral changes as shown in Figure 3. According to a preliminary analysis of the products the electrolysis leads to a ligand exchange:

$$2\ Ni^{II}(BABA)(MNT) \rightarrow Ni^{II}(BABA)_2^{2+} + Ni^{II}(MNT)_2^{2-}$$

The electrochemistry of Ni(BABA)(MNT) has been investigated recently (64). The first reduction occurs reversibly at $E_0' = -0.7$ V vs. SCE. However, the oxidation is irreversible ($E_{P/2} = 0.8$ V). For the related complex Ni(o-phen)$(S_2C_2Ph_2)$ it was shown that the cation Ni(o-phen)$(S_2C_2Ph_2)^+$ generated by photooxidation in halocarbon solvents undergoes a facile ligand exchange to yield the symmetric complexes Ni(o-phen)$_2^{2+}$ and Ni$(S_2C_2Ph_2)_2$ (65). According to these considerations the ac electrolysis can be rationalized by the following reaction scheme:

$Ni(BABA)(MNT) + e^- \rightarrow Ni(BABA)(MNT)^-$ cathodic cycle

$Ni(BABA)(MNT) - e^- \rightarrow Ni(BABA)(MNT)^+$ anodic cycle

$2\ Ni(BABA)(MNT)^+ \rightarrow Ni(BABA)_2^{2+} + Ni(MNT)_2$ ligand exchange

$2\ Ni(BABA)(MNT)^- + Ni(MNT)_2 \rightarrow 2\ Ni(BABA)(MNT) + Ni(MNT)_2^{2-}$ electron transfer

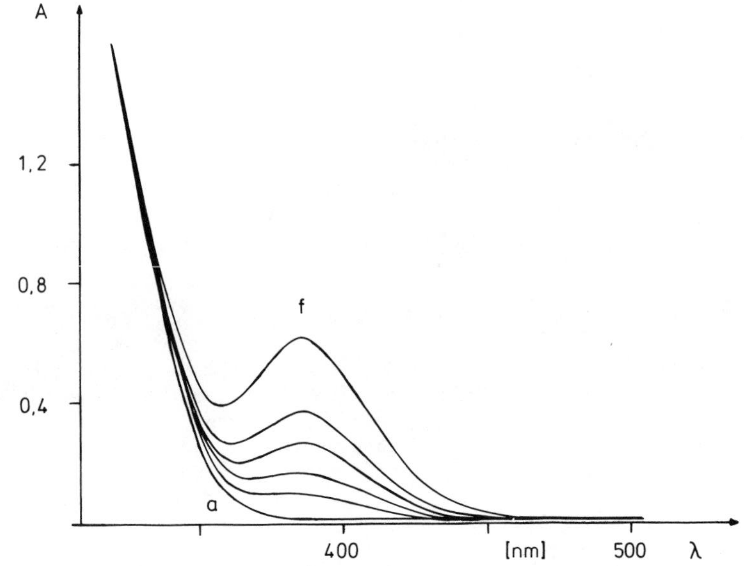

Figure 2. Spectral changes during ac electrolysis of 6.5×10^{-4} M $Cr(CO)_6$ in acetonitrile/0.05 m Bu_4NBF_4 at (a) 0 and (f) 300-min electrolysis time at 2.5 V/10 Hz and 5 mA, 1-cm cell.

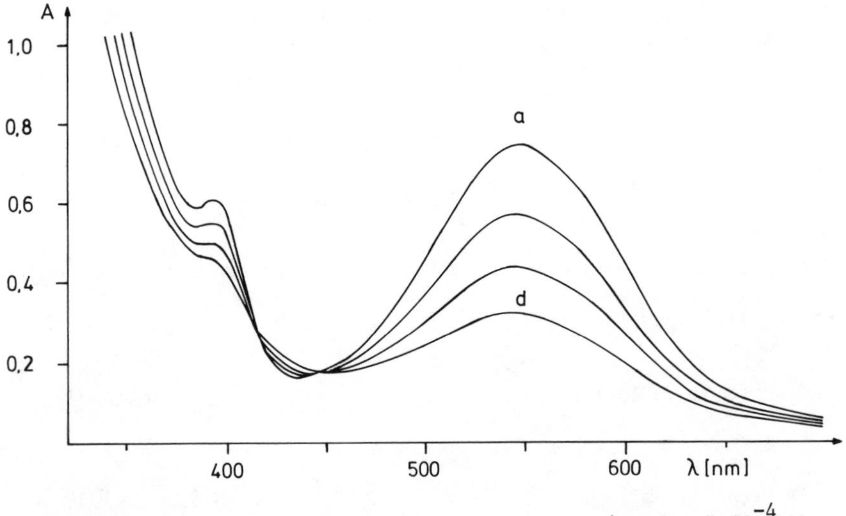

Figure 3. Spectral changes during ac electrolysis of 1.5×10^{-4} M Ni(BABA)(MNT) in acetonitrile/0.1 M Bu_4NBF_4 at (a) 0 and (d) 30-min electrolysis time at 3 V/20 Hz and 40 mA, 1-cm cell.

The ligand exchange produces $Ni(MNT)_2$ which is not stable but a strong oxidant (66). It oxidizes apparently the reducing anion $Ni(BABA)(MNT)^-$ in two subsequent electron transfer steps.

Reactions Related to the AC Electrolysis

There are other reactions of transition metal complexes which are relevant to our observations on the ac electrolysis. Recently, new mechanisms of ligand substitution reactions have been reported which are characterized by electron transfer reactions as key steps although the overall reactions are not redox processes, e.g.,

$$ML + e^- \rightarrow ML^-$$

$$ML^- + L' \rightarrow ML'^- + L$$

$$ML'^- - e^- \rightarrow ML'$$

overall: $$ML + L' \rightarrow ML' + L$$

The substitutionally labile complex may be generated not only by reduction but by oxidation as well. An immediate relationship of such a reaction to the ac electrolysis proceeding without generation of excited states can be recognized. The initial production of the substitutionally labile oxidation state of ML can be achieved electrochemically (67-76), chemically (75-77) or photochemically (78). In the electrochemical experiments reduction or oxidation was accomplished by a direct current. In most cases these processes are catalytic chain reactions with Faradaic efficiencies much larger than unity. Electrochemical substitution of $M(CO)_6$ with M = Cr, Mo, W was carried out by cathodic reduction to $M(CO)_6^-$ which dissociates immediately to yield $M(CO)_5^-$. Upon anodic reoxidation at the other electrode coordinatively unsaturated $M(CO)_5$ is formed and stabilized by addition of a ligand L to give $M(CO)_5L$ (68).

Photochemical substitution via a labile oxidation state may occur by excited-state electron transfer. If the metal complex has a long-lived excited state, it can undergo an electron exchange with a reductant or oxididant in a bimolecular reaction. The labile reduced or oxidized complex thus produced is susceptible to a ligand substitution. A catalytic chain reaction takes place when the substituted complex in the labile oxidation state undergoes a further electron exchange with another unsubstituted complex. The chain terminates by back electron transfer between the labile oxidation state and the external redox partner which was generated initially. The cation $Re(o\text{-}phen)(CO)_3(CH_3CN)^+$ undergoes this new type of photosubstitution (78). The occurrence of a chain reaction was confirmed by the quantum yields which were as large as $\phi = 24$ depending on the experimental conditions. Of course, the efficiency of the usual photosubstitutions which originate from LF excited states of metal complexes do not exceed unity.

Conclusion

The use of ac electrolysis in all its variations is certainly an interesting and valuable technique for study of the mechanism of electron transfer reactions. The generation of a short-lived redox pair as chemical intermediates is an important feature of the ac electrolysis. In the future it may even be developed to synthetic applications irrespective of the mechanistic details. In some cases it could be a convenient alternative to photochemical reactions. In other cases it represents a new reaction type which has no precedent.

Acknowledgments

We thank Professor Andreas Merz for helpful discussions. Financial support of this work by the Deutsche Forschungsgemeinschaft and the Fonds der Chemischen Industrie is gratefully acknowledged.

Literature Cited

1. Balzani, V.; Bolletta, F.; Gandolfi, M. T.; Maestri, M. Top. Curr. Chem. 1978, 75, 1.
2. Meyer, T. J. Acc. Chem. Res. 1978, 11, 94.
3. Sutin, N.; Creutz, C. Adv. Chem. Ser. 1978, 168, 1.
4. Schuster, G. B.; Schmidt, S. P. Adv. Phys. Org. Chem. 1982, 18, 187.
5. Lyttle, F. E.; Hercules, D. M. Photochem. Photobiol. 1971, 13, 123.
6. Martin, J. E.; Hart, E. J.; Adamson, A. W.; Halpern, J. J. Am. Chem. Soc. 1972, 94, 9238.
7. Gafney, H. D.; Adamson, A. W. J. Chem. Ed. 1975, 52, 480.
8. Jonah, C. D.; Matheson, M. S.; Meisel, D. J. Am. Chem. Soc. 1978, 100, 1449.
9. Bolletta, F.; Rossi, A.; Balzani, V. Inorg. Chim. Acta 1981, 53, L23.
10. Vogler, A.; El-Sayed, L.; Jones, R. G.; Namnath, J.; Adamson, A. W. Inorg. Chim. Acta 1981, 53, L35.
11. Balzani, V.; Bolletta, F. J. Photochem. 1981, 17, 479.
12. Bolletta, F.; Balzani, V. J. Am. Chem. Soc. 1982, 104, 4250.
13. Rubinstein, I.; Bard, A. J. J. Am. Chem. Soc. 1981, 103, 512.
14. Schuster, G. B. Acc. Chem. Res. 1979, 12, 336.
15. Vogler, A.; Kunkely, H. Angew. Chem. Int. Ed. Engl. 1981, 20, 469.
16. Faulkner, L. R.; Bard, A. J. In "Electroanalytical Chemistry"; Bard, A. J., Ed.; Marcel Dekker Inc.: New York, 1977; Vol. 10, p. 1.
17. Faulkner, L. R.; Glass, R. S. In "Chemical and Biological Generation of Excited States"; Adam, W.; Cilento, G., Eds.; Academic Press, New York, 1982; chapter 6 and references cited therein.
18. Park, S.-M.; Tryk, D. A. Rev. Chem. Intermediates 1981, 4, 43.
19. Pragst, F. Z. Chem. 1978, 18, 41.
20. Tokel, N. E.; Bard, A. J. J. Am. Chem. Soc. 1972, 94, 2862.
21. Tokel-Takvoryan, N. E.; Hemingway, R. E.; Bard, A. J. J. Am. Chem. Soc. 1973, 95, 6582.
22. Chang, M. M.; Saji, T.; Bard, A. J. J. Am. Chem. Soc. 1977, 99, 5399.
23. Wallace, W. L.; Bard, A. J. J. Phys. Chem. 1979, 83, 1350.

24. Rubinstein, I.; Bard, A. J. J. Am. Chem. Soc. 1980, 102, 6641.
25. Luttmer, J. D.; Bard, A. J. J. Phys. Chem. 1981, 85, 1155.
26. Rubinstein, I.; Bard, A. J. J. Am. Chem. Soc. 1981, 103, 5007.
27. Bolletta, F.; Ciano, M.; Balzani, V.; Serpone, N. Inorg. Chim. Acta 1982, 62, 207.
28. Glass, R. S.; Faulkner, L. R. J. Phys. Chem. 1981, 85, 1160.
29. Itoh, K.; Honda, K. Chem. Lett. 1979, 99.
30. Abruña, H. D.; Bard, A. J. J. Am. Chem. Soc. 1982, 104, 2641.
31. Gonzales-Velasco, J.; Rubinstein, I.; Crutchley, R. J.; Lever, A. B. P.; Bard, A. J. Inorg. Chem. 1983, 22, 822.
32. Luong, J. C.; Nadjo, L.; Wrighton, M. S. J. Am. Chem. Soc. 1978, 100, 5790.
33. Tokel-Takvoryan, N. E.; Bard, A. J. Chem. Phys. Lett. 1974, 25, 235.
34. Park, S. M.; Bard, A. J. Chem. Phys. Lett. 1976, 38, 257.
35. The thermal generation of reactive excited states ("photochemistry without light") has been reported before (36).
36. White, E. H.; Miano, J. D.; Watkins, C. J.; Breaux, E. J. Angew. Chem. Int. Ed. Engl. 1974, 13, 229 and references cited therein.
37. Vogler, A.; Kunkely, H. Angew. Chem. Int. Ed. Engl. 1984, 23, 316.
38. Fordyce, W. A.; Brummer, J. G.; Crosby, G. A. J. Am. Chem. Soc. 1981, 103, 512.
39. Rice, S. F.; Gray, H. B. J. Am. Chem. Soc. 1983, 105, 4571.
40. Che, C.-M.; Butler, L. G.; Gray, H. B.; Crooks, R. M.; Woodruff, W. H. J. Am. Chem. Soc. 1983, 105, 5492.
41. Heuer, W. B.; Totten, M. D.; Rodman, G. S.; Hebert, E. J.; Tracy, H. J.; Nagle, J. K. J. Am. Chem. Soc. 1984, 106, 1163.
42. Che, C.-M.; Butler, L. G.; Gray, H. B. J. Am. Chem. Soc. 1981, 103, 7796.
43. Che, C.-M.; Atherton, S. J.; Butler, L. G.; Gray, H. B. J. Am. Chem. Soc. 1984, 106, 5143.
44. Che, C.-M.; Herbstein, F. H.; Schaefer, W. P.; Marsh, R. E.; Gray, H. B. J. Am. Chem. Soc. 1983, 105, 4604.
45. Bryan, S. A.; Dickson, M. K.; Roundhill, D. M. J. Am. Chem. Soc. 1984, 106, 1882.
46. Che, C.-M.; Schaefer, W. P.; Gray, H. B.; Dickson, M. K.; Stein, P. B.; Roundhill, D. M. J. Am. Chem. Soc. 1982, 104, 4253.
47. Stein, P.; Dickson, M. K.; Roundhill, D. M. J. Am. Chem. Soc. 1983, 105, 3489.
48. Nocera, D. G.; Gray, H. B. J. Am. Chem. Soc. 1984, 106, 824.
49. Schäffl, S.; Kunkely, H.; Vogler, A., unpublished results.
50. Kalyanasundaram, K. Coord, Chem. Rev. 1982, 46, 159.
51. Watts, R. J. J. Chem. Ed. 1983, 60, 834.
52. Van Houten, J.; Watts, R. J. Inorg. Chem. 1978, 17, 3381.
53. Hoggard, P. E.; Porter, G. B. J. Am. Chem. Soc. 1978, 100, 1457.
54. Gleria, M.; Minto, F.; Beggiato, G.; Bortolus, P. J. Chem. Soc., Chem. Comm. 1978, 285.
55. Durham, B.; Caspar, J. V.; Nagle, J. K.; Meyer, T. J. J. Am. Chem. Soc. 1982, 104, 4803.
56. Allen, G. H.; White, R. P.; Rillema, D. P.; Meyer, T. J. J. Am. Chem. Soc. 1984, 106, 2613.
57. Caspar, J. V.; Meyer, T. J. Inorg. Chem. 1983, 22, 2444.

58. Caspar, J. V.; Meyer, T. J. J. Am. Chem. Soc. 1983, 105, 5583.
59. Kunkely, H.; Merz. A.; Vogler, A. J. Am. Chem. Soc. 1983, 105, 7241.
60. Wrighton, M. S.; Morse, D. L.; Pdungsap, L. J. Am. Chem. Soc. 1975, 97, 2073.
61. Cheim, C. U.; Wang, H. C.; Szwarc, M.; Bard, A. J.; Itaya, K. J. Am. Chem. Soc. 1980, 102, 3100.
62. Picket, C. J.; Pletcher, D. J. J. Chem. Soc., Dalton Trans. 1976, 749.
63. Schäffl, S.; Vogler, A., unpublished results.
64. Vogler, A.; Kunkely, H.; Hlavatsch, J.; Merz, A. Inorg. Chem. 1984, 23, 506.
65. Vogler, A.; Kunkely, H. Angew. Chem. Int. Ed. Engl. 1981, 20, 386.
66. Davison, A.; Edelstein, N.; Holm, R. H.; Maki, A. H. Inorg. Chem. 1963, 2, 1227.
67. Bezems, G. J.; Rieger, P. H.; Visco, S. J. Chem. Soc., Chem. Comm. 1981, 265.
68. Grobe, J.; Zimmermann, H. Z. Naturforsch. 1981, 36b, 301.
69. Tanaka, K.; U-eda, K.; Tanaka, T. J. Inorg. Nucl. Chem. 1981, 43, 2029.
70. Hershberger, J. W.; Kochi, J. K. J. Chem. Soc., Chem. Comm. 1982, 212.
71. Hershberger, J. W.; Klingler, R. J.; Kochi, J. K. J. Am. Chem. Soc. 1982, 104, 3034.
72. Darchen, A.; Mahe, C.; Patin, H. J. Chem. Soc., Chem. Comm. 1982, 243.
73. Miholová, D.; Vlček, A. A. J. Organometal. Chem. 1982, 240, 413.
74. Hershberger, J. W.; Amatore, C.; Kochi, J. K. J. Organometal. Chem. 1983, 250, 345.
75. Hershberger, J. W.; Klingler, R. J.; Kochi, J. K. J. Am. Chem. Soc. 1983, 105, 51.
76. Zizelman, P. M.; Amatore, C.; Kochi, J. K. J. Am. Chem. Soc. 1984, 106, 3771.
77. Harrison, J. J. J. Am. Chem. Soc. 1984, 106, 1487.
78. Summers, D. P.; Luong, J. C.; Wrighton, M. S. J. Am. Chem. Soc. 1981, 103, 5238.

RECEIVED November 8, 1985

10

Surface-Enhanced Raman Spectroscopy

Use in the Detection of Adsorbed Reactants and Reaction Intermediates at Electrodes

M. J. Weaver, P. Gao, D. Gosztola, M. L. Patterson, and M. A. Tadayyoni

Department of Chemistry, Purdue University, West Lafayette, IN 47907

The application of surface-enhanced Raman spectroscopy (SERS) for monitoring redox and other processes at metal-solution interfaces is illustrated by means of some recent results obtained in our laboratory. The detection of adsorbed species present at outer- as well as inner-sphere reaction sites is noted. The influence of surface interaction effects on the SER spectra of adsorbed redox couples is discussed with a view towards utilizing the frequency-potential dependence of oxidation-state sensitive vibrational modes as a criterion of reactant-surface electronic coupling effects. Illustrative data are presented for $Ru(NH_3)_6^{3+/2+}$ adsorbed electrostatically to chloride-coated silver, and $Fe(CN)_6^{3-/4-}$ bound to gold electrodes; the latter couple appears to be valence delocalized under some conditions. The use of coupled SERS-rotating disk voltammetry measurements to examine the kinetics and mechanisms of irreversible and multistep electrochemical reactions is also discussed. Examples given are the outer- and inner-sphere one-electron reductions of Co(III) and Cr(III) complexes at silver, and the oxidation of carbon monoxide and iodide at gold electrodes.

There has recently been much activity in developing molecular spectroscopic probes of electrochemical interfaces, as for other types of heterogeneous systems. The ultimate objectives of these efforts include not only the characterization of adsorbate molecular structure interactions under equilibrium conditions, but also the extraction of mechanistic and kinetic information from spectral detection of reactive adsorbates.

Two problems are inherent, however, in applying such techniques to electrochemical interfaces. Firstly, the extremely small quantity of material present within the molecular thin interfacial region can present severe challenges in analytical detection. Secondly, this difficulty is exacerbated by the usual need for the incoming and outgoing photons to traverse the bulk solution, so that under normal

0097–6156/86/0307–0135$06.00/0

circumstances the resulting spectra reflects the composition of the solution rather than that of the interface. The first difficulty is gradually being overcome by the continuing development in high-power sources (e.g., lasers) and detection systems. The second difficulty can be minimized by the use of thin-layer electrochemical cells, although at the expense of flexibility in electrode design and the resulting inapplicability of some electrochemical techniques.

Great interest arose in the surface science community upon the discovery and realization of surface-enhanced Raman scattering (SERS). (1) Under certain conditions, adsorbates at metal surfaces exhibit strikingly (ca. 10^4-10^7 fold) more intense Raman scattering than in bulk media. The physical phenomenon (or phenomena) responsible for SERS is incompletely understood as yet, (2) even though the major research effort in the area has been devoted to its elucidation. The virtues of SERS as an *in situ* surface molecular probe were recognized at the outset since the high enhancement factors enable absolute Raman spectra for interfacial species to readily be obtained even in the presence of high bulk concentrations using conventional electrochemical cells. (1,2) This technique therefore surmounts both the difficulties of spectral detection noted above.

A drawback which has limited somewhat the practical applications of SERS to surface chemistry is that satisfactory enhancement can apparently only be obtained at relatively few metal surfaces, most prominently silver, copper, and gold, under conditions where the surface is mildly roughened. These metals, especially gold, are nevertheless of importance in electrochemistry in view of their strongly adsorptive and electrocatalytic properties in aqueous media.

Our interest in SERS stemmed from our research activities concerned with establishing connections between the molecular structure of electrode interfaces and electrochemical reactivity. A current objective of our group is to employ SERS as a molecular probe of adsorbate-surface interactions to systems of relevance to electrochemical processes, and to examine the interfacial molecular changes brought about by electrochemical reactions. The combination of SERS and conventional electrochemical techniques can in principle yield a detailed picture of interfacial processes since the latter provides a sensitive monitor of the electron transfer and electronic redistributions associated with the surface molecular changes probed by the former. Although few such applications of SERS have been reported so far the approaches appear to have considerable promise.

In this conference paper, we discuss some recent electrochemical SERS results obtained in our laboratory, both for simple electron-transfer reactions and more complex multistep processes, with the objective of illustrating the types of molecular and dynamical information that can be extracted from this approach. The majority of results discussed here involve the gold-aqueous interface. While most SERS studies to date have utilized silver surfaces, we have recently formulated a simple pretreatment procedure for gold that yields surfaces displaying unusually stable as well as intense SERS with red laser excitation. (3) Gold is a particularly tractable surface for electrochemistry since it provides a wide polarizable window even in aqueous media and has strongly adsorptive properties, enabling a variety of catalytic electrooxidation as well as electroreduction processes to be examined. Emphasis will be placed on a conceptually based discussion of representative results along with implications for future studies; experimental and other details can be found in the various original papers cited.

Molecular Generality of SERS; Outer- and Inner-sphere Adsorbates

Similarly to electron-transfer reactions in homogeneous solution, it is useful to distinguish between inner- and outer-sphere pathways, referring to precursor states where the reactant does, or does not, penetrate the "coordination layer" of solvent molecules adjacent to the metal surface.(4) In order to utilize SERS to examine redox processes at electrodes, it would clearly be desirable to detect outer- as well as inner-sphere adsorbates. The detection of the latter type, where the adsorbate is bound directly to the metal surface, is clearly facilitated by the high interfacial (even monolayer) concentrations that are often encountered. Much lower surface concentrations of outer-sphere adsorbates are usually anticipated since their adsorption is determined primarily by coulombic forces. The detection of the latter adsorbates using SERS might be further exacerbated by small surface Raman scattering cross sections since most SERS theories predict that the surface enhancement factors diminish sharply as the adsorbate-surface separation increases and degree of interaction decreases.(2)

Indeed, under most circumstances SERS appears to probe preferentially inner-sphere adsorbates. Nevertheless, outer-sphere ionic adsorbates can also be detected at silver and gold electrodes by coating the metal surface with suitable charged or dipolar chemisorbed species, thus providing a strong electrostatic attraction and generating extremely high surface concentrations of outer-sphere ions.(3,5,6) (This procedure is closely analogous to the use of multicharged ions of opposite charge so to generate detectable concentrations of ion pairs that form precursor states for homogeneous outer-sphere electron transfer reactions(7).)

Both outer-sphere cationic and anionic species yield detectable SER spectra under these conditions. Monolayers of chloride and bromide anions have been used to generate SER spectra for cationic hexaammine and pentaammine transition-metal complexes,(5,6) and thiourea-coated surfaces yield spectra for a number of anionic species, including oxyanions and hexocyano complexes.(3) As might be anticipated, the SERS frequencies and bandshapes for such outer-sphere adsorbates, including metal-ligand and intraligand vibrational modes, are essentially unaltered from the bulk-phase Raman values, indicating that the adsorbate-surface interactions are weak. (5,6) Nevertheless, the surface enhancement factors for several outer-sphere metal complexes are comparable to (within ca. 2-3 fold of) those involving the same vibrational modes for closely related inner-sphere adsorbed complexes, bound to the surface via thiocyanate bridging ligands.(5b) This result indicates that adsorbate-surface binding can have only a relatively small influence upon the degree of surface Raman enhancement.

Surface Interaction Effects Upon Adsorbed Redox Couples: Valence-Trapped Versus Valence-Delocalized Cases

A central question in electrochemical electron transfer is the manner and extent to which the interface may modify the reaction energetics, both with respect to the stability of the precursor and successor states and the shape of the electron-transfer barrier itself.(8) For outer-sphere reactions, the reactant-electrode interactions are expected to be sufficiently weak and nonspecific so that the electron-transfer barrier is largely unaltered by the proximity of the metal surface. For inner-sphere processes, on the other hand, the degree of reactant-surface electronic coupling may be sufficient so to distort the intersecting reactant and product potential-energy surfaces, thereby altering (and usually lowering) the barrier height.(8) In extreme cases, the two potential-energy surfaces may become entirely merged, whereupon the Franck-Condon barrier is eliminated and the electron will be delocalized between the donor and acceptor sites on the redox center and the metal surface even on the vibrational time scale. This latter case may be termed "Class III" behavior by analogy with the Robin-Day classification for bulk-phase mixed-valence complexes,(9) with the weak-coupling limit termed "Class I" and the intermediate case of strong electronic coupling, yet valence trapped, being labeled "Class II". (Although the metal surface itself does not acquire specific oxidation states, the analogy with bulk "mixed valence" systems is appropriate given that electron transfer occurs from and to a "donor or acceptor site" on the metal surface in the vicinity of the adsorbed redox couple.) Electron-delocalization effects for adsorbed species, primarily monoatomic adsorbed anions and cations, have previously been discussed in terms of the concept of partial charge transfer,(10) but this issue has received little attention for *bona fide* adsorbed redox couples.

We have examined a number of adsorbed transition-metal redox couples using SERS and conventional electrochemistry in order to detect distinct redox states using SERS and to ascertain the degree to which these states are perturbed upon adsorption. Several likely outer-sphere couples have been examined; these include $Os(NH_3)_5py^{3+/2+}$ (py = pyridine) at chloride-coated silver,(6) $Ru(NH_3)_5py^{3+\ 2+}$ and $Ru(NH_3)_6^{3+/2+}$ at chloride-coated silver and gold,(5a,11) and $Fe(CN)_6^{3-/4-}$ at thiourea-coated gold.(3) Characteristic features of these systems are the presence of potential-independent vibrational modes characteristic of the oxidized and reduced forms that occur at potentials positive and negative, respectively, of the formal potential, E_f, of the couple, with progressive displacement of bands due to one redox form by these due to the other as the potential is altered within the vicinity of E_f. The SERS frequencies and bandshapes of both oxidized and reduced adsorbate forms are essentially coincident with the bulk Raman spectral features.(3,5a,6)

It is of interest to examine quantitatively such potential-dependent redox equilibria as determined by SERS in comparison with that obtained by conventional electrochemistry. Figure 1 shows such data determined for $Ru(NH_3)_6^{3+/2+}$ at chloride-coated silver. The solid curves denote the surface concentrations of the Ru(III) and Ru(II) forms as a function of electrode potential, normalized to values at -100 and -500 mV vs SCE. These are determined by integrating cyclic voltammograms for this system obtained under conditions [very dilute (50 µM) $Ru(NH_3)_6^{3+}$, rapid (50 V sec^{-1}) sweep rate] so that the faradaic current arises entirely from initially adsorbed, rather than from diffusing, reactant (cf. ref. 6b). The dashed curves denote the corresponding potential-dependent normalized Ru(III) and Ru(II) surface concentrations, obtained from the integrated intensities of the 500 cm^{-1} and 460 cm^{-1} SERS bands associated with the symmetric Ru(III)-NH_3 and Ru(II)-NH_3 vibrational modes.(5a)

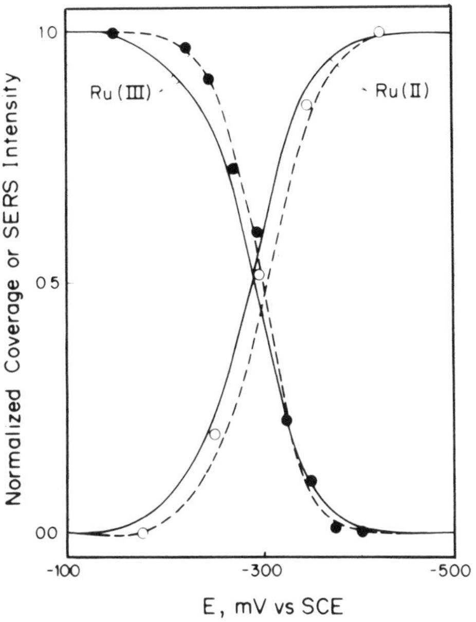

Figure 1. Plots of the relative surface concentrations (solid curves) and relative peak intensities (points, dashed curve) of the 500 cm^{-1} [Ru(III)-NH_3 stretch] and 460 cm^{-1} [Ru(II)-NH_3 stretch] as a function of electrode potential for $Ru(NH_3)_6^{3+/2+}$ electrostatically adsorbed at chloride-coated silver. Ru(III) and Ru(II) surface concentrations determined by integrating cyclic voltammogram obtained for conditions [50 mM $Ru(NH_3)_6^{3+}$ in 0.1 M KCl, 50 V sec^{-1} sweep rate] where faradaic response dominated by adsorbed redox couple. Both surface concentration and SERS intensities for Ru(III) and Ru(II) normalized to values at -100 and -500 mV vs SCE. SER spectra obtained using 647.1 irradiation.

The SERS and electrochemical data are clearly in close agreement, indicating that the former technique is indeed sensing the preponderant adsorbate detected by electrochemical means. Further, the effective formal potential of the adsorbed $Ru(NH_3)_6^{3+/2+}$ couple, E_f^s, can be extracted from the intersection point of the Ru(III) and Ru(II) curves, yielding $E_f^s = -300 \pm 10$ mV vs SCE. Interestingly, this value is about 120 mV more negative than the corresponding bulk formal potential, $E_f = -180$ mV vs SCE. This shift can be understood simply in terms of the electrostatic potential, ϕ_r, at the reaction site for the adsorbed redox couple since[5a] $(E_f^s - E_f) = \phi_r = -120$ mV. This ϕ_r value is comparable to that determined at the outer Helmholtz plane for these experimental conditions using the simple Gouy-Chapman model,(12) providing further evidence that the redox couple is excluded from the inner layer region containing the adsorbed chlorine layer. Similar results are also obtained from $Ru(NH_3)_6^{3+/2+}$ adsorbed at halide-coated gold electrodes.(11) These data together with the identical frequencies and bandshapes for the adsorbed and bulk-phase species suggest that adsorbed $Ru(NH_3)_6^{3+/2+}$ exhibits Class I behavior.

Significant redox center-surface interactions are anticipated to be induced upon binding transition-metal redox centers to the metal surface via small coordinated ligands. Under these circumstances the vibrational frequencies of oxidation-state sensitive modes are expected to be altered as a result of electronic coupling effects. For very strong coupling, such that Class III behavior is approached, the discrete pairs of vibrational frequencies for the oxidized and reduced forms should merge into those reflecting an effective "hybrid" oxidation state. Such behavior has been observed in some bulk-phase mixed-valence systems.(9b) For electrochemical systems, behavior diagnostic of such coupling effects might be obtained from the merging of the vibrational frequencies associated with the oxidized and reduced forms observed in the vicinity of the formal potential.

We have recently been examining various inner-sphere adsorbed redox couples with the aim of deducing if such electron delocalization effects are indeed encountered. One type of candidate system is transition-metal cyano complexes. Some representative data for the $Fe(CN)_6^{3-/4-}$ couple adsorbed at gold are presented in Figs. 2 and 3. The former is a plot of the peak frequency of the SERS C-N stretching mode, ν_{CN}, against electrode potential for gold in 1 m$\underline{M}$ $Fe(CN)_6^{3-}$ with 0.1 $\underline{M}$ MCl + 0.01 $\underline{M}$ $HClO_4$ supporting electrolyte, where M = Na^+, K^+, and Cs^+. Figure 3 shows a representative potential-dependent set of SER spectra obtained for M = K^+. Intense and broad ν_{CN} SERS bands are obtained, indicative of strong adsorption of $Fe(CN)_6^{3-/4-}$ to the gold surface via one or more cyano bridging ligands. A strong dependence of the spectra at a given potential upon the nature of the supporting electrolyte cation is observed, not surprisingly in view of the multicharged anionic nature of the $Fe(CN)_6^{3-/4-}$ couple and the extensive ion pairing expected within the double layer with the cationic diffuse-layer charge.

More surprisingly, however, are the striking variations observed in the potential dependence of ν_{CN} as the cation is altered. For the heavier alkali cations, K^+, Rb^+, and Cs^+, one and sometimes two ν_{CN} bands were commonly observed at each potential, whose frequencies shifted to lower values as the potential becomes less positive in the vicinity of the $Fe(CN)_6^{3-/4-}$ formal potential, 200 mV vs SCE. (This is

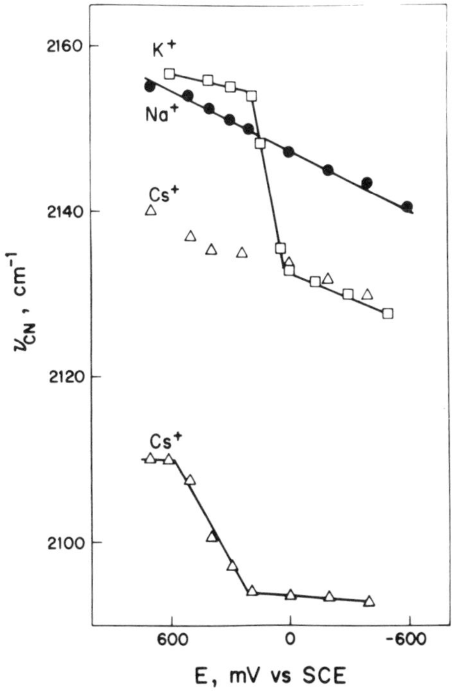

Figure 2. Frequencies of SER C-N stretching mode, ν_{CN}, plotted against electrode potential for $Fe(CN_6{}^{3-/4-}$ adsorbed at gold electrode in supporting electrolytes containing various alkali metal cations. Solutions were 1 m$\underline{M}$ $Fe(CN)_6{}^{3-}$ or $Fe(CN)_6{}^{4-}$ with 0.1 $\underline{M}$ MCl + 0.01 $\underline{M}$ $HClO_4$, where M = Na^+, K^+, or Cs^+ as indicated. Laser excitation wavelength was 647.1 nm.

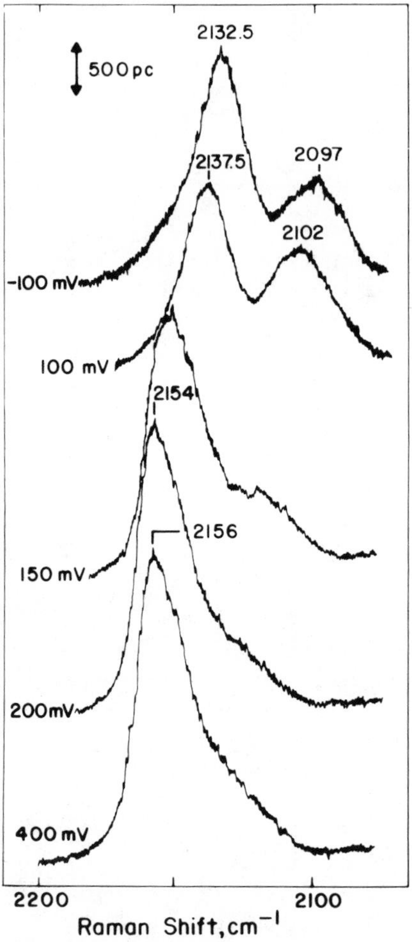

Figure 3. Representative set of SER spectra in C-N stretch region for $Fe(CN)_6^{3-/4-}$ adsorbed at gold in 0.1 M KCl + 0.01M $HClO_4$ (See caption to Fig. 2).

illustrated for M = K^+ in Figure 3.) Although the ν_{CN} band was sufficiently broad (FWHM ~ 30-40 cm^{-1}) so that separate SERS bands attributable to adsorbed $Fe(CN)_6^{3-}$ and $Fe(CN)_6^{4-}$ could not be clearly resolved, the sharp transition region seen in the ν_{CN}-E plots around E ~ 200 mV is indicative of valence-trapped behavior; thus distinct oxidized and reduced forms appear to be present on the vibrational time scale (~ 10^{-14} sec). These potential-dependent SERS bands occur at significantly different frequencies to those for bulk-phase $Fe(CN)_6^{3-}$ and $Fe(CN)_6^{4-}$ (ca. 2140 and 2095 cm^{-1}).

Employing lighter supporting electrolyte cations, M = Li^+, Na^+, however, yields a major ν_{CN} SERS band whose frequency varies linearly with potential over the entire available potential region, -600 to +700 mV, the plots exhibiting no discernable break in the vicinity of the $Fe(CN)_6^{3-/4-}$ formal potential. (The data are shown for Na^+ in figure 2.) This suggests that the redox center engages in sufficiently strong electronic coupling with the metal surface so that the separate identity of the Fe(III) and Fe(II) adsorbed redox states is diminished or even lost entirely; i.e., valence delocalization occurs. The very small difference in Fe-C bond lengths between $Fe(CN)_6^{3-}$ and $Fe(CN)_6^{4-}$ make this system a likely candidate for Class III behavior if cyano-bridged adsorption can induce strong iron-surface electronic coupling since inner-shell distortional contributions to the electron-transfer barrier should be essentially absent. A possible explanation of the markedly different behavior with light and heavy alkali cations is that the latter, having smaller *hydrated* radii, have a greater tendency to form ion pairs with the adsorbed complex , and to a differing extent with $Fe(CN)_6^{3-}$ than $Fe(CN)_6^{4-}$. This may give rise to a structural difference between the oxidized and reduced adsorbate forms and hence induce a Franck-Condon barrier to electron transfer. Such ion pairing effects are liable to be smaller with the lighter Li^+ and Na^+ cations, thereby rationalizing the onset of Class III behavior in the presence of these cations.

Such conclusions are, however, extremely speculative at present. The $Fe(CN)_6^{3-/4-}$ system suffers from a number of complications, including the possibility of surface reactions to form Prussian Blue-like cyano-bridged chains. Details of these results will be given in a forthcoming publication. (14) This brief summary is included here in order to illustrate how information on electron delocalization effects for adsorbed redox couples might be obtained, at least in principle, by examining potential-dependent SERS frequencies.

Irreversible and Multistep Electrochemical Reactions

The foregoing has been concerned with the application of SERS to gain information on surface electronic coupling effects for simple adsorbed redox couples that are reversible in the electrochemical as well as chemical sense, that is, exhibit Nernstian potential-dependent responses on the electrochemical time scale. As noted in the introduction, a major hoped-for application of SERS to electrochemical processes is to gain surface molecular information regarding the kinetics and mechanisms of multiple-step electrode reactions, including the identification of reactive surface intermediates.

For electrochemically reversible systems, the interfacial composition and structure will depend only on the applied potential since the system is entirely under thermodynamic control. For irreversible reactions, on the other hand, the progress of the reaction and hence the surface composition can be determined additionally by other factors, such as potential scan rate, mass transport conditions, etc. Two types of conditions can be envisaged for the examination of such reactions using SERS. The first and simplest approach involves steady-state conditions, most conveniently using rotating-disk voltammetry. The second involves utilizing potential perturbation methods, such as linear sweep voltammetry. The latter has a clear advantage for kinetic measurements, and in favorable cases allows the contributions to reaction from initially adsorbed reactant to be separated from those arising from diffusing solution species. However, it has the disadvantage of demanding relatively rapid time-resolved SERS measurements. Although these are feasible, especially by using optical multichannel detectors, we have chosen to employ the former steady-state approach for most studies so far in view of its simplicity.

We initially employed coupled SERS-rotating disk voltammetry (RDV) to examine the irreversible reductions of $Co(NH_3)_6{}^{3+}$, $Cr(NH_3)_5NCS^{2+}$, and $Cr(NH_3)_5Br^{2+}$ at silver electrodes.(15) These systems were selected since each occurs via a rate-determining one-electron step, the first and the last two reactions occuring via outer-sphere and inner-sphere (anion-bridged) mechanisms, respectively, followed by very rapid aquation of the divalent metal products.(15) The $Co(NH_3)_6{}^{3+}$ yields easily measurable SERS in chloride media (*vide supra*), most conveniently probed via the A_{1g} Co(III)-NH_3 mode at 515 cm^{-1}. Adsorbed $Cr(NH_3)_5Br^{2+}$ and $Cr(NH_3)_5NCS^{2+}$ can be probed via the Cr(III)-NH_3 mode, or for the latter additionally via the ν_{CN} ligand mode.(15) The approach involves monitoring the potential-dependent SERS intensities and frequencies of these modes for the adsorbed reactant at a rotating disk electrode on the rising part of the voltammetric wave, i.e., in a potential region where the kinetics are under mixed mass transfer-hererogeneous electron transfer control.

For the outer-sphere $Co(NH_3)_6{}^{3+}$ reduction, the SERS and current-potential data are closely compatible in that the SERS intensities drop sharply at potentials towards the top of the voltammetric wave where the overall interfacial reactant concentration must decrease to zero. Some discrepancies between the SERS and electrochemical data were seen for the inner-sphere $Cr(NH_3)_5Br^{2+}$ and $Cr(NH_3)_5NCS^{2+}$ reductions, in that the SERS intensities decrease sharply to zero at potentials closer to the foot of the voltammetric wave. This indicates that the inner-sphere reactant bound to SERS-active sites is reduced at significantly lower overpotentials than is the preponderant adsorbate.(15) This suggests that SERS-active surface sites might display unusual electrocatalytic activity in some cases.

Of wider interest, however, is the examination of multiple-step reactions, since SERS should yield fresh insights concerning the molecular identity of the adsorbed intermediates. An important example of such a reaction that has been studied on a number of noble metal electrodes, including gold, is the electrooxidation of carbon monoxide to carbon dioxide.(16) We have recently studied SERS of CO adsorbed at gold electrodes as a function of potential. In acidic and neutral perchlorate electrolytes containing saturated (1 mM) CO, at potentials well negative of where CO electrooxidation commences, a SERS C-O stretching mode, ν_{CO}, is obtained in the frequency region

2080-2110 cm^{-1}, downshifted from the bulk-phase value, 2138 cm^{-1}. The frequency increases with increasing positive potential although the intensity and bandshape remains essentially unaltered. A plot of the ν_{CO} frequency-potential dependence is given in Figure 4. This potential dependence is commonly observed for vibrational bands associated with functional groups interacting strongly with the metal surface, and can be rationalized either in terms of progressive changes in adsorbate-surface σ and π bonding contributions (18a) or to the influence of the inner-layer electric field upon the C-O bond polarization.(18b) The ν_{CO} frequency in the vicinity of the potential of zero charge at gold, 2105 cm^{-1}, is close to that found for CO adsorbed at gold-gas interfaces, and suggests that the carbon is bound to a single surface gold atom.(19)

Altering the potential to progressively more positive values within a few hundred millivolts of where CO electrooxidation commences (as signaled by the passage of detectable anodic current) yields sharp increases in the ν_{CO} frequency and marked intensity decreases along with band broadening. Returning the potential to more negative values exposes these changes as being partly irreversible, the band remaining broad and retaining higher frequencies than those recorded prior to the positive potential excursion. (These frequency changes are traced as the dashed arrowed lines on the ν_{CO} frequency-potential plot in Figure 3.) Similar ν_{CO} frequency increases have been observed at gold-gas interfaces either in the presence of adsorbed oxygen or at incompletely reduced surfaces.(20) It is therefore reasonable to ascribe the higher frequency, 2130-2150 cm^{-1}, features to CO bound to "oxidized" sites, perhaps with some surface reconstruction to form Au(I), and/or bound in the vicinity of adsorbed oxygen or hydroxide. Such spectra were obtained even in strongly acidic media at potentials well negative of where extensive gold surface oxidation commences. The "oxidized" sites are possibly those at which sustained CO electrooxidation occurs at high anodic overpotentials.

The formation of these sites could be responsible for the progressive catalyst deactivation which is observed for sustained CO electrooxidation at gold and other metals.(16b,c) Such sites could be associated with the formation of more tightly bound, and presumably less reactive, oxygen species that are required for CO oxidation. Interestingly, recent combined kinetic/*in situ* infrared studies of the deactivation of CO oxidation by O_2 at Ru- and Rh-SiO_2 catalysts also exposed a connection between the appearance of higher frequency ν_{CO} bands and the occurence of catalyst deactivation.(21)

We have also recently examined the electrooxidation of iodide at gold using the combined SERS-RDV approach.(22) The system was chosen as a simple example of a multistep process where the reaction products (iodine and/or triodide) as well as the reactant and any intermediates should be strongly adsorbed. This reaction has been studied extensively using conventional electrochemical techniques, yet the reaction mechanism remains in doubt.(23) At potentials well negative of the I^-/I_2 formal potential, iodide yields a pair of SERS bands at gold at 124 and 158 cm^{-1}, associated with adsorbed I^--surface vibrations. As the potential is made more positive, these bands are supplemented and eventually supplanted by bands at 110, 145, and 160-175 cm^{-1}, the latter two being especially intense. The new SERS bands are assigned to higher polyiodides and molecular iodine. Although the 110 and 145 cm^{-1} polyiodide bands appear at potentials significantly prior to the

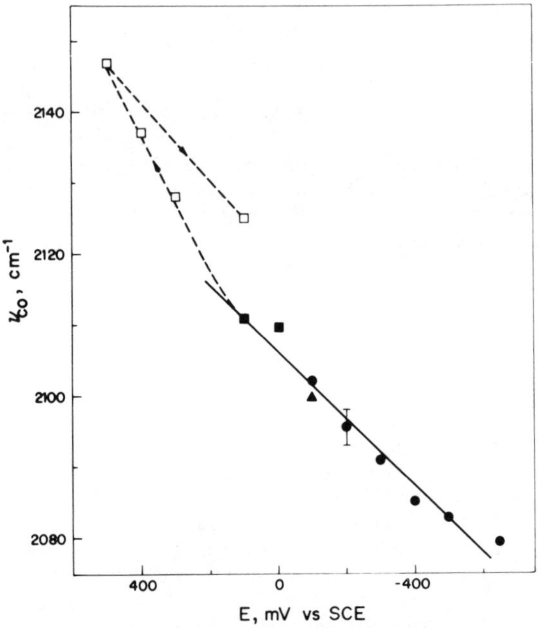

Figure 4. Plot of peak frequency for C-O stretching mode, ν_{CO}, of adsorbed carbon monoxide against electrode potential at gold electrodes, for CO-saturated (ca. 1m$\underline{M}$) solutions. Electrolytes were: (■,□) 0.1 $\underline{M}$ $HClO_4$, (●) 0.1 $\underline{M}$ $NaClO_4$, and (▲) 0.1 $\underline{M}$ KF. The arrowed dashed curves represent the sequential peak frequencies obtained upon potential excursions from 100 mV to 500 mV and return, into the region where CO electrooxidation, occurs. The solid straight line, drawn through the points obtained at potentials (⩽ 100 mV) where adsorbed CO is stable towards electrooxidation, has a slope of about 50 cm^{-1} V^{-1}.

onset of detectable faradaic current flow, the magnitude of the current correlates well with the intensity of the iodine SERS band. This indicates that iodine, rather than triodide, is the major *surface* species formed in the overall reaction, even though triodide is the major *solution* species detected under some conditions. (23)

An interesting facet of this system is the behavior of adsorbed iodide. This anion can irreversibly adsorb at gold, as evidenced by the survival of the characteristic SERS bands after transfer of the electrode following rinsing to a cell containing only 0.1 M $HClO_4$ supporting electrolyte. These bands also survive positive potential excursions as far as 1.2 V vs SCE after transfer, indicating that the irreversibly adsorbed iodide is electro-inactive. A similar conclusion for iodide adsorbed at platinum has been made some time ago on the basis of conventional electrochemical data. (23) This finding shows that adsorbed iodide cannot provide a viable reaction intermediate for iodide electrooxidation. However, it probably still plays a key role in the electrooxidation mechanism by providing binding sites for iodine and polyiodide species formed as reaction intermediates, as evidenced by the loss of 124 cm^{-1} SERS band for iodide-containing solutions at potentials where faradaic current flows. (22)

Concluding remarks

Although such applications of SERS are still in their infancy, this approach appears to have considerable promise for obtaining molecular structural information for adsorbed reactants and reaction intermediates in electrochemical systems. Related developments might also be envisaged for monitoring catalytic processes at metal-gas interfaces, where SERS also offers the crucial advantage of immunity from bulk-phase interferences even at the high gas pressures of interest in heterogeneous catalytic systems. Another promising, although virtually unexplored, area of application of SERS is for monitoring photochemical processes at metal surfaces. Surface-enhanced photochemical degradation of some aromatic molecules has recently been demonstrated at metal-gas interfaces using SERS. (24)

The related technique of *in situ* surface infrared spectroscopy is also starting to be employed to study electrochemical processes. (25) Although the difficulties from bulk spectral interference can be severe, the infrared probe has the advantage that it can be employed on a much wider variety of surfaces than SERS. It seems likely that considerable development of both surface Raman and infrared probes with regard to their application to the characterization of reactive electrochemical systems will occur in the near future.

Acknowledgments

The support of this work by the Air Force Office of Scientific Research, the Office of Naval Research, Eastman Kodak Co., and the NSF Materials Research Laboratory at Purdue is gratefully acknowledged.

Literature Cited

1. (a) Fleishman, M,; Hendra, P. J.; McQuillan, A.J. Chem. Phys. Lett., 1974, 26, 163. (b) Jeanmaire, D. L.; Van Duyne, R. P. J. Electroanal. Chem., 1977, 84, 1.
2. For example, (a) Furtak, T. E.; Reyes, J., Surf. Sci., 93, 382. (b) Chang, R. K.; Furtak, T. E. (eds) "Surface-Enhanced Raman Scattering", Plenum, New York, 1982. (c) Chang, R.K.; Laube, B. L., CRC Crit. Rev. Solid State Mat. Sci., 1984, 12, 1.
3. Gao, P.; Patterson, M. L.; Tadayyoni, M. A.; Weaver, M. J., Langmuir, 1985, 1, 173.
4. Weaver, M. J.; Satterberg, T. L., J. Phys. Chem., 1977, 81, 1772.
5. (a) Tadayyoni, M. A.; Farquharson, S.; Weaver, M. J., J. Chem. Phys., 1984, 80, 1363; (b) Weaver, M. J.; Farquharson, S.; Tadayyoni, M.A.; J. Chem. Phys., 1985, 82, 4867
6. (a) Farquharson, S.; Guyer, K. L.; Lay, P. A.; Magnuson, R. H.; Weaver, M. J.; J. Am. Chem. Soc., 1984, 106, 5123; (b) Farquharson, S.; Weaver, M. J.; Lay, P. A.; Magnuson, R. H.; Taube, H.; J. Am. Chem. Soc., 1983, 105, 3350.
7. Gaswick, D.; Haim, A.; J. Am. Chem. Soc., 1974, 93, 7347; Miralles, A. J.; Armstrong, R. E.; Haim, A.; J. Am. Chem. Soc., 1977, 99, 1416
8. For example, Barr, S. W.; Guyer, K. L.; Li, T. T-T.; Liu, H. Y.; Weaver, M. J.; J. Electrochem. Soc., 1984, 131, 1626.
9. (a) Robin, M. B.; Day, P.; Adv. Inorg. Chem. Radiochem., 1967, 10, 247; (b) For a recent review see Creutz, C.; Prog. Inorg. Chem., 1983, 30, 1.
10. For example, (a) Lorenz, W.; Salie, G.; J. Electroanal, Chem., 1977, 80, 1; (b) Frumkin, A. N.; Damaskin, B. B.; Petrii, O. A.; Z. Phys. Chem. Leipzig, 1975, 256, 728; (c) Schultze, J. W.; Koppitz, F. D.; Electrochim. Acta, 1976, 21, 327.
11. (a) Tadayyoni, M. A.; Ph.D. Thesis, Purdue University, 1984; (b) Gao, P.; unpublished results.
12. Tadayyoni, M. A.; Weaver, M. J.; J. Electroanal. Chem., 1985, 187, 283.
13. Brunschwig, B. S.; Creutz, C.; MaCartney, D. H.; Sham, T-K.; Sutin, N.; Far. Disc. Chem. Soc., 1982, 74, 113.
14. Gao, P.; Weaver, M. J.; manuscript in preparation.
15. Farquharson, S.; Milner, D.; Tadayyoni, M. A.; Weaver, M. J.; J. Electroanal. Chem., 1984, 178, 143.
16. For example, (a) Roberts, J. L.; Sawyer, D. T.; Electrochim. Acta. 1965, 10, 989; (b) Gibbs, T. K.; McCallum, C.; Pletcher, D.; Electrochim. Acta., 1977, 22, 525; (c) Farrugia, T. R.; Fredlein, R. A.; Aust. J. Chem., 1984, 37, 2415.
17. Tadayyoni, M. A.; Weaver, M. J.; Langmuir, in press.
18. (a) Russell, J. W.; Overend, J.; Scanlon, K.; Severson, M.; Bewick, A.; J. Phys. Chem., 1982, 86, 3066; (b) Lambert, D. K.; Solid State Comm., 1984, 51, 297.
19. Sheppard, N.; Nguyen, T. T.; in "Advances in Infrared and Raman Spectroscopy", Clark, R. J. H.; Hester, R. E., eds, Heyden, London, Vol. 5, 1978, p. 67.
20. Yates, D. J. C.; J. Coll. Interface Sci., 1979, 29, 194.
21. Kiss, J. T.; Gonzalez, R. D.; J. Phys. Chem., 1984, 88, 892, 898.
22. Tadayyoni, M. A.; Gao, P.; Weaver, M. J.; J. Electroanal. Chem., in press.

23. For example, (a) Hubbard, A. T.; Osteryoung, R. A.; Anson, F.C.; Anal. Chem., 1966, 38, 692; (b) Toren, E. C.; Driscoll, C. P.; Anal. Chem., 1966, 38, 873; (c) Johnson, D. C.; J. Electrochem. Soc., 1972, 119, 331.
24. Goncher, G. M.; Parsons, C. A.; Harris, C. B.; J. Phys. Chem., 1984, 88, 4200.
25. For example, Bewick, A.; in "Proc. of Symposium on the Chemistry and Physics of Electrocatalysis", McIntyre, J. D. E.; Weaver, M. J.; Yeager, E. B., eds, The Electrochemical Society, Pennington, N. J., 1984, p. 301.

RECEIVED December 2, 1985

11

Thermal and Photoinduced Long Distance Electron Transfer in Proteins and in Model Systems

George McLendon, John R. Miller[1], Ken Simolo, Karen Taylor, A. Grant Mauk[2], and Ann M. English[3]

Department of Chemistry, University of Rochester, Rochester, NY 14627

All biological energy (and, thus, all fossil energy) is ultimately derived from a series of basic electron transfer reactions, starting with the primary charge separation in photosynthesis. The subsequent energy flow proceeds through a series of subsequent redox reactions, largely involving metallo-proteins in which the energy of reduction is coupled to proton transport and manufacture of ATP for biosyntheses. (Fig 1).

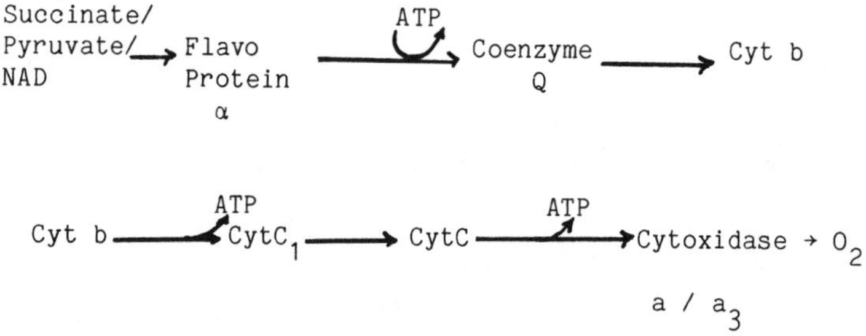

Figure 1. Mitochondral electron transport chain.

Despite the obvious importance of such redox reactions, until recently such reactions remained rather poorly characterised, and poorly understood.

---

[1] Current address: Chemistry Division, Argonne National Labs, Argonne, IL 60438
[2] Current address: Department of Biochemistry, School of Medicine, University of British Columbia, Vancouver, British Columbia, V6T 1Y6 Canada
[3] Current address: Enzymology Research Group, Departments of Chemistry and Biology, Concordia University, Montreal, Quebec, H3G 1M8 Canada

0097-6156/86/0307-0150$06.00/0

Within the past few years, however, rapid advances have occurred in several key areas, including: 1) electron transfer theory[1-3] 2) experiments on model reactions (eg: electron transfer at long, fixed distance),[4-6] 3) experimental techniques for monitoring rapid biological electron transfer,[7-10] and 4) structural characterization of the redox proteins themselves,[11-15] including detailed models for the protein-protein complexes within which electron transfer occurs. As a result of these advances, rapid experimental progress in protein redox chemistry is occurring.

In this paper we focus on two prototypic protein redox couples, cytochrome (c)c/cytochrome b_5 (b_5) and c/cytochrome c peroxidase (ccp) which have been the subjects of detailed studies in our labs.

Theories of Protein-Protein Electron Transfer. The sketch presented here is necessarily brief. Details of the theory can be found in several recent reviews.[1-3] Like any reaction, the rate constant for electron transfer can be written as the product of a prefactor times an activation barrier:

$$k = A \exp^{-\Delta G^*/k_BT}$$

An unusual feature of biological electron transfer reactions is that the reactive prosthetic groups are held at fixed distances within a protein matrix and these distances are greater than the collision distance. Thus, the prefactor for biological electron transfer is not related to a simple collision frequency. Instead, A is governed by the overlap of donor and acceptor wavefunctions, (Fig 2), since such overlap determines the probability of electron exchange between the donor and acceptor. When this overlap is large (ie: electronic interaction energy $H_{AB} > 100\ cm^{-1}$) when reactants attain the correct transition state nuclear configuration will proceed to products, with a probability $\kappa=1$. Such reactions are called adiabatic. If the overlap is small, then the appropriate nuclear configuration can be reached many times without net reaction, and the reaction is called nonadiabatic. As detailed elsewhere, the overlap will generally decrease exponentially with increasing donor-acceptor distances: $A \propto \exp - (\alpha R)$ where the parameter α may depend on the nature of the donor and acceptor, the donor ionization potential, as well as the nature of the "stuff" (eg: protein matrix) in between the donor and acceptors. For many reactions, α has been experimentally found to be $\alpha \cong 1.2\pm0.2$[4,6] $Å^{-1}$

We now turn to the activation energy. The most widely used approach to relate ΔG^* to structural parameters of the reactants is due to Marcus[3] (fig 3). In essence, Marcus theory states that the activation free energy ΔG^*, is determined by the balance between the reorganization energy, λ, and the free energy of reaction, ΔG. The reorganization energy λ can be understood as the energy required for a vertical transition between the energy minimum of the reactant surface and the product surface at the same nuclear coordinate. In such a picture, λ essentially defines the Franck Condon factor for vibrational overlap of reactant and product states. The total λ is made up of internal (λ_i) and medium (λ_s) displacements: $\lambda = \lambda_i + \lambda_s$. An energy λ_i is required since, in general, bond lengths (and angles) will change between an oxidized and reduced molecule. Thus, λi can be modeled by a harmonic oscillator treatment. If $\hbar\omega >> kT$, then reaction along this vibrational coordinate requires nuclear tunneling. Semiclassical and quantum mechanical treatments have been

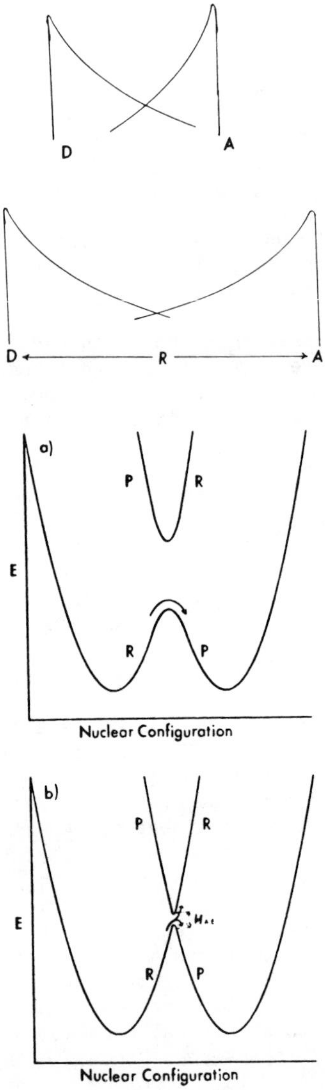

Figure 2. Top: Note that for simple (Slater) wavefunctions, the overlap between donor (D) and acceptor (A) decreases exponentially as distance (R) increases.
Bottom: this overlap can equivalently be viewed as an interaction energy, H_{AB}, between reactant and product surfaces, leading to an avoided crossing. (a) When H_{AB} is large (>100 cm^{-1}) the reaction remains on the lower surface, and the reaction is "adiabatic". (b) When H_{AB} is small, some trajectories may cross to the upper "R" surface and return to the reactant well without making products.

developed for such cases, and have been applied to low temperature data for the photosynthetic reaction centers. A second important energy, λs, arises from repolarization of the medium around a developing charge. A classical continuum treatment suggests

$$\lambda s = \Delta e^2 \left(\frac{1}{D_{op}} - \frac{1}{D_s} \right) \left(\frac{1}{r}\right) \text{ (for r >> collisional)}$$

where D_{op} and D_s are the optical and static dielectric constants and r is the distance between donor and acceptor. The continuum treatment has been questioned recently.[15,16]

For reactions in polymers (like proteins) whether the static dielectic constant is the appropriate parameter is also questionable. Recent data for electron transfer in dry lexan films (D_s ~ 2.4) suggest λ_s ~ 1.0 V,[6c] while classical calculations require λ_s < 0.3 V.

Thus, in order to understand biological electron transfer in a theoretical context, we wish to characterize the parameters which control the prefactor, A, through the distance dependence (the "α" parameter) and characterize the reorganization energy, λ.

The Cytochrome c/cytochrome b5 Complex: Structural Features. One important feature of the c/b5 system is the detailed structural information which is available for these proteins. The structure of cytc is known at ~ 1.5Å resolution in both the oxidized and reduced states.[11a] EXAFS studies have also been reported, which show no observable changes in metal coordination geometry on oxidation/reduction.[11b] The cytochrome b5 structure has also been solved at high resolution by Matthews et al.[12]

Based on these studies and the known chemical properties of these cytochromes, in 1976 Salemme proposed a novel model for the strong noncovalent c/b5 complex.[13] This model is graphically shown in fig. 4. Key features include an electrostatic binding region in which several lysine residues on cytc align with appropriate acidic amino acids on cytochrome b5 to form strong "salt-links".

In the proposed complex structure, the hemes are separated by a nearest contact distance of ~ 8 Å, (16 Å Fe-Fe distance) and are predicted to be in parallel planes.

A variety of subsequent studies have examined several features of this model, and generally support its predictions. For example, the c(III)/b5(III) binding constant, K_M, is quite sensitive to ionic strength, decreasing from K_M~$3X10^7 M^{-1}$ at μ = 0 M to K_M~$10^4 M^{-1}$ at μ = 10 mM, confirming the importance of ionic interactions in stabilizing the complex.[17] Some specific residues involved in this interaction have been identified by the NMR studies of Moore.[18] In agreement with the model, resonances for Lys 13, Lys 72, & Phe 82 are selectively broadened.

Energy transfer measurements by McLendon et al[19] suggest a Zn-Fe distance of ca. 17 Å, in good agreement with Salemme's prediction. Thus the general structural features of the model appear to be well founded.

Electron Transfer Kinetics in the cytc/cyt b5 Complex. Pulse radiolysis techniques have been used to investigate the reaction sequence

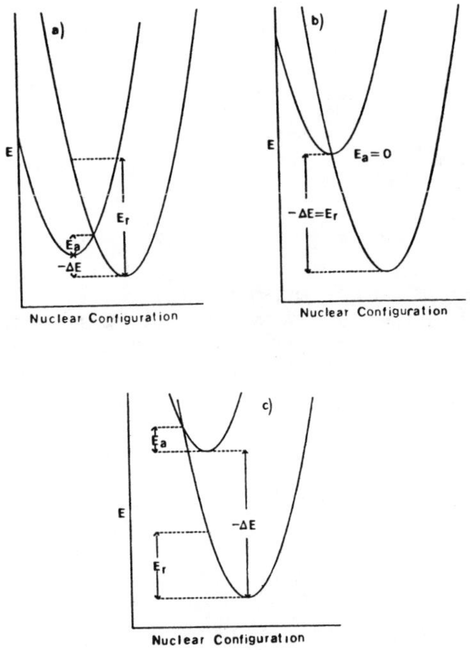

Figure 3. Schematic of Marcus theory (zero interaction). Key: (a) activated process, $\Delta G < \lambda$; (b) activationless, $\Delta G = \lambda$; (c) inverted, $\Delta G > \lambda$

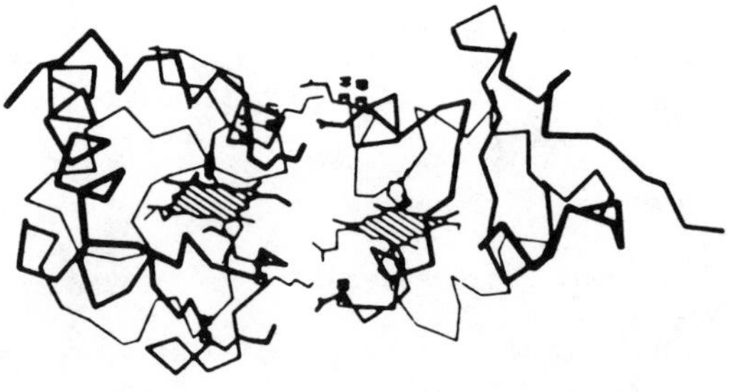

Figure 4. Computer model of the cyt c/cyt b_5 complex (courtesy Tom Poulos, Genex).

$$\text{Fe(III) b5/Fe(III) c} \xrightarrow[k\sim 10^{9}M^{-1}s^{-1}]{e^{-}(aq)} \text{Fe(II) b5 \ Fe(III) c}$$

$$\text{Fe(II) b5/Fe}^{III}\text{c} \longrightarrow \text{Fe(III) b5/Fe(II)c} \quad k = 1.6 \pm .4 \times 10^{3}\ s^{-1}$$

At low ionic strength, ($\mu \leq 10^{-3}$ M) where the complex is fully formed, the transfer rate is first order and is independent of the concentrations of [b_5], [c] or [e^-aq]. As shown in fig 5, the decay of initially produced Fe(II) b5 --> Fe(III)b5 measured at 428 nm is kinetically coupled to the conversion Fe(III) c --> Fe(II) c measured at 416 nm.[10a] At high ionic strength (μ=100 mM Phosphate), where the complex is broken up, we observe a second order rate which increases linearly with increasing [c]. The second order rate constant so obtained is $k_{bi} = 4X10^{7}M^{-1}s^{-1}$ which agrees well with independent stopped flow measurements under identical conditions.[10b]

Species Dependence. We have found that this intramolecular rate is rather sensitive to small perturbations. For example, when the primary sequence of cytc is altered, the rate constant can change by over a factor of 4: $k_{chicken} = 4000s^{-1}$ $k_{cow} = 6000s^{-1}$ $k_{horse} = 1800s^{-1}$ $k_{tuna} = 2400s^{-1}$ $k_{yeast} = 1200s^{-1}$

Similar examples of the dependence of intramolecular transfer rates on protein primary structure are found in reactions in the cytc/cytc peroxidase system.[20]

Metal Substitution: Photoinduced Electron Transfer. The data for the Fe^{II} b5/Fe^{III} c reaction, by themselves, are insufficient to establish either the reorganization energy λ, or the exchange energy H_{AB}, for the protein couple. As one approach to this problem, we prepared and characterized derivatives of cytochrome c in which Fe is replaced by metals with very different reactivity (eg: Zn(II)). Since Znc, and the analogous free base, H_2porphc, have filled d shells, they have relatively long lived excited states, which can serve as strong reducing agents: $E°_{(^3Znc^*/(Znc)^{\cdot+})} \cong 0.8$ V $\tau_{Znc} \cong 10$ msec, $E°_{(^3porphc^*)/porphc)^{+}} \cong 0.4$ V, $\tau_{porphc} \cong 2$ msec).[21]

We, and others, have shown that these derivatives maintain the same structure as native cytochrome c, as judged by circular dichroism, high resolution magnetic resonance, and by binding to Fe(III) cyt b5, or other physiological partners.[10a] Thus, by combining pulse-radiolysis studies of Fec/Feb5 or $porphc^{\cdot-}$/Feb5 with photolysis studies of metal substituted cytochromes it is possible to vary the reaction exothermicity from $\Delta G \cong 0.2$ eV to $\Delta G \cong 1.1$eV. The results of flash photolysis studies with porphc/b5 and Znc/b5 are shown in fig 6. Data from all derivatives are summarised in figure 7, as a plot of k_{et} vs. ΔG. Considering the uncertainties in rates which may result from small conformational differences the data can be adequately described using simple Marcus theory (solid line). This result suggests a total reorganization energy for reaction of $\lambda \cong 0.7$ eV. It remains unknown how such reorganization energy might be partitioned between λi and λs. Preliminary measurements of the temperature dependence of the reaction Fe^{II} b5/Fe^{III} c suggest Eact ~ 4

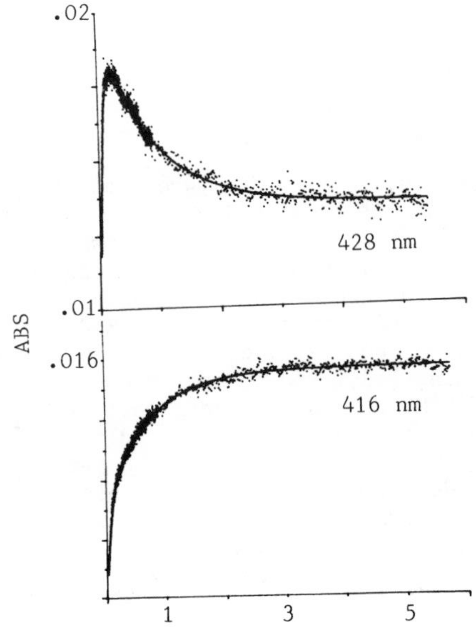

Figure 5. Intramolecular electron transfer kinetics in the c/b_5 complex. Key: top, decay of fe(II)b_5 (428 nm); bottom, growth of Fe(II)c (416 nm).

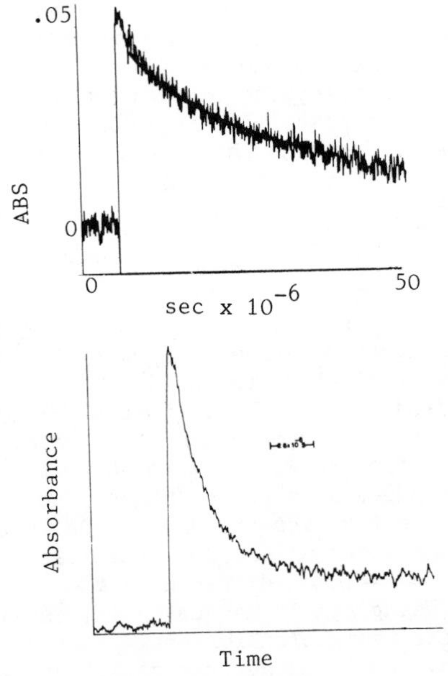

Figure 6. Top: quenching of ^{3}(porph cytc) by Fe(III) cyt b_5 (10 μm each, pH 7, 1mMpi) (460 nm); bottom: quenching of ^{3}(Zn cytc)* by Fe(III) cyt b_5 (460 nm).

kcal/Mol (fig 8) which is somewhat higher than the value of 3.6 kcal/Mol calculated from Marcus theory assuming $\lambda = 0.7$ eV as in fig 7. However, the observed value might contain some contribution from conformational changes which affect orbital overlap, and studies are ongoing.

The Marcus type fit shown in figure 7 implicitly assumes that the primary effect of metal substitution is to change the exothermicity, while holding constant both the reorganization energy and the donor acceptor electronic coupling.

We have already noted that extensive conformational studies suggest that the metal substituted cytochromes c (eg: Zncytc, porphcytc) are essentially isostructural with Fecytc. Thus we expect that for the Zn, Fe, and H_2 porphyrin cyto chromes metal-porphyrin dependent variations in structure will not greatly affect the general trend observed for the dependence of rate on ΔG.

The question of metal dependent effects on electronic coupling requires detailed examination. There are two possible sources for such effects

First, as the donor binding energy decreases from Fe^{II} to Zn^{II}, the electronic mixing, expressed in the dependence of rate on distance $k \propto \exp^{-}(\alpha R)$ should change. In the simplest barrier tunneling theory[2] $\alpha \cong (IP_{donor} - IP_{medium})^{1/2}$ For this theory, then, α depends strongly on IP donor. In more detailed theories, based on superexchange,[4c,6c] α depends weakly (roughly logarithmically) on IP donor. The available experimental data for the dependence of α on IP support the superexchange model[6]: α depends weakly on IP. Thus, we expect the shape of the rate vs. ΔG plot will be minimally affected by changes in the "α" parameter among the different porphyrins.

A second type of variation in electronic coupling could occur. It is possible that in the Fe system, the donor wavefunction is highy localized at the iron, whereas in the excited state Zn porphyrin, the electron is clearly widely delocalized around the ring If this were true, then the "effective" distance for the Fe reaction would be ca: 4 Å longer than for the * Zn reaction, corresponding to a hundred fold rate difference.

However, the available evidence strongly suggests that extensive delocalization occurs between the Fe center and the porphyrin π system. For example, combined NMR studies and theoretical calculations of Fe(III) porphyrins suggest the spin is extensively delocalized into the π system. ($\phi_{MO} = 0.7\phi_{Fe} + 0.3\phi_{porph}$).[23] Furthermore, excess spin density, and associated electron density, in cytochrome c is directed at the heme edge from which electron transfer would occur, reflecting an anisotropic interaction of the d xz, d yz orbitals with the axial methionine.[23d,e] If we take the conservative estimates of 10% electron density at the reactive edge of the Fe system, and ~25% electron density at the edge for the more delocalized Zn system, then a maximum "normalization factor" of about two fold in the electronic frequency factor can be estimated to correct for differential wavefunction overlap in the Fecyt vs. Zncyt system.

We conclude that the basic trend of increasing rate with increasing ΔG in the c/b5 system primarily reflects a Franck Condon term rather than an electronic term. However, since small rate differences may be physiologically significant, "tuning" of the electronic factor is certainly worthy of further study.

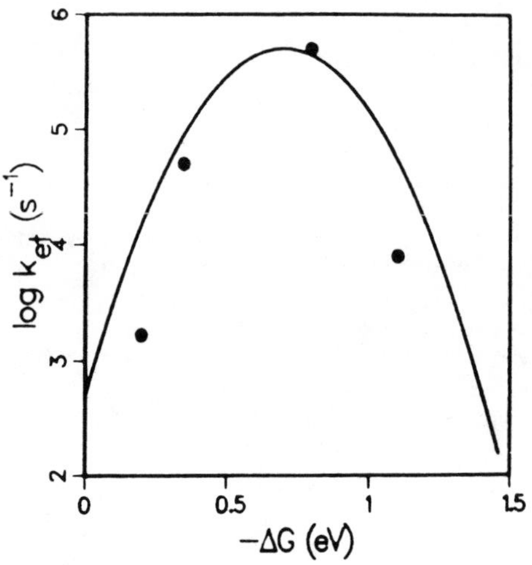

Figure 7. Plot of ln (intracomplex) electron transfer rate (10 μm each, pH 7, 1mM phosphate) vs ΔG.

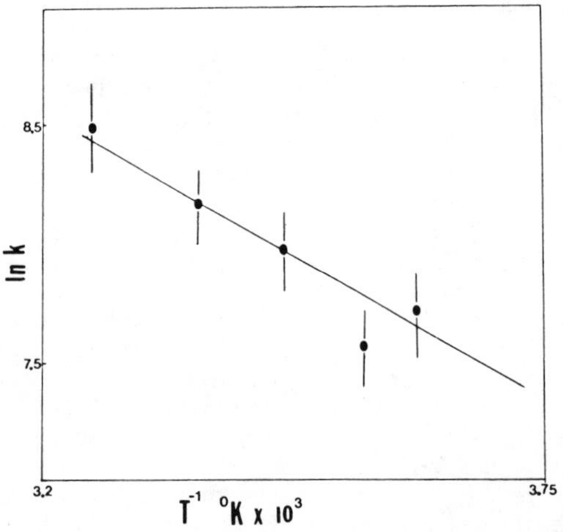

Figure 8. Arrhenius plot for data in Figure 5.

A Second Example: Cytochrome c Peroxidase/cytochrome c. Cytochrome c peroxidase (ccp) catalyzes reduction of H_2O_2 in yeast, with cytochrome c providing the reducing equivalents:

$$Fe^{III}ccp + H_2O_2 \rightarrow Fe^{IV}\ ccp^{o+}\ (ES) + H_2O$$

$$ES + (2)\ Fe(II)cytc \rightarrow (2)\ Fe(III)\ cytc + Fe(III)ccp$$

Detailed crystallographic structures are available for ccp, and c, and a detailed structural proposal exists for the noncovalent ccp/cytc complex,[14] as shown in figure 9. In this complex, the hemes are roughly parallel with a closest approach distance of ca 16 Å, (24 Å Fe-Fe). A wide variety of physical evidence, including chemical cross linking studies, supports this proposed structure.[21]

Thus, the interesting initial questions for relating structure to activity are: what are the rates of reaction, and how do these rates depend on reaction exothermicity, metal site structure, and protein structure. The c/ccp system is well suited for addressing such questions. Fe ccp can be produced in a variety of reactive states including Fe(II) (high spin), Fe(II) (low spin), Fe(III) (high spin), Fe(III) (low spin), and Fe(IV), and metal substitution to introduce Zn, Mn and metals is facile.

Similarly, a variety of metallocytochrome c derivatives can be produced, eg: Fe(II) cytc, Fe(III) cytc, Zncytc, porph cytc.

$Fe^{II}ccp/Fe^{III}cytc$

The reaction $Fe^{II}ccp/Fe^{III}cytc \rightarrow Fe^{III}ccp/Fe^{II}cytc$ proceeds with $\Delta E \cong 0.4V$. The reaction has been monitored both by pulse radiolysis, and by simple mixing of $Fe^{II}ccp + Fe^{III}cytc$, with equivalent results: $k = 0.25 \pm 0.07\ s^{-1}$ (figure 10) It is interesting that a dependence of rate on the primary structure of the protein is observed: (at constant ΔG) for horse cytc/ccp(yeast) $k = 0.25\ s^{-1}$ but for yeast cytc/(yeast) ccp $k = 4\ s^{-1}$ and for tuna cytc/yeast ccp $k \cong 0.1\ s^{-1}$, even though the general three dimensional structures are essentially identical for horse, tuna and yeast cytochromes c. These determinations disprove an earlier suggestion[24] based on modulated excitation spectroscopy, that $k \sim 10^3 s^{-1}$. Clearly the rate is slow, but does this slow rate reflect λ or H_{AB}^2?

The comparative rate study of $porphc/Fe^{III}ccp \rightarrow porphc/Fe^{II}ccp$ ($\Delta E \sim 0.9V$ $k=100\ s^{-1}$) suggests a high reorganization energy ($\lambda \sim 2V$) for this couple. It is likely that much of this reorganization energy is an inner sphere reorganization, reflecting the coordination change between the high spin 6 coordinate Fe(III)/5 coordinate Fe(II) couple.

As already noted, it is possible to compare the reactivity of another oxidation state of ccp via the reaction Fe(IV) ccp + cytc(red) → Fe(III) ccp + cyt c (ox).

For Fe(II)cytc $\Delta E_{reaction} \sim 1V$ and $k \cong 10^3\ s^{-1}$ while for Zn(II)cytc, $\Delta E_{reaction} \cong 0.4V$, $k = 2 \pm 0.4\ s^{-1}$ and for H_2 porph cytc $\Delta E_{reaction} = 0.1V$ $k \cong 4 \pm 0.4 \times 10^{-3}\ s^{-1}$. We again see a strong dependence of rate in driving force, consistent with $\lambda \sim 1.4V$.

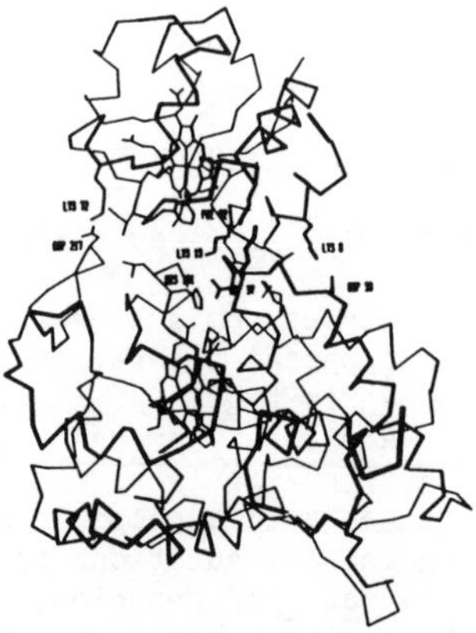

Figure 9. Computer model of cytc/ccp complex. (courtesy Tom Poulos).

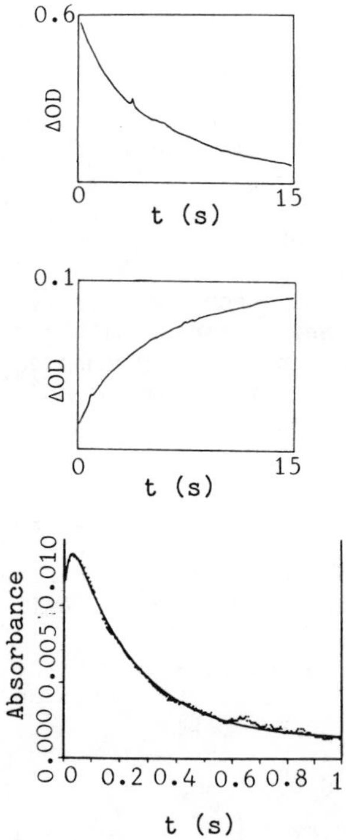

Figure 10. Reaction of horse cytc Fe(III)/Fe(II)ccp top 436 nm (ccp). Reaction of horse cytc Fe(III)/Fe(II)ccp middle 416 nm (cytc). Reaction of yeast cytc Fe(III)/Fe(II)ccp bottom 436 nm.

Thus, the picture obtained from the cytc/ccp system is quite ~~similar to that in the cytc/cytb5/b$_5$ system: λ for protein-protein~~ electron transfer appears to be large.

<u>Comparisons With Other Intramolecular Protein Redox Reactions</u>. Although work remains limited, progress in this area is indeed rapid, and many new results have been reported within the last year. Experiments can be divided into two basic categories: transfer between physiological partners (eg: cyt b5/cytc; cytc/cytc peroxidase (ccp), cyt b5/Hb), and transfer between two groups in the same protein ranging from Hoffman's studies of α Znporph/βFeporph Hb,[8] to studies of Ru substituted proteins (cytc, azurin or myoglobin) by Gray and Iseid.[7] Currently available results are summarized in Table I.

Several points arise from this compilation. First of all, for the protein-protein redox couples, the dependences of rate on ΔG, and/or temperature, are <u>generally</u> consistent with moderately large reorganization energies (0.8V < λ < 2.0 V). Thus the moderately large value of λ inferred for the c/b5 couple appears likely to be a common phenomenon rather than an anomaly. Recent model reactions (not protein) in low dielectric media like isooctane[5] or polycarbonate[6c] show relatively large experimental solvent reorganization energies, ranging from λ = 0.6 - 1.0V while classical dielectric continuum theory predicts λ = 0 - 0.3V. The reasons for the failure of continuum theory are not understood, but may reflect the high local electric fields which occur within several angstroms of an injected charge.

In this context, values of λ ≧ 1V for proteins do not seem anomalous, but are quite in line with observations of simple redox reactions in low dielectric media. We also note that the rates observed for the protein complexes at <u>optimal</u> exothermicity are similar to those observed at optimal ΔG and similar distances in small molecules.[4-6]

A second point of interest is that the λ values inferred for high spin Fe(III) complexes (eg: Hb) are much larger than those seen for low spin systems (eg: cytc). The slow reactions seen for high spin Fe heme proteins likely reflect a large internal reorganization energy associated with the reaction from 5 coordinate Fe(II) to 6 coordinate Fe(III).

A final point is that electron transfer reactions of the Ru modified proteins are generally slow when compared with the protein-protein reactions or with simple model systems. The reasons for this discrepancy are far from clear. Likely explanations include an underestimate of λ (the date of Iseid,[7b] and some data of Winkler et al.[7a] suggest λ ≧ 0.8V), and an underestimate of the appropriate distance for electron transfer. If transfer proceeds from the heme directly to the Ru atom, rather than via the imidazole ligand, then the larger distance listed in the Table is appropriate. Some evidence for this possibility comes from studies of a model ruthenium-porphyrin system.[25] At high exothermicity, a rate constant of $k \leq 10^7\ s^{-1}$ is observed for electron transfer from the porphyrin to Ru. While rapid, this rate is far slower than expected for an adiabatic system ($k \geq 10^{11}\ s^{-1}$). We infer, therefore, that adiabaticity in this system is precluded by the poor overlap of the donor and acceptor orbitals reflecting in part the highly metal

Table I. Some Rates of Intramolecular Biological Electron Transfer Reactions

System	E°	R(Å)	$k(s^{-1})$	ref
$Fe^{II}cytb_5/Fe^{III}cytc$	0.2	8[a](16)	$1.5X10^3$	b
$^3Zncytc^*/Fe^{III}b_5$	0.8	8	$3X10^5$	
$^3H_2porfc^*/Fe^{III}b_5$	0.4	8	$1X10^4$	
$Zncytb_5/Fe^{III}cytc$	1.1	8	$4X10^3$	
$H_2porfc^-/Fe^{III}b_5$	1.1	8		
$Fe^{III}b_5/Fe^{III}Hb$	0.05	8(16)	0.01	c
$ZnHb/Fe^{III}b_5$	0.9	8	$8X10^3$	d
$Znb_5/Fe^{III}Hb$	1.0	8	$2X10^3$	c
α Zn β Fe^{III} Hb	0.9	20	100	e
$Fe^{IV}ccp(ES)/Fe^{II}cytc$	0.9	16(24)	$2X10^3$	f
$ccp(ES)/Zn^{II}cytc$	0.4	16	2	g
$ccp(ES)/H_2porfcytc$	0.1	16	$4X10^{-3}$	g
$Fe^{II}ccp/Fe^{III}cytc$	0.4	16	0.25	h
$Fe^{III}ccp/porfcytc^-$	1.0	16	180	h
ccpES/protein radical (TRP)	0.1	?	0.1	i
$^3Znccp^*/Fe^{III}c$	0.8	16	200	i'
$Fe^{II}cytc_1/Fe^{III}cytc$	0.1	?	50[j]	j
$Fe^{II}cytc$/cyt oxidase	0.1	12(20)	>700	k
cyt c_3 (D. Gigas)	0.06	4	$>10^4$	l
flavo cytc 553 (flavin→heme)			10^3	m
Ruthenium substituted proteins:				
$(NH_3)_5Ru$ His 33 cytc	0.15	11(17)	40	n
$(NH_3)_5Ru$ His azurin	0.2	10	2.5	o
$(NH_3)_5Ru$ His Mb	0.05	13(18)	0.02	p
$(NH_3)_5Ru$ His ZnMb	~0.8	13	$2X10^3$	q

a Number given is closest edge-edge distance ((Fe-Fe) center-center distance)

b G. McLendon, J. R. Miller J. Am. Chem. Soc. (1985) in press.

c G. McLendon, J. R. Miller, A. G. Mauk, to be published.

d K. Simolo, A. G. Mauk, M. Mauk, G. McLendon J. Am. Chem. Soc. (1984) 5013.

e S. Peterson-Kennedy, J. McGourty, B. Hoffman J. Am. Chem. Soc. (1984) 5010.

f T. Yonetani, J. Biol. Chem. (1965) 240, 4503.
g K. Taylor, G. McLendon, J. Am. Chem. Soc. (submitted).
h K. Taylor, E. Chaung, A. English, J. Miller, G. McLendon Proc. Nat. Acad. Sci. in press.
i P. Ho, G. Sutoris, N. Wang, E. Margolish, B. Hoffman J. Am. Chem. Soc. (1985) 107, 1070, B. Hoffman, P. Ho, Soloman, K. Kang, E. Margolish Biochem (1984) 23, 4122.
j Evidence suggests the rate is limited by a conformational change:
k M. Cusanovich, G. Tollin Biochem (1982) 21 3122.
l H. Sautos, I. Moura, J. Legall, A. Xavier Eur J. Bioch 141, 283 (1984).
m M. Cusanovich, G. Tollin Biochem (1982) 21 842.
n D. Nocera, J. Winkler, K. Yocum, E. Bordignon, H. Gray J. Am. Chem. Soc. (1984) 106 5145, S. Iseid, C. Kuehn, G. Worosila J. Am. Chem. Soc. (1984) 106, 1722.
o N. Kostic, R. Margolish, C. Che, H. Gray J. Am. Chem. Soc. (1983) 105 7765.
p R. Crutchley, W. Ellis, H. Gray J. Am. Chem. Soc., in press.
q A. Axup, R. Crutchley, H. Gray, to be published.

localized redox orbital on Ru. If a similar situation occurs in the Ru proteins, then the slow rates observed from the proteins compared with model systems become far less "mysterious".

Summary

Perhaps two key points are emerging from the frenzied activity in protein electron transfer. First, it is clear that rapid electron transfer can occur over long distances in proteins. Furthermore, the rates of these reactions are generally consistent with expectations based on simple small molecule reactions. In particular it appears that proteins, like other polymers, may provide relatively high reorganization energies for electron transfer.

Second, although general consistency with theory is found, it is clear that the precise nature of the reactants and the "stuff in between" can significantly modulate the observed rates by many orders of magnitude. Such modulations could be quite significant in a physiological context, and provide a rich area for future theoretical and experimental investigation.

Acknowledgments

This work was supported by grants from the NSF and NIH (to G.M.), D.O.E. (to J.R.M.) and the Canadian Research Council (To A.G.M. and A.E.) Additional fellowship support was provided by Sherman Clarke fellowships to K.T. and K.S.

We are pleased to acknowledge the help and encouragement of Profs. Harry Gray, Dan Nocera, Jay Winkler, and Rudy Marcus, and the experimental assistance of Eddy Cheng, Marcia Mauk, and Bill Mooney.

Literature Cited

(1) Newton, M. D.; Sutin, N. Ann. Rev. Phys. 1984, 35, 437.
(2) McLendon, G.; Guarr, T.; McGuire, M.; Simolo, K.; Strauch, S.; Taylor, K.; Coord. Chem. Revs. 1985 64, 113.
(3) Marcus R. A.; Sutin, N. Biochem Biophys. Acta,, in press.
(4) a. Miller, J. R. Science 1975, 189, 221.
b. Miller, J. R.; Beitz, J. V.; Huddleston, R. K. J. Am. Chem. Soc. 1984, 106, 5057
(5) a. Miller, J. R.; Beitz, J. V. J. Chem. Phys. 1981, 74, 6746.
b. Miller, J. R.; Calcaterra, L. T.; Closs, G. L. J. Am. Chem. Soc. 1984, 106, 3047.
(6) a. Guarr, T.; McGuire, M.; Strauch, S.; McLendon, G. ibid. 1983, 105, 616.
b. Strauch, S.; McLendon, G.; McGuire, M.; Guarr, T. J. Phys. Chem. 1983, 87, 3579.
c. Guarr, T.; McGuire M.; McLendon, G. J. Am. Chem. Soc., in press.
(7) a. Nocera, D. G.; Winkler, J. R.; Yocum, K. M.; Bordignon, E.; Gray, H. B. ibid. 1984, 106, 5145.
b. Kostic, N. M.; Margalit, R.; Che C.-M.; Gray, H. B. ibid. 1983, 105, 7765.
c. Crutchley, R. J.; Ellis, W. R.; Gray, H. B. ibid., in press.
d. Isied, S. S.; Kuehn, C.; Worosila, G. ibid. 1984, 106, 1722.
(8) a. McGourty, J. L.; Blough, N. V.; Hoffman, B. M. ibid. 1983, 105, 4470.
b. Peterson-Kennedy, S. E.; McGourty, J. L.; Hoffman, B. M. ibid. 1984, 106, 5010.
(9) Simolo, K. P.; McLendon, G.; Mauk, M. R.; Mauk, A. G. ibid. 1984, 106, 5012.
(10) a. McLendon, G.; Miller, J. R. ibid., in press.
b. Strittmatter, P. in "Rapid Mixing and Sampling Techniques in Biochemistry"; Chance, B.; Eisenhardt, R. H.; Gibson, Q. H.; Lonberg-Holm, K. K., Ed.; Academic: New York, 1964; pp 71-85.
(11) a. Takano, T.; Dickerson, R. E. Proc. Natl. Acad. Sci. U.S.A., 1980, 77, 6371.
b. Korszun, Z. R.; Moffat, K.; Frank, K.; Cusanovich, M. A. Biochemistry, 1982, 21, 2253.
(12) Matthews, F. S.; Czerwinski, E. W.; Argos, P. in "The Porphyrins", Dolphin, D. Ed.; Academic: New York, 1979; Vol VII, pp. 107-147.
(13) Salemme, F. R. J. Mol. Biol., 1976, 102, 563.
(14) For a review see; Poulos, T. L.; Finzel, B. C. Peptide and Protein Revs., 1984, 4, 115.
(15) Weaver, M. J.; Gennett, T. Chem. Phys. Lett., 1985, 113, 213.
(16) a. Calef, D. F.; Wolynes, P. G. J. Phys. Chem., 1983, 87, 3387.
b. van der Zwan, G.; Hynes, J. T. J. Chem. Phys., 1982, 76, 2993.
(17) Mauk, M. R.; Reid, L. S.; Mauk, A. G. Biochemistry, 1982, 21, 1843.
(18) Eley, C. G. S.; Moore, G. R. Biochemistry, 1983, 215, 11.
(19) McLendon, G.; Winkler, J. R.; Nocera, D. G.; Mauk, M. R.; Mauk, A. G.; Gray, H. B. J. Am. Chem. Soc., 1985, 107, 739.

(20) Ho, P. S.; Sutoris, C.; Liang, N.; Margoliash, E.; Hoffman, B. M. ibid., 1985, 107, 1070.
(21) Cheung, E.; Taylor, K.; Kornblatt, J. A.; English, A. M.; McLendon, G.; Miller, J. R. Proc. Natl. Acad. Sci. U.S.A., in press.
(22) Magner, E.; McLendon, G., to be published.
(23) a. Shulman, R. R.; Glarum, S. H.; Karplus, M. J. Mol. Biol., 1971, 57, 93.
b. Wütrich, K.; Baumann, R. Helv. Chim. Acta, 1973, 56, 585.
c. Zerner, M.; Gouterman, M.; Kobayashi, H. Theor. Chim. Acta, 1966, 6, 363.
d. Smith, M.; McLendon, G. J. Am. Chem. Soc., 1981, 103, 4912.
e. Senn, H.; Keller, R. M.; Wüthrich, K. Biochem. Biophys. Res. Comm., 1980, 92, 1362.
(24) Potasek, M. J. Science, 1978, 201, 151.
(25) a. Franco, C.; McLendon, G., Inorg. Chem., 1984, 23, 2370.
b. Franco, C.; McLendon, G., Proc. NATO Workshop on Photocatalysis Reidel Press (1985).

RECEIVED November 8, 1985

12

Photochemistry of Dinuclear d^8–d^8 Iridium and Platinum Complexes

Janet L. Marshall, Albert E. Stiegman, and Harry B. Gray

Arthur Amos Noyes Laboratory, California Institute of Technology, Pasadena, CA 91125

The long-lived ($^3B_2$ and $^3A_{2u}$) excited states of the d^8-d^8 dimers $[Ir(\mu\text{-}pz)(COD)]_2$ and $Pt_2(pop)_4^{4-}$, respectively, undergo a variety of photochemical reactions (pzH is pyrazole; COD is 1,5-cyclooctadiene; pop is $P_2O_5H_2^{2-}$, bridging pyrophosphite). Electron transfer reactions to one-electron acceptors such as pyridinium cations or substituted benzophenones are quite facile with acceptors that have reduction potentials as negative as -2.0 V vs. NHE. With halocarbon acceptors, d^7-d^7 oxidative addition products are obtained. Several organic substrates with relatively weak C-H bonds react with the triplet ($d\sigma^*p\sigma$) excited state of $Pt_2(pop)_4^{4-}$ by H-atom transfer.

An electronic excited state of a metal complex is both a stronger reductant and oxidant than the ground state. Therefore, complexes with relatively long-lived excited states can participate in intermolecular electron transfer reactions that are uphill for the corresponding ground state species. Such excited state electron transfer reactions often play key roles in multistep schemes for the conversion of light to chemical energy (1).

The redox potentials of an excited state can be estimated from its spectroscopic energy (0-0 transition, $E_{0\text{-}0}(M/M^*)$) and the oxidation and reduction potentials of the ground state:

$$E^\circ(M^+/{*M}) = E^\circ(M^+/M) - E_{0\text{-}0}(M/M{*M}) \quad (1)$$

$$E^\circ({*M}/M^-) = E^\circ(M/M^-) + E_{0\text{-}0}(M/{*M}) \quad (2)$$

Excited state potentials can also be estimated from kinetic studies of electron transfer quenching reactions involving a series of acceptors and/or donors with varying potentials. By applying electron transfer theory to the quenching step, in conjunction with the predicted dependence of the quenching rate constant on ΔG° for the electron transfer reaction, estimates for the redox potentials may be obtained (2). These approaches have been used successfully in the evaluation of the redox properties of several metal complexes,

0097–6156/86/0307–0166$06.00/0

including *$[Ru(bpy)_3]^{2+}$, *$[Cr(bpy)_3]^{2+}$, *$[Ir(Me_2phen)_2Cl_2]^+$, and *$[Re_2Cl_8]^{2-}$ (3-11).

Our interest in this area concerns the electron transfer reactivity of dimeric d^8-d^8 complexes of platinum(II) and Ir(I). Spectroscopic studies in our laboratory indicate that the "face-to-face" dimers of D_{4h} symmetry exemplified by $Pt_2(pop)_4^{4-}$ (pop is $P_2O_5H_2^{2-}$, bridging pyrophosphite) 1 have low-lying triplet ($^3A_{2u}$) and singlet ($^1A_{2u}$) emissive excited states derived from the $(d\sigma)^2(d\sigma^*)^1(p\sigma)^1$ electronic configuration (12-16). Recently, we have observed fluorescence and phosphorescence from similar excited states ($^1B_2$ and $^3B_2$) in lower symmetry (C_{2v}) pyrazolyl-bridged iridium(I) dimers such as $[Ir(\mu\text{-pz})(COD)]_2$ (pzH is pyrazole, COD is 1,5-cyclooctadiene 2 (17). From spectroscopic and electrochemical studies, the $^3B_2(d\sigma^*\text{-}p\sigma)$ excited state of 2 is predicted to be a powerful reductant with E° ($Ir_2^+/^3Ir_2^*$) estimated to be ca. -1.81 V vs. SSCE in CH_3CN. As expected, the rather long-lived (250 ns (CH_3CN, 22 ± 2°C)) $^3B_2(d\sigma^*p\sigma)$ excited state readily reduces methyl viologen (MV^{2+}) at a diffusion-limiting rate ($k_q' = 1.6 \times 10^{10}\ M^{-1}s^{-1}$; E° ($MV^{2+}/MV^+$) = -0.45 V vs. SSCE, CH_3CN (18)). (Figure 1).

We have confirmed that this state is a powerful reductant by an investigation of the electron transfer quenching of $^3[Ir(\mu\text{-pz})(COD)]_2^*$ by a series of pyridinium acceptors with varying reduction potentials (Figure 2; Table I). For acceptors with reduction potentials of ∿ -1.5 to -1.9 V (vs. SSCE, CH_3CN), the quenching rate constants range from $8 \times 10^8\ M^{-1}s^{-1}$ to $1 \times 10^6\ M^{-1}s^{-1}$. The important point is that, as predicted (3), the rate of electron transfer decreases markedly as the reduction potential of the acceptor approaches the value previously estimated for E° ($Ir_2^+/^3Ir_2^*$) (17).

While the $^3A_{2u}$ ($d\sigma^*p\sigma$) excited state reduction potential of $Pt_2(pop)_4^{4-}$ cannot be calculated accurately due to the irreversibility of the ground state electrochemistry, it can be estimated from quenching experiments. In aqueous solution, this excited state readily reduces one-electron acceptors such as $[Os(NH_3)_5Cl]^{2+}$ and nicotinamide. From these data, E° ($Pt_2^{3-}/^3Pt_2^{4-*}$) is estimated to be ca. -1 V vs. NHE in aqueous solution (13). The excited state reduction potential decreases significantly in nonaqueous solvents, presumably due to the decrease in ground state potential resulting from placing a tetraanion in low dielectric solvents. Quenching of $^3[Pt_2(pop)_4^{4-}]^*$ in acetonitrile by a series of substituted benzophenone acceptors suggests that the excited state redox potential E° ($Pt_2^{3-}/^3Pt_2^{4-*}$) is ∿ -2 V vs. NHE. Therefore, in nonaqueous solvents, the $^3A_{2u}(d\sigma^*p\sigma)$ excited state should readily reduce a variety of substrates (19).

Both 1 and 2 luminesce from singlet and triplet ($d\sigma^*p\sigma$) excited states. In both cases, the triplet states are long-lived and, as just discussed, strong reducing agents. The singlet excited states, however, have lifetimes of < 10 ps (20), and recent experiments strongly suggest that the $^1B_2$ ($d\sigma^*p\sigma$) state of 2 does not participate in intermolecular electron transfer reactions (21).

Although the conversion of light energy to chemical energy via the electron transfer reactivity of $^3[Ir(\mu\text{-pz})(COD)]_2^*$ is rather facile, the photochemical products rapidly return to starting materials because the back electron transfer reactions are highly exothermic. For example, back electron transfer between the oxidized iridium

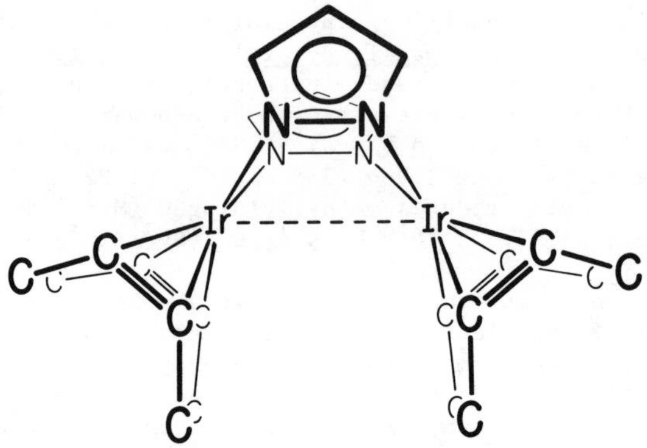

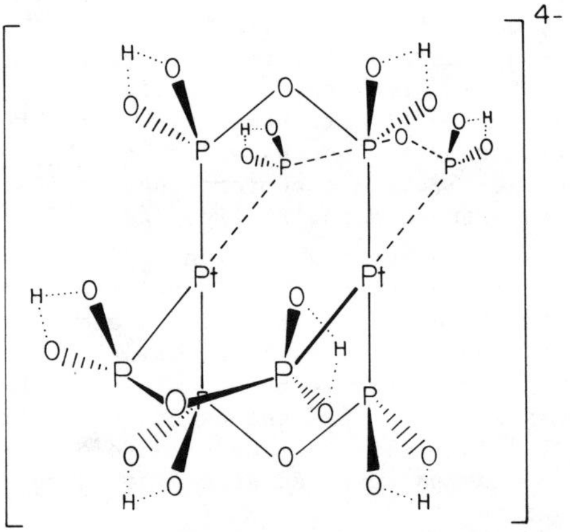

Figure 1. Diagrams of $[Ir(\mu\text{-}pz)(COD)]_2$ and $Pt_2(pop)_4^{4-}$.

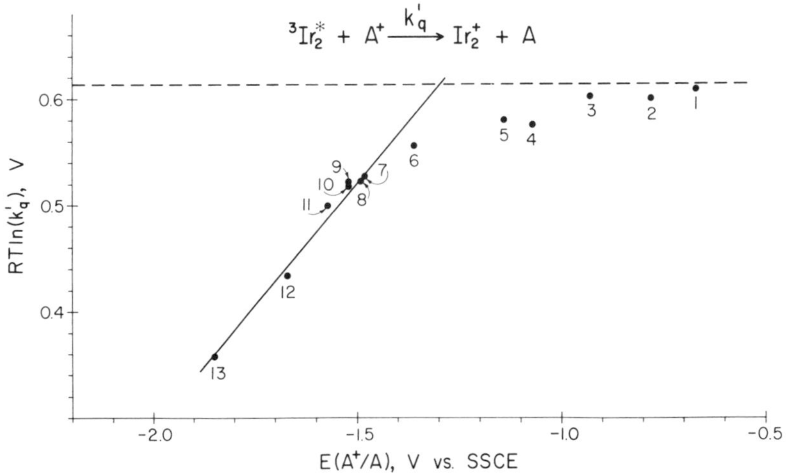

Figure 2. Plot of RTlnk_q' (V) vs. E(A^+/A) (V) for the electron transfer quenching of $^3[Ir(\mu\text{-pz})(COD)]_2^*$ in CH_3CN (μ = 0.1 M $[(n\text{-}C_4H_9)_4NPF_6]$) at 22 ± 2°C. The solid line that fits points 7-13 has a slope of 0.5. Data and the quencher numbering scheme are given in Table I.

Table I. Data for electron transfer quenching of $^{3}[Ir(\mu\text{-}pz)(COD)]_2^*$ by alkylated quenchers

$(PF_6)^-$

	R	R'	$E(A^+/A)$, V vs. SSCE[a]	k_q', $M^{-1}s^{-1}$[b]
1.	CH_3	4-CN	-0.67	2.0×10^{10}
2.	CH_3	$4\text{-}CO_2CH_3$	-0.78	1.3×10^{10}
3.	C_2H_5	$4\text{-}CONH_2$	-0.93	1.5×10^{10}
4.	$C_6H_5CH_2$	$3\text{-}CONH_2$	-1.07	5.5×10^{9}
5.	CH_3	$3\text{-}CONH_2$	-1.14	6.2×10^{9}
6.	C_2H_5	----	-1.36	2.5×10^{9}
7.	CH_3	$2\text{-}OCH_3$	-1.48	8.1×10^{8}
8.	CH_3	$4\text{-}CH_3$	-1.49	6.6×10^{8}
9.	C_2H_5	$4\text{-}C(CH_3)_3$	-1.52	6.5×10^{8}
10.	CH_3	$2,6\text{-}(CH_3)_2$	-1.52	5.8×10^{8}
11.	CH_3	$2,3,6\text{-}(CH_3)_3$	-1.57	2.9×10^{8}
12.	CH_3	$2,4,6\text{-}(CH_3)_3$	-1.67	2.2×10^{7}
13.	CH_3	$2,6\text{-}(CH_3)_2\text{-}4\text{-}OCH_3$	-1.85	1.1×10^{6}

a. For quenchers 1-3, $E(A^+/A) = E_{1/2}(A^+/A)$. For quenchers 4-13, the reductions are irreversible; therefore, the values of $E(A^+/A)$ are the cathodic peak potentials, $E_p(A^+/A)$, measured at a constant scan rate (200 mV/s). Both $E_{1/2}(A^+/A)$ and $E_p(A^+/A)$ were measured by cyclic voltammetry (CH_3CN, $\mu = 0.1$ M $[(n\text{-}C_4H_9)_4NPF_6]$, $22 \pm 2°C$).

b. The rate constants (k_q') are corrected for diffusion effects.

dimer and reduced methyl viologen can be monitored by flash photolysis and the rate constant is near that of the diffusion limit:

$$[Ir(\mu\text{-}pz)(COD)]_2^+ + MV^+ \xrightarrow{k_b} [Ir(\mu\text{-}pz)(COD)]_2 + MV^{2+} \qquad (3)$$

Similarly, the back electron transfer reactions involving alkylated pyridinium acceptors are very rapid and no net photochemistry is observed.

To utilize the strong reducing power of the $^3(d\sigma^*p\sigma)$ excited states of the platinum and iridium dimers, the nonproductive back electron transfer reactions need to be inhibited. An effective way to accomplish this is to use acceptors that are thermally unstable after the initial electron transfer. Reduction of alkyl halides has been shown to lead to short-lived radical anions $RX^{\cdot-}$, which rapidly decompose to give R· and X^- ($k_d = 3 \times 10^7\ s^{-1}$ for $CH_3Cl^{\cdot-}$, $k_d >$ $3 \times 10^8\ s^{-1}$ for $CH_3Br^{\cdot-}$) (22). The triplet excited states of 1 and 2 are capable of reducing a number of halocarbons, and this one-electron reduction step leads to net photochemistry because of radical anion fragmentation (23).

Excitation of the $^1A_{1g} \rightarrow {}^1A_{2u}$, $^3A_{2u}$ (D_{4h}) ($^1A_1 \rightarrow {}^1B_2$, $^3B_2$ (C_{2v})) electronic transitions of $Pt_2(pop)_4^{4-}$ and $[Ir(\mu\text{-}pz)(COD)]_2$, respectively, is followed in each case by a rapid reaction with 1,2-dichloroethane (DCE) to form a d^7-d^7 dihalide dimer ($Pt_2(pop)_4(Cl)_2^{4-}$ or $[Ir(\mu\text{-}pz)(COD)(Cl)]_2$) and ethylene:

$$Pt_2(pop)_4^{4-} + ClCH_2CH_2Cl \xrightarrow[\lambda \geq 350\ nm]{h\nu} Cl\text{-}Pt\text{—}Pt\text{-}Cl + C_2H_4 \qquad (4)$$

$$[Ir(\mu\text{-}pz)(COD)]_2 + ClCH_2CH_2Cl \xrightarrow[\lambda \geq 450\ nm]{h\nu} Cl\text{-}Ir\text{—}Ir\text{-}Cl + C_2H_4 \qquad (5)$$

This binuclear photooxidative addition reaction is general for a number of halocarbons (Figure 3). While DCE and 1,2-dibromoethane react cleanly to give the dihalide metal dimers and ethylene, substrates such as bromobenzene or methylene chloride react through an alkyl or aryl intermediate. This intermediate reacts further to yield the dihalide d^7-d^7 metal complexes.

A mechanism that accounts for the oxidative addition of halocarbons has been proposed for the two d^8-d^8 dimers (Figure 4) (23). The mechanism involves the oxidative quenching of the triplet excited state of the metal dimer as the primary photoprocess. This gives a radical anion species that dissociates a halide, thereby producing an organic radical. The dissociated halide adds to the partially oxidized metal dimer to form a mixed valence Ir^I-Ir^{II}-X or Pt^{II}-Pt^{III}-X intermediate. This intermediate reacts further with the remaining organic radical (presumably in a second, thermal electron transfer step) to form the final d^7-d^7 dihalide dimer.

This mechanism is supported by the product distribution found for the photochemical reactions with 1,2-bromochloroethane. At high concentrations of this substrate, the resulting platinum and iridium dimers are the dibromide species, $Pt_2(pop)_4(Br)_2^{4-}$ and [Ir(μ-pz)(COD)

$\underline{[Ir(\mu\text{-}pz)(COD)]_2}$

Ir_2 + XCH_2CH_2X $\xrightarrow{h\nu}$ $X\text{-}Ir\!-\!Ir\text{-}X$ + C_2H_4
      (X = Cl, Br)

Ir_2 + CH_2Cl_2 $\xrightarrow{h\nu}$ $ClCH_2\text{-}Ir\!-\!Ir\text{-}Cl$ $\longrightarrow$ $Cl\text{-}Ir\!-\!Ir\text{-}Cl$

$\underline{Pt_2(pop)_4^{4-}}$ (non-aqueous solvents)

Pt_2 + XCH_2CH_2X $\xrightarrow{h\nu}$ $X\text{-}Pt\!-\!Pt\text{-}X$ + C_2H_4
      (X = Cl, Br)

Pt_2 + $2C_6H_5X$ $\xrightarrow{h\nu}$ $C_6H_5Pt\!-\!Pt\text{-}X$ $\longrightarrow$ $X\text{-}Pt\!-\!Pt\text{-}X$ + biphenyl
      (X = Cl, Br)

$CH_2Cl_2 \xrightarrow{h\nu} ClCH_2\text{-}Pt\!-\!Pt\text{-}Cl$
Pt_2 +   or   or $\longrightarrow$ $Cl\text{-}Pt\!-\!Pt\text{-}Cl$
$CR_2Cl_2 \xrightarrow{h\nu} ClCR_2\text{-}Pt\!-\!Pt\text{-}Cl$

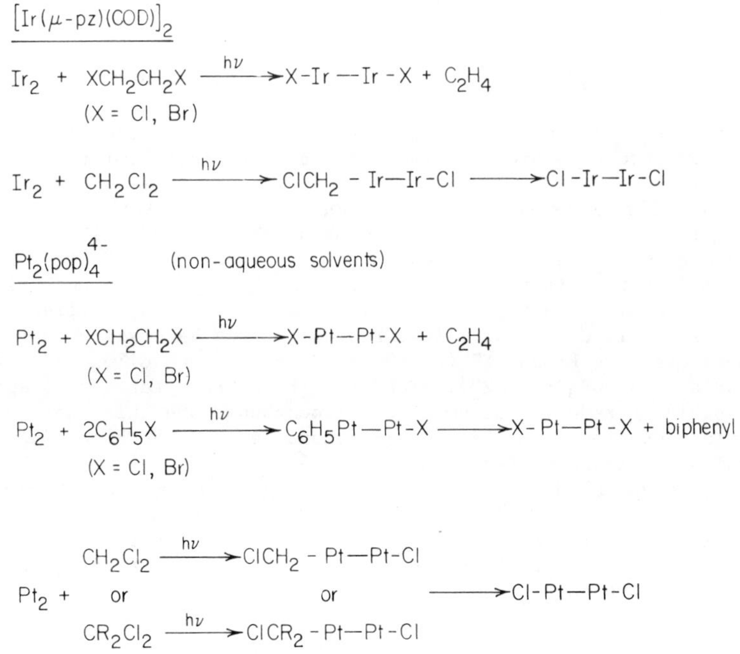

Figure 3.   Reactivity of $^3[Ir(\mu\text{-}pz)(COD)]_2^*$ and $^3[Pt_2(pop)_4^{4-}]_2^*$
with halocarbons.

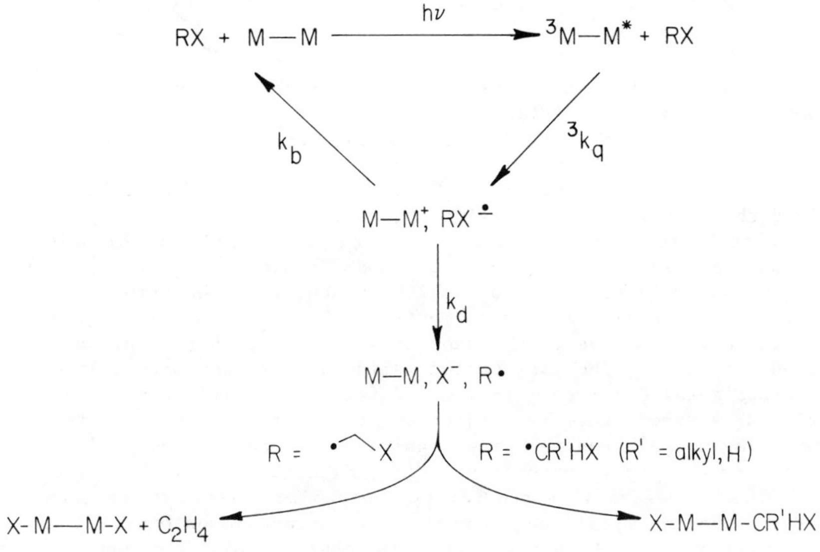

Figure 4.   Proposed general mechanistic scheme for halocarbon
photooxidative addition to binuclear Ir(I) and Pt(II).

$(Br)]_2$, exclusively. Low concentrations of 1,2-bromochloroethane, however, yield the mixed halide metal dimers $Pt_2(pop)_4(Br)(Cl)^{4-}$ and $Ir_2(\mu\text{-}pz)_2(COD)_2(Br)(Cl)$. This result is predicted by the proposed mechanism (Figure 5). Photolysis results in formation of $Pt_2(pop)_4(Br)^{4-}$ or $Ir_2(\mu\text{-}pz)_2(COD)_2(Br)$ as intermeditaes. The intermediate can react with another bromochloroethane molecule, as it does when the latter species is in high concentration, to yield the dibromide dimer or it can react with the chloroethane radical to yield the mixed halide metal species. The latter pathway becomes competitive at low halocarbon concentrations. In general, the oxidative addition of halocarbons is typical of the photochemistry arising from electron transfer from d^8-d^8 metal dimers with the final product being the stable d^7-d^7 metal-metal bonded dimers (24-25).

Bimolecular quenching of the excited states of metal complexes generally involves electron transfer or energy transfer processes (1). Recently, however, $Pt_2(pop)_4{}^{4-}$ has been found to undergo a photochemical reaction involving atom abstraction as a primary photoprocess (26). The reaction involves the catalytic conversion of isopropanol to acetone:

$$Pt_2(pop)_4{}^{4-} + (CH_3)_2CHOH \xrightarrow{h\nu} Pt_2(pop)_4{}^{4-} + H_2 + (CH_3)_2CO \quad (6)$$

While $^3[Pt_2(pop)_4{}^{4-}]^*$ is a strong reductant, it is not sufficiently reducing to transfer one electron to an alcohol (27). Extraction of a hydrogen in a primary photoprocess would produce an isopropyl radical that could undergo disproportionation to yield acetone (28):

$$2(CH_3)_2\dot{C}OH \longrightarrow (CH_3)_2CHOH + (CH_3)_2CO \quad (7)$$

or further react with a metal dimer (29).

$$^3[Pt_2(pop)_4{}^{4-}]^* + (CH_3)_2\dot{C}OH \longrightarrow Pt_2(pop)_4(H)^{4-} + (CH_3)_2CO \quad (8)$$

Our kinetic investigation of this reaction provides compelling evidence for an atom abstraction mechanism. Quenching of the triplet excited state of 1 by various alcohols occurs only when an α-hydrogen is present (19). No quenching occurs with t-butanol or triphenylcarbinol. Furthermore, completely deuterated isopropanol yields a kinetic isotope effect of 1.5 (Table II).

Table II. Data for the quenching of $^3[Pt_2(pop)_4{}^{4-}]^*$ by organic substrates in CH_3CN solution at 22 ± 2°C.

Quencher	$k_q(M^{-1}s^{-1})$
$(CH_3)_2CHOH$	10^3
$(C_6H_5)CH_3$	10^4
$(C_6H_5)_2CHOH$	10^5

k_q(2-propanol)/k_q(2-propanol(d-8)) = 1.5.

Hydrocarbons with relatively weak C-H bonds also react with $^3[Pt_2(pop)_4{}^{4-}]^*$ by H-atom transfer. For example, toluene can be photocatalytically converted to bibenzyl by abstraction of a methyl

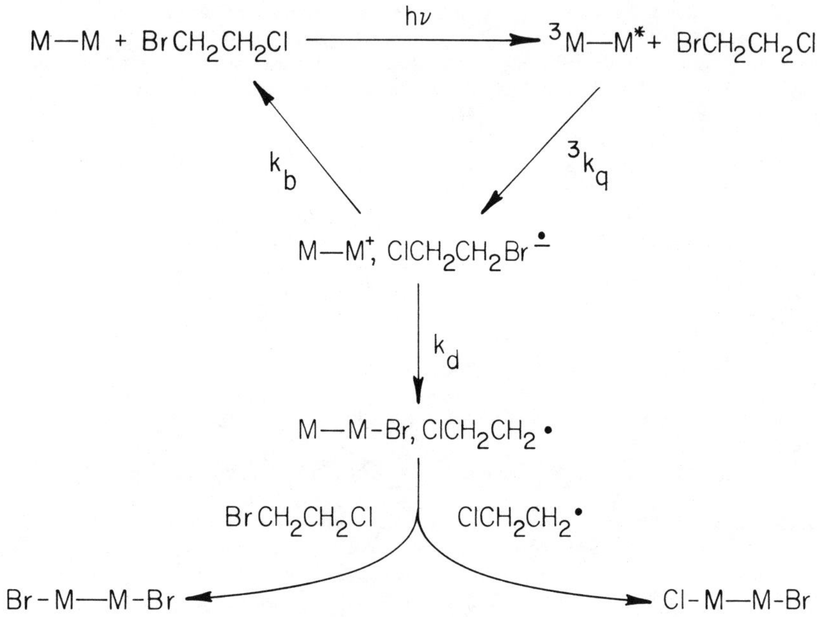

Figure 5. Proposed pathways for $BrCH_2CH_2Cl$ photooxidative addition to binuclear Ir(I) and Pt(II).

hydrogen and coupling of the radical products:

$$Pt_2(pop)_4^{4-} + C_6H_5\text{-}CH_3 \xrightarrow{h\nu} Pt_2(pop)_4^{4-} + C_6H_5\text{-}CH_2\text{-}CH_2\text{-}C_6H_5 + H_2 \quad (9)$$

An intermediate has been observed for the reaction in alcohols. Narrow band irradiation (370 nm) of $Pt_2(pop)_4^{4-}$ in isopropanol results in the disappearance of $Pt_2(pop)_4^{4-}$ and the appearance of an intermediate with a maximum at 313 nm. This intermediate will thermally back-react ($t_{1/2}$ = 85 min) by first-order kinetics to reform $Pt_2(pop)_4^{4-}$ or will convert back immediately upon 313 nm irradiation. Long-term photolysis at low energy (370 nm) does not produce acetone, while equation (6) occurs readily with broadband irradiation.

The exact nature of the intermediate has not been determined; however, its absorption spectrum strongly resembles that of a d^7-d^7 (metal-metal-bonded) dimer. One possible structure would be that derived from the oxidative addition of isopropanol: $(CH_3)_2C(H)$-O-Pt-Pt-H. Another possibility is a dihydrido species, H-Pt-Pt-H. Irradiation of either of these intermediates could yield organic products and regenerate $Pt_2(pop)_4^{4-}$.

In summary, the triplet ($d\sigma^*p\sigma$) excited states of the d^8-d^8 metal dimers $[Ir(\mu\text{-}pz)(COD)]_2$ and $Pt_2(pop)_4^{4-}$ undergo a variety of photochemical reactions. Electron transfer to one-electron quenchers such as pyridinium cations or halocarbons readily occurs with acceptors that have reduction potentials as negative as -2.0 V. With the latter reagents, net two-electron, photoinduced electron transfer yields d^7-d^7 oxidative addition products. Additionally, the triplet ($d\sigma^*p\sigma$) excited state of $Pt_2(pop)_4^{4-}$ apparently is able to react by extracting a hydrogen atom from a C-H bond of an organic substrate.

Acknowledgment. J. L. M. thanks the Sun Co. for a graduate fellowship. This research was supported by National Science Foundation Grant CHE84-19828.

Literature Cited

1. Balzani, V.; Bolletta, F.; Gandolfi, M. T.; Maestri, M. Top. Curr. Chem. 1978, 75, 1-64.
2. Meyer, T. G. Prog. Inorg. Chem. 1938, 30, 389-441 and references therein.
3. Bock, C. R.; Connor, J. A.; Gutierrez, A. R.; Meyer, T. J.; Whitten, D. G.; Sullivan, B. P.; Nagle, J. K. J. Am. Chem. Soc. 1979, 101, 4815-4824.
4. Bock, C. R.; Whitten, D. G.; Meyer, T. J. J. Am. Chem. Soc. 1975, 97, 2909-2911.
5. Lin, C. T.; Bottcher, W.; Chou, M.; Creutz, C.; Sutin, N. J. Am. Chem. Soc. 1976, 98, 6536-6544.
6. Toma, H. E.; Creutz, C. Inorg. Chem. 1977, 16, 545-550.
7. Bock, C. R.; Connor, J. A.; Gutierrez, A. R.; Meyer, T. J.; Whitten, D. G.; Sullivan, B. P.; Nagle, J. K. Chem. Phys. Lett. 1975, 61, 522-525.
8. Ballardini, R.; Varani, G.; Indelli, M. T.; Scandola, F.; Balzani, V. J. Am. Chem. Soc. 1978, 100, 7219-7223.
9. Sutin, N. J. Photochem. 1979, 10, 19-40.
10. Sutin, N.; Creutz, C. Pure and Appl. Chem. 1980, 52, 2717-2738.

11. Nocera, D. G.; Gray, H. B. J. Am. Chem. Soc. 1981, 103, 7439-7350.
12. The $Pt_2(pop)_4^{4-}$ anion can be made with a variety of counterions. Spectroscopic and aqueous photochemical studies were performed with the barium and potassium salts, respectively. The tetrabutylammonium salt is required for photochemistry in nonaqueous solvents.
13. Che, C.-M.; Butler, L. G.; Gray, H. B. J. Am. Chem. Soc. 1981, 103, 7796-7797.
14. Fordyce, W. A.; Brummer, J. G.; Crosby, G. A. J. Am. Chem. Soc. 1981, 103, 7061-7064.
15. Rice, S. F.; Gray, H. B. J. Am. Chem. Soc. 1983, 105, 4571-4575.
16. Parker, W. L.; Crosby, G. A. Chem. Phys. Lett. 1984, 105, 544-546.
17. Marshall, J. L.; Stobart, S. R.; Gray, H. B. J. Am. Chem. Soc. 1984, 106, 3027-3029.
18. Bock, C. R.; Meyer, T. J.; Whitten, D. G. J. Am. Chem. Soc. 1974, 96, 4710-4712.
19. Heinrichs, M. A.; Stiegman, A. E.; Gray, H. B. unpublished results.
20. Miskowski, V.; Stiegman, A. E.; Gray, H. B. unpublished results.
21. Winkler, J. R.; Marshall, J. L.; Netzel, T. L.; Gray, H. B. J. Am. Chem. Soc. submitted.
22. Kochi, J. K. "Organometallic Mechanisms and Catalysis"; Academic Press: New York, 1978; Chapter 7.
23. Caspar, J. V.; Gray, H. B. J. Am. Chem. Soc. 1984, 106, 3029-3030.
24. Fukuzumi, S.; Nishizawa, N.; Tanaka, T. Bull. Chem. Soc. Jpn. 1983, 56, 709-714.
25. Fukuzumi, S.; Hironaka, K.; Nishizawa, N.; Tanaka, T. Bull. Chem. Soc. Jpn. 1983, 56, 2220-2227.
26. Roundhill, D. M. J. Am. Chem. Soc. 1985, 107, 4354-4356.
27. Bard, A. J.; Lind, H. "Encyclopedia of Electrochemistry of the Elements"; Marcel Dekker: New York, 1973; Vol. XI, p. 181.
28. Hay, J. M. "Reactive Free Radicals"; Academic Press: New York, 1974; pp. 96-102.
29. Nonhebel, D. C.; Walton, J. C. "Free Radical Chemistry"; Cambridge University Press, 1974; p. 19.

RECEIVED November 8, 1985

13

Photochemical Production of Reactive Organometallics for Synthesis and Catalysis

William C. Trogler

Department of Chemistry, D-006, University of California at San Diego, La Jolla, CA 92093

Photochemical reactions of unsaturated metallacycles and of platinum and palladium complexes that contain chelating oxalate yield intermediates with two reactive sites that are either ligand or metal centered. Unsaturated metallacycles exhibit low lying π^* excited states that can also function as photoreceptors to promote ligand dissociation elsewhere in the molecule. A strong coupling model for excited state reactivity of metal carbonyls is presented. Reactions of photogenerated PtL_2 and PdL_2 fragments (L = trialkylphosphine) are summarized along with methods of preparing silica attached photocatalysts.

Synthetically useful photochemical reactions of organotransition metal complexes can be classified according to Scheme I.

Scheme I

1) Ligand Photodissociation (1):

$$Cr(CO)_6 \xrightarrow{h\nu} Cr(CO)_5 + CO$$

2) Homolysis of Metal Ligand Bond (2):

$$CoMe([14]aneN_4)OH_2^{2+} \xrightarrow{h\nu} Co([14]aneN_4)^{2+} + CH_3 + H_2O$$

3) Photochemical Homolysis of a Metal-Metal Bond (3):

$$Mn_2(CO)_{10} \xrightarrow{h\nu} 2Mn(CO)_5$$

or

$$Pd_2(CNCH_3)_6^{2+} \xrightarrow{h\nu} 2\ Pd(CNCH_3)_3^{+}$$

0097–6156/86/0307–0177$06.00/0

4) Photooxidation (4):

$$Fe(\eta\text{-}C_5H_5)_2 \xrightarrow[CCl_4]{h\nu} [Fe(\eta\text{-}C_5H_5)_2]^+Cl^-$$

5) Photochemical Reductive Elimination of H_2 (5):

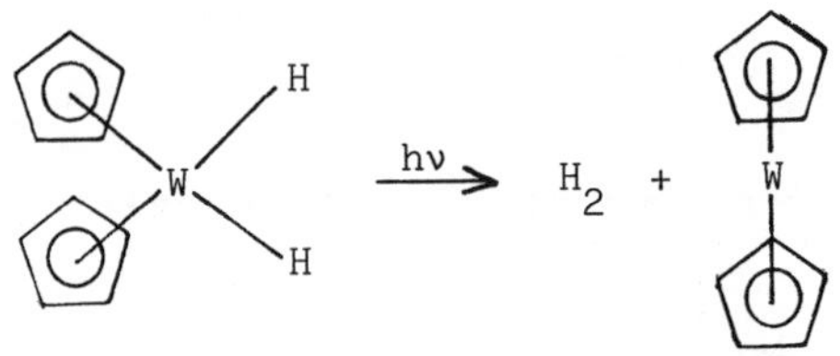

18e 16e

These reactions generate a single reactive site and occur via 15e, 16e, 17e, or 18e intermediates (6-8). One of our goals was to examine the photochemical behavior of complexes that contain an unsaturated chelate chromophore. Photofragmentation of these systems might lead to two reactive centers, either on the ligand or metal. This could produce intermediates that exhibit novel chemistry.

Metallacyclopentadiene, Metalladiazabutadiene, and Metallatetraazadiene Photochemistry

Consider the series of metallacycles A-C. These unsaturated ring systems were expected to show low lying electronic transitions

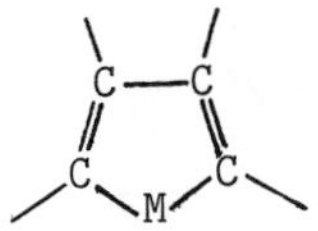

metallacyclopentadiene

A

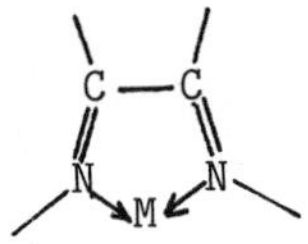

metalladiazabutadiene

B

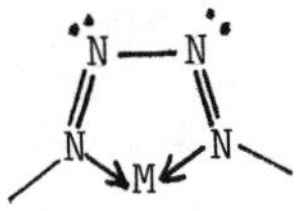

metallatetraazabutadiene

C

because of the unsaturated metal-ligand π system. The photochemistry of $CpCo[C_4Ph_4][PPh_3]$ (9), where Cp = $\eta\text{-}C_5H_5$ and Ph = C_6H_5, in benzene solvent is summarized in Equation 1. In the absence of O_2 phosphine dissociation was shown to yield a 16e intermediate that

$$CpCo[C_4Ph_4](PPh_3) \xrightarrow{h\nu} CpCo(\eta\text{-}C_4Ph_4) \quad (1)$$

$$CpCo[C_4Ph_4](PPh_3) \xrightarrow[O_2]{h\nu} CpCo[\eta^4\text{-}OC(Ph)C(Ph)C(Ph)C(Ph)O]$$

rearranged to the η^4-tetraphenyl(cyclobutadiene) complex. In the presence of O_2, stereospecific oxidation of the tetraphenyl metallacycle occurred to yield Z-dibenzoylstilbene. Single crystal X-ray diffraction (9) showed that this ligand bound to CO in an η^4-eneone fashion.

Several tetraazabutadiene complexes that contained the CpCo fragment were synthesized (10,11), Equation 2. All the derivatives

$$CpCo(CO)_2 + 2N_3R \longrightarrow CpCo[N(R)NNN(R)] \quad (2)$$

$$R = CH_3{=}Me,\ C_6H_5,\ C_6F_5,\ 2,6\text{-}Me_2C_6H_3,\ 2,4\text{-}F_2C_6H_3$$

were intensely colored and calculations (SCF-Xα-DV) (12) of the model complex CpCo[N(H)NNN(H)], Figure 1, showed the presence of a low-lying 30a' metallacycle π* orbital. Strong π back bonding to the tetraazabutadiene ligand was evidenced by short Co-N bond lengths and a single short N-N bond length in the structure of $CpCo[N(C_6F_5)NNN(C_6F_5)]$, Figure 2. A further indication of the strong π acceptor character of the N_4R_2 ligand was the formation (13) of stable 19e anions on electrochemical or chemical (Na/Hg) reduction. That a delocalized metallacycle π orbital was the acceptor orbital was suggested by the large variation in reduction potentials on changing the substituent as well as the Co hyperfine splitting in the EPR spectra (Table I, ~60% cobalt character). These results

Table I. Reduction Potentials vs. NHE in CH_3CN and EPR Spectral Data in THF Solution for $(\eta\text{-}C_5H_5)Co(1,4\text{-}R_2N_4)$ Complexes.

R	$E^{o'}$,V	g_{iso}	$A_{iso}(Co)$,G
CH_3	-1.53	2.055	57.9
$2,6\text{-}(CH_3)_2C_6H_3$	-1.31	2.061	56.3
C_6H_5	-1.01	2.078	50.0
$2,4\text{-}F_2C_6H_3$	-0.97	2.070	51.6
C_6F_5	-0.70	2.066	51.7

contrasted with those for complexes that contain ring System A where little π back-bonding to the ligand is observed (14).

Irradiation of the neutral R = Me derivative led to slow decomposition; however, the aryl derivatives extruded N_2 on photolysis ($\Phi = 10^{-3}$-10^{-4}, λ = 313 nm) to form a series of benzoquinone diimine complexes (10,11) in yields of 65-90%, Equations 3-5. Because this reaction had no precedent, and because C-F and C-C bond cleavage was unknown in the organic photochemistry of nitrenes, the structure of the perfluorophenyl photoproduct was verified (15) by crystallography. Metrical parameters of the structure are consistent with strong Co-N π bonding in these product metallacycles. Benzoquinone diimines do not appear to be as good π acceptors as tetraazabutadienes judging by their more negative (13) (by 0.3 to 0.6 V) reduction potentials. The mechanism of fragmentation-rearrangement that

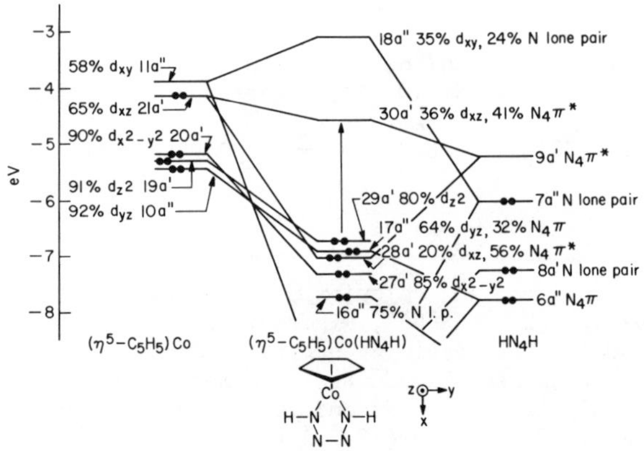

Figure 1. Orbital energy diagram from SCC-Xα-DV calculations of the CpCo and H-N=N-N=N-H fragments as well as the CpCo(1,4-H_2N_4) molecule. Reproduced from Ref. 12. Copyright 1982, American Chemical Society.

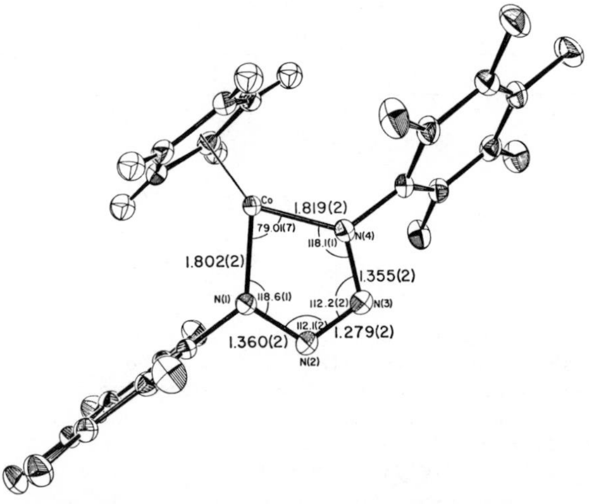

Figure 2. ORTEP (50% ellipsoids) of CpCo[1,4-$(C_6F_5)_2N_4$] with selected bond distances and angles. Reproduced from Ref. 12. Copyright 1982, American Chemical Society.

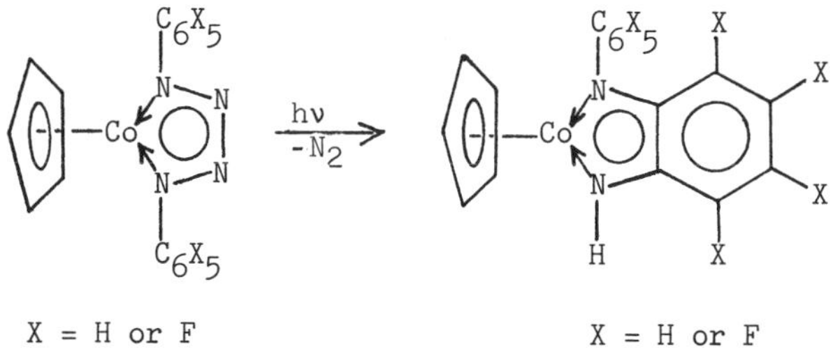
C_6X_5
N
N
Co
N
N
C_6X_5
hν
$-N_2$
C_6X_5
X
X
Co
N
N
H
X
X
X = H or F
X = H or F

(3)

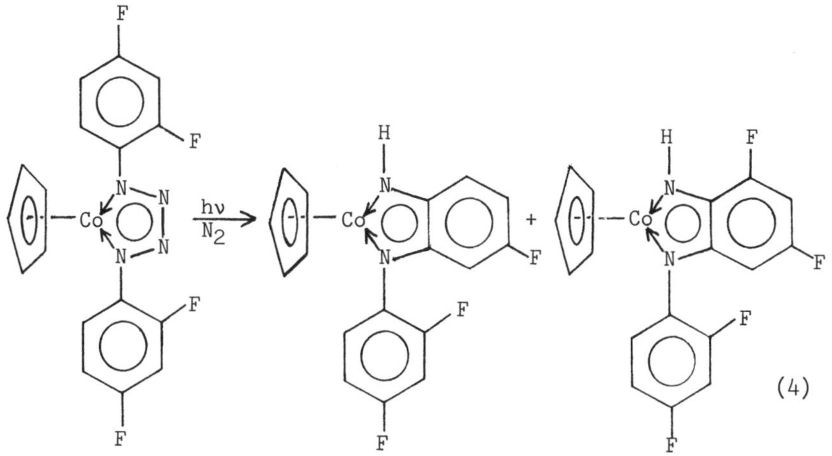
F
F
N
N
Co
N
N
F
F
hν
N_2
H
N
Co
N
F
F
F
+
H
F
N
Co
N
F
F
F

(4)

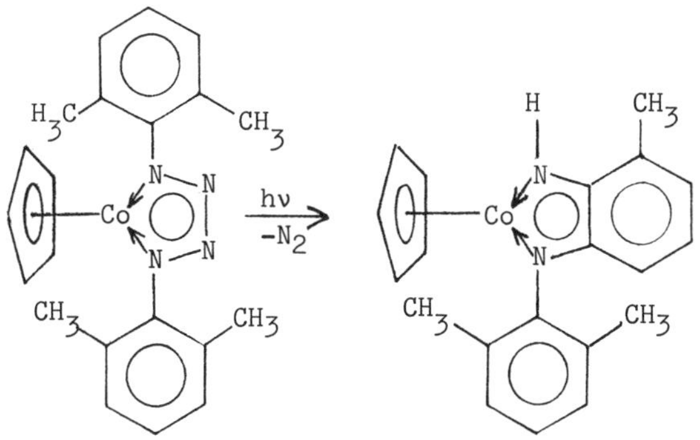
H_3C
CH_3
N
N
Co
N
N
CH_3
CH_3
hν
$-N_2$
H
CH_3
N
Co
N
CH_3
CH_3

(5)

we favor is shown in Equation 6. The final step of the mechanism,

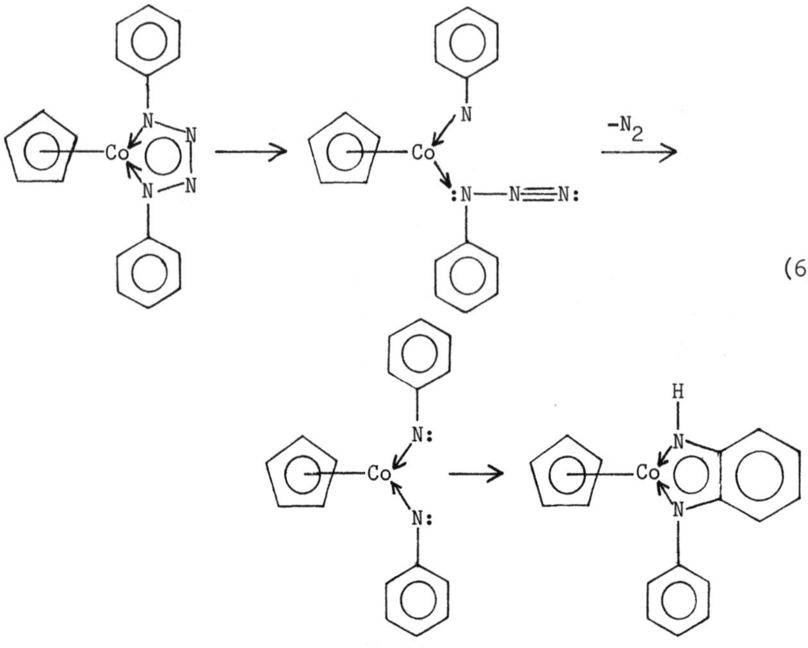

(6)

attack at an ortho ring position, may occur by a radical displacement mechanism (16); for the 2,6-$Me_2C_6H_3$ derivative the eliminated methyl group led to formation of methane (11).

Photodissociation of CO From Tricarbonyliron Diazabutadienes and Tetraazabutadienes

The isolobal behavior (17) of the CpCo and $Fe(CO)_3$ fragments is known. Both $Fe(CO)_3[1,4\text{-}R_2N_4]$ and $Fe(CO)_3[N(R)C(R')C(R')N(R)]$ complexes can be prepared and both are photoactive. Molecular orbital calculations (18) show that strong back-bonding occurs from the $Fe(CO)_3$ fragment to the tetraazabutadiene ligand π system just as for the CpCo derivative. The similarity between the average CO stretching frequency in $Fe(CO)_3[1,4\text{-}Me_2N_4]$ and $Fe(CO)_5$ suggests (18) that the π acceptor ability of a tetraazabutadiene chelate compares with that of two CO's. Diazabutadiene complexes, whose carbonyl IR stretching frequencies lie 35-40 cm^{-1} to lower energy (19), are weaker π acceptors. Therefore, the relative back-bonding ability of the metallacycles, **C** > **B** > **A**, parallels the electronegativity of the ring atoms.

Tetraazabutadiene (18) and diazabutadiene (20) complexes containing the $Fe(CO)_3$ moiety exhibit intense visible absorptions attributed to transitions from Fe d orbitals to a low-lying metallacycle π* orbital. Although the excited state does not directly involve Fe-CO bonding orbitals, efficient CO substitution (21) occurs in the presence of neutral ligands, Equations 7, 8, and 9. The ability to generate coordinately unsaturated iron centers with visible light in

$$Fe(CO)_3[1,4\text{-}Me_2N_4] + L \xrightarrow{h\nu} Fe(CO)_2L[1,4\text{-}Me_2N_4] + CO \quad (7)$$

$$L = NC_5H_5,\ P(Me_3)_3,\ PPh_3,\ P(\underline{c}\text{-}C_6H_{11})_3,\ P(OPh)_3,\ C_2H_4$$

$$Fe(CO)_3[N(Ph)C(Me)C(Me)N(Ph)] \xrightarrow[PPh_3]{h\nu} \quad (8)$$

$$Fe(CO)_2(PPh_3)[N(Ph)C(Me)C(Me)N(Ph)] + CO$$

$$P(OMe)_3 + Fe(CO)[P(OMe)_3]_2[1,4\text{-}Me_2N_4] \xrightarrow{h\nu} \quad (9)$$

$$CO + Fe[P(OMe_3)]_3[1,4\text{-}Me_2N_4]$$

the presence of other UV photosensitive complexes (e.g. $Fe(CO)_5$) permits condensation reactions (21) such as Equation 10.

$$Fe(CO)_3[1,4\text{-}(CH_3)_2N_4] + Fe(CO)_5 \xrightarrow[\text{irradiation}]{\text{visible}} \quad (10)$$

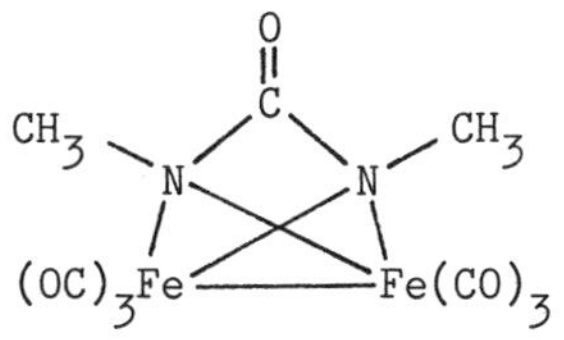

There is an advantage to these photodissociation reactions that produce a 16e intermediate because thermal reactions of these complexes proceed by associative pathways (19,22). The low energy associative thermal path is attributed (19,22) to the ability of the heterodiene ligands to accept an electron pair in the S_{N2} transition state. Mechanistic differences between thermal and photochemical CO replacements can lead to reactivity differences because of different steric requirements of the intermediates, e.g., Equation 11.

$$Fe(CO)_3[N(\underline{t}\text{-}C_4H_9)C(CH_3)C(CH_3)N(\underline{t}\text{-}C_4H_9)] \xrightarrow[P(CH_3)_3]{\Delta} Fe(CO)_3[P(CH_3)_3]_2$$

$$\downarrow h\nu,\ P(CH_3)_3 \quad (11)$$

$$Fe(CO)_2[P(CH_3)_3][N(\underline{t}\text{-}C_4H_9)C(CH_3)C(CH_3)N(\underline{t}\text{-}C_4H_9)]$$

Steric crowding in the thermal S_{N2} transition state for Equation 11 favors loss of the bulky diazabutadiene ligand, while under photochemical conditions simple substitution occurs.

Recently Kokkes, Stufkens, and Oskam (23) have questioned whether CO dissociation occurs in the metallacycle systems we studied. Their evidence against dissociation was that sterically hindered derivatives they examined, $Fe(CO)_3[NRCHCHNR]$, where R = 2,6-i-(C_3H_7)-C_6H_3, c-C_6H_{11}, 4-Me(C_6H_4), t-Bu = t-C_4H_9, and $CH[CH(CH_3)_2]_2$, did not photodecompose in solution (in the absence of ligands). They failed to notice our observation (21) (the first sentence under the heading Photochemical Reactions) that "visible light photolysis of $Fe(CO)_3$-[1,4-Me_2N_4] (in hexanes, cyclohexane, benzene, CH_2Cl_2, THF, or CH_3CN) results in the loss of CO to yield an unstable species". Even if their statement were true for $Fe(CO)_3[1,4-(CH_3)_2N_4]$ it is doubtful whether a photodecomposition criterion for photodissociation is meaningful. For example, $Cr(CO)_6$ does not decompose efficiently when irradiated in pure hydrocarbon solvents (in the absence of ligands) because of rapid reverse binding (24) of dissociated CO. Iron pentacarbonyl exhibits efficient photodecomposition (in the absence of ligands) because the bimolecular reaction between $Fe(CO)_5$ and photogenerated $Fe(CO)_4$ yields insoluble $Fe_2(CO)_9$. Sterically unhindered $Fe(CO)_3[1,4-Me_2N_4]$ mimics the behavior of $Fe(CO)_5$ in forming a cluster on irradiation in the absence of ligands. Furthermore selective photodissociation of CO from the tetraazabutadiene complex produces a coordinatively unsaturated species that reacts (like photogenerated $Fe(CO)_4$) with $Fe(CO)_5$ to form a dimer, Equation 10.

Therefore, we attribute the photostability of the complexes studied by Kokkes et al., to the reversible process of Equation 12.

$$Fe(CO)_3(DAB) \underset{}{\overset{h\nu}{\rightleftharpoons}} Fe(CO)_2(DAB) + CO \qquad (12)$$

DAB = diazabutadiene chelate

The $Fe(CO)_2(DAB)$ species should resist binuclear decomposition pathways because of steric hindrance from the bulky substituents on the DAB ligand. In addition coordinately unsaturated species may be further stabilized by weak coordination to benzene solvent employed in the photochemical studies.

The alternative mechanism to CO dissociation, proposed by Stufkens (23) for the DAB complexes, is not consistent with the difference between thermal and photochemical reaction products, Equation 11. In solution Kokkes et al. propose that one end of the DAB chelate dissociates on photolysis. If this were the case it would be difficult to understand why the photochemical reaction (where the DAB ligand is half attached) leads only to CO displacement, while the associative thermal reaction leads only to DAB displacement. Consider the mechanism, Equation 13, established (19) for thermal loss of DAB. The key to DAB loss is formation of the monodentate species **D** of Equation 13. This intermediate is identical to that proposed by Kokkes et al. (23) for photochemical CO replacement. According to their mechanism, Equation 14, the same species **D**, forms in a two step process and would therefore be thermally equilibrated. Thus the alternative mechanism is not consistent with thermal chemistry of these systems.

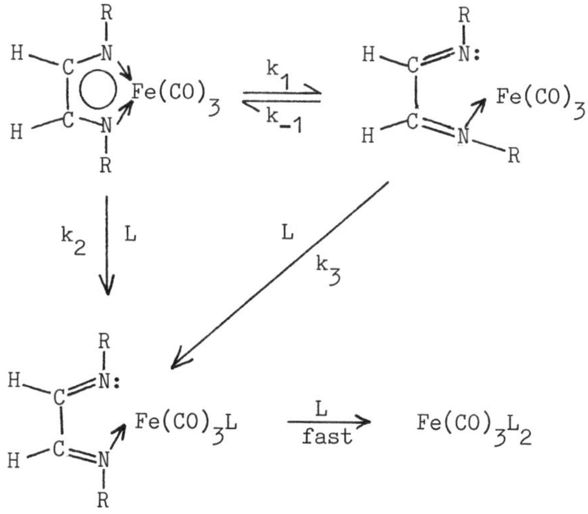

D

(13)

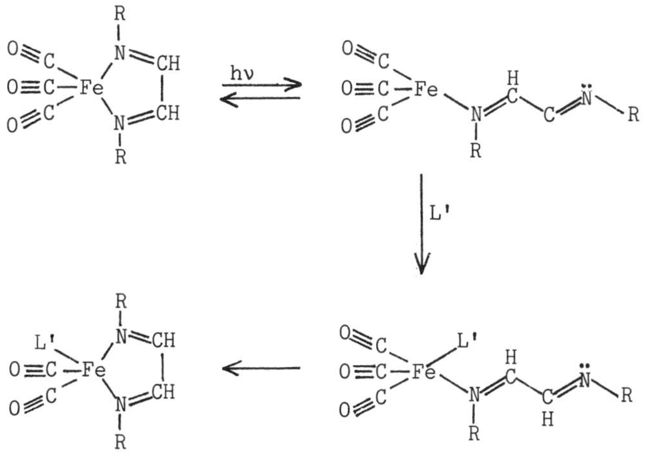

(14)

D

Carrying the analogy between the photochemistry of $Fe(CO)_5$ and $Fe(CO)_3[1,4\text{-}Me_2N_4]$ one step further we note that both compounds (25,26) behave as photoassisted olefin hydrosilation and isomerization catalysts. One distinction between the two catalyst systems is the latter (26) operates effectively with long wavelength radiation, Table II. Hydrosilation activity requires continuous photolysis;

olefin isomerization activity remains during dark reactions after catalyst generation. Further study of these catalytic reactions is needed.

Table II. Photocatalytic Reactions of $Fe(CO)[1,4-(CH_3)_2N_4]$ with Olefins and Trialkylsilanes[a]

Olefin (M)	Silane (M)	Fe, M	Irrad. (min)	% Conv.	Products (%)
Ethylene (0.12)	$HSiEt_3$ (0.42)	0.01	80	65	$SiEt_4$ (98) $Et_3SiCH{=}CH_2$ (2)
Ethylene (0.12)	$HSiMe_3$ (0.36)	0.01	80	75	$EtSiMe_3$ (90)
1-Pentene (2)	$HSiMe_3$ (2)	0.005	90[b]	> 95	Pentene isomers and Pentane (~85)
					Pentylsilane and Pentenylsilanes (~15)
cis-2-Pentene (8)	$HSiMe_3$ (0.8)	0.02	58[c]	> 95	Pentene isomers (no hydrosilation products)

[a] Reactions at 25°C in benzene (or neat) using a total solution volume of 0.3 mL. The reactions were monitored by proton NMR and for the first three entries, the products were analyzed by GC-mass spectrometry. A 200W mercury-xenon arc lamp was used for the irradiations together with Corning 3-74 (λ > 400 nm, first two entries) or 0-52 (λ > 340 nm, last entry) filters. No thermal reactions were observed prior to photolysis.

[b] The last 40% of the reaction took place during 10 h in the dark. Continued photolysis for 265 min gave no change in the NMR spectrum.

[c] Part of the reaction took place, after irradiation, during 20 h in the dark.

[d] Mostly trans-2-pentene and < 5% 1-pentene.

Strong Coupling Model For Organometallic Photoreactions

We noted (21) that the quantum yield for photosubstitution of CO in $Fe(CO)_3[1,4\text{-}Me_2N_4]$, $Fe(CO)_2(PPh_3)[1,4\text{-}Me_2N_4]$, and $Fe(CO)_3$[PhNC(Me)C-(Me)NPh] (e.g. Figure 3) increased in an exponential fashion with increasing excitation energy. There was no correlation with absorption spectral features. The high quantum efficiency for CO substitu- at long wavelengths was unexpected because Xα calculations for iron tricarbonyl tetraazabutadiene complexes (18) and MO calculations for diazabutadiene analogues (20) suggest that the lowest excited states do not alter metal-CO bonding. Resonance Raman spectra of the diazabutadiene complexes (27) support this conclusion. To rationalize the observations we suggested (21) a strong coupling model for excited state reactivity.

Most discussions (28-32) of inorganic photochemical reactions have focused on the specific nature of excited states and correlations with photoreactivity. This assumes the weak coupling model (33) for excited state reactivity. This approach will be successful when the excited state that precedes the photoreaction has a long lifetime or is localized (e.g., MLCT, $\sigma \rightarrow \sigma^*$, ...). Frequently one finds flat Φ vs λ profiles and reactivity from the lowest excited state in the weak coupling limit. There is another limit that should be considered. If the excited state is short lived, not well localized, and if photochemistry competes with vibrational deactivation of an excited state then a strong coupling (33) (i.e., strong coupling between the initially prepared vibronic state and the dissociation continuum) model may be more appropriate.

We introduced the premise that a constant fraction of the excitation energy is available for M-CO dissociative processes. A quasi-statistical (34) partitioning of excitation energy would be favored by 1) dense manifolds of vibronic levels, 2) lack of symmetry selection rules on nonradiative decay pathways, and 3) delocalized excited electronic states that do not couple strongly with any single vibration mode; we qualify the last condition by noting that even localized excitations can lead (35) to "statistical" behavior. The word "statistical" is not used in a strict thermal sense, because partitioning of excitation energy depends on the specifics of intramolecular vibronic coupling of the initially prepared state.

Experimental manifestations of strong coupling that are expected include the following: 1) quantum yields for photodissociative pathways that depend on the amount by which the excitation energy exceeds the thermodynamic threshold for bond breaking; 2) multiple reaction pathways that become available at higher excitation energies; 3) structure sensitivity to reaction quantum yields because energy flow relies on the vibrational modes that initially receive the energy and how they couple to other modes; 4) bonding character of the excited state becomes irrelevant.

Photoreactions of the metallacycles discussed (21) show linear plots of $\ln\Phi_{CO}$ vs excitation energy before limiting quantum yields are reached. There was a correlation between the donor atom set about Fe and quantum yields. Thus $Fe(CO)_2(PPh_3)[1,4\text{-}Me_2N_4]$ and $Fe(CO)_3[1,4\text{-}Me_2N_4]$ have similar absorption spectra, but quite different quantum yields for CO substitution. Absorption spectra of $Fe(CO)_3[1,4\text{-}Me_2N_4]$ and $Fe(CO)_3[PhNC(Me)C(Me)NPh]$ are different; however, both compounds possess the same donor atom set and exhibit similar quantum yields for CO loss. It is also noteworthy that isoelectronic $CpCo[1,4\text{-}R_2N_4]$ complexes that do not contain an easily dissociable group photofragment by N_2 loss (10,11). There was a correlation between the mode of photochemical decomposition of the $Fe(CO)_3$ and CpCo tetraazabutadiene complexes and the lowest energy fragment in their electron impact mass spectra (11). For these reasons we favor a strong coupling description of the photoreactions of these compounds where light excitation is rapidly converted into vibrational energy that then results in bond breaking as governed by energetic and statistical considerations. It should be noted that there are other explanations (36) for wavelength dependencies of quantum yields.

Recent studies (37-39) of the gas phase photochemistry of $Fe(CO)_5$ and $Cr(CO)_6$ by time resolved IR methods shows that ejection of two CO's per incident photon occurs as the excitation energy increases. That is consistent with the strong coupling model proposed. It also appears (40) that few of the ejected CO groups are vibrationally excited. We speculate that the energy gap between high frequency CO vibrations and low frequency M-C or M-M vibrations in simple metal carbonyls (e.g., $Cr(CO)_6$ or $Mn_2(CO)_{10}$) traps excited state energy in the M-C and M-M vibrational modes. This could explain the high quantum efficiencies for CO dissociation or M-M bond homolysis in such compounds. This may also be why quantum efficiencies for CO dissociation in substituted carbonyls (41) decrease markedly. Introduction of ligands with low frequency vibrational modes provides a sink for vibrational excitation energy and perhaps a better path for energy migration to the surroundings in condensed phases.

Photochemistry of Oxalate and Dithiooxalate Complexes of Nickel, Palladium, and Platinum

Photooxidation of coordinated oxalate has been known since the earliest studies of transition metal photochemistry (42). In these reactions oxalate ligand is photooxidized to CO_2, and up to two metal centers are reduced by one electron (e.g. ferrioxalate). We wondered whether the oxalate ligand could be a two-electron photoreductant, by simultaneous or rapid sequential electron transfer, with metals prone to 2e redox processes. Application of this concept to 16e square planar d^8 complexes, Equation 15, was attractive because it should produce solvated 14e metal complexes that are inorganic analogues of

$$\underset{16e}{M(C_2O_4)L_2} \xrightarrow{h\nu} 2CO_2 + \underset{14e}{ML_2} \qquad M = Ni, Pd, Pt \tag{15}$$

carbenes. The reports (43,44) that platinum(0) complexes could be isolated by irradiating $Pt(C_2O_4)(PPh_3)_2$ and that rhodium(I) species were obtained by irradiating $Rh(C_2O_4)Cl(py)_3$ suggested that this process might work. Because 14e PtL_2 fragments can be made thermally (45,46) when L is a bulky phosphine [e.g., PCy_3 or $P(\underline{t}\text{-Bu})_3$], we examined (47,48) the photochemistry of sterically unhindered complexes. The photoreactivity of $Pt(C_2O_4)(PEt_3)_2$, Et = C_2H_5, is summarized in Scheme II. Photochemical conversions are high and few side products (e.g., Figure 4) form. All the reactions suggest formation of a reactive $Pt(PEt_3)_2$ fragment that can be trapped as a platinum(0) species or combined with oxidative addition substrates to yield platinum(II) compounds.

This chemistry has been extended to produce palladium(0) intermediates (48). Much of the chemistry is similar to that of the Pt analogues except that the palladium(0) complexes are more unstable and difficult to isolate. A reaction characteristic of palladium is the addition of allyl compounds to form cationic allyl complexes, Equation 16. This has been postulated (49) as a key step in the mechanism for $Pd(diphos)_2$ catalyzed reactions of allyl compounds.

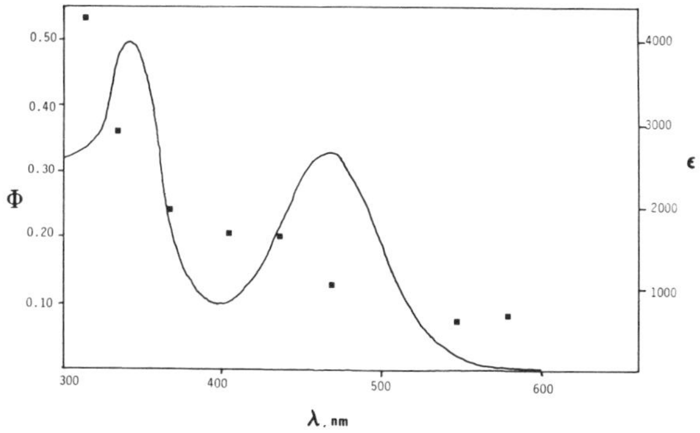

Figure 3. Electronic absorption spectrum of $Fe(CO)_3(1,4\text{-}Me_2N_4)$ and quantum yields for photosubstitution of CO by PPh_3. Reproduced from Ref. 21. Copyright 1981, American Chemical Society.

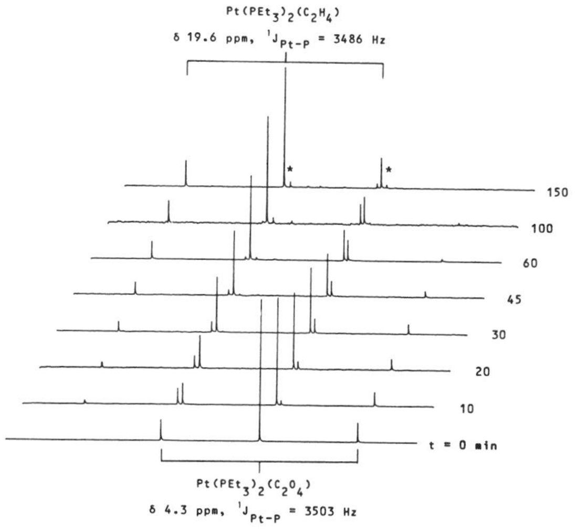

Figure 4. Successive $^{31}P\{^1H\}$NMR spectra (109 MHz) showing the photochemical conversion under 1 atm ethylene of $Pt(PEt_3)_2(C_2O_4)$ to $Pt(PEt_3)_2(C_2H_4)$. Irradiation times are shown at right. The symmetrically disposed satellite peaks result from those molecules that contain ^{195}Pt (33.8% abundance, $I = \frac{1}{2}$). Signals marked by an asterisk in the final spectrum are unidentified side products, which form at long irradiation times. Reproduced from Ref. 48. Copyright 1985, American Chemical Society.

Scheme II

Reactions of $Pt(C_2O_4)(PEt_3)_2$ On Ultraviolet Irradiation

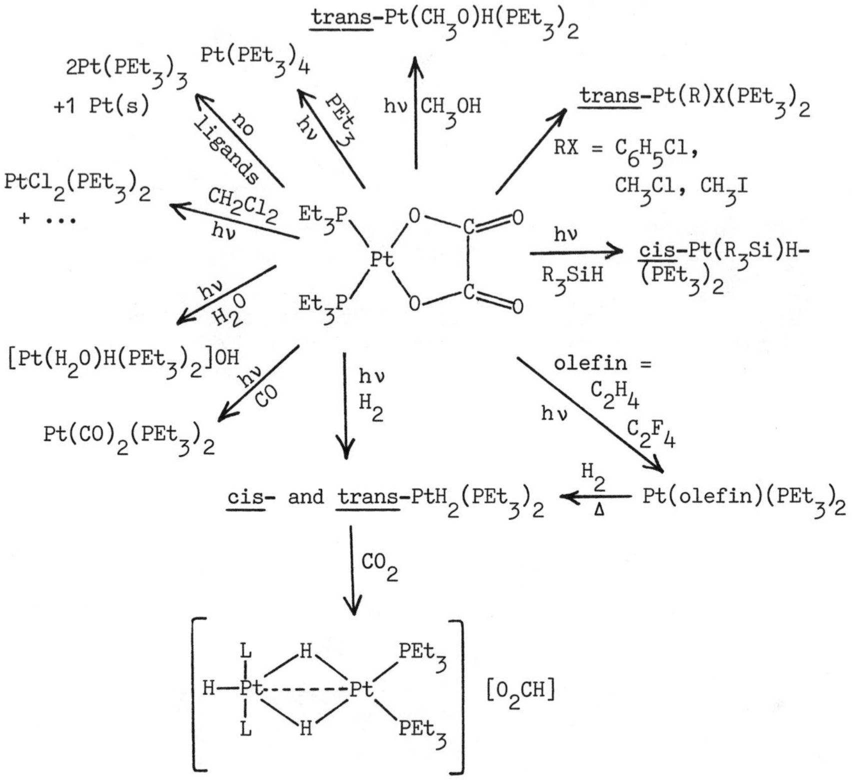

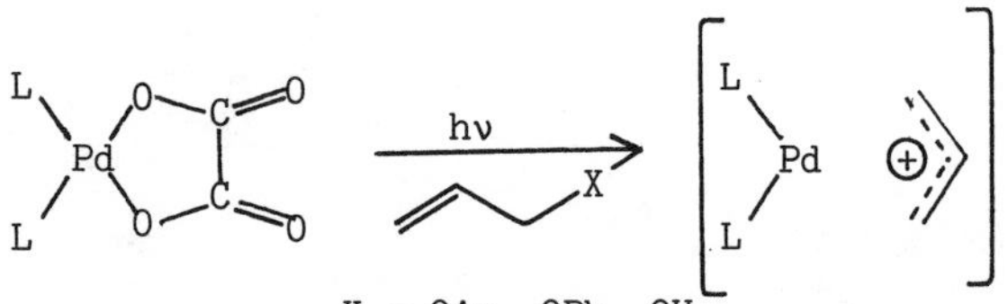 (16)

X = OAc, OPh, OH
OEt, Cl
L_2 = diphos or $[P(\underline{n}\text{-Bu})_3]_2$

Presumably Pd(diphos) is generated in the catalytic cycle by decomposition or ligand dissociation from the bis(diphos) complex. The reactivity of photogenerated PdL_2 differs from PtL_2 since the latter species does not add allyl substrates cleanly. Photochemical routes to PtL_2 and PdL_2 species, that contain less sterically hindered phosphines, complements thermal chemistry known (45,46,50) for bulky ML_2 species (M = Pd and Pt, L = PCy_3 or $P(\underline{t}\text{-}Bu)_3$). High reactivity of the photogenerated species is illustrated by the following comparison (49,51,52).

$$\text{Ph-Cl} + \text{Pt}(C_2O_4)(PEt_3)_2 \xrightarrow[15^\circ C]{h\nu} \underline{\text{trans}}\text{-PtClPh}(PEt_3)_2$$

$$\text{Ph-Cl} + \text{Pt}(PEt_3)_3 \xrightarrow[110^\circ C]{} \underline{\text{trans}}\text{-PtClPh}(PEt_3)_2$$

$$\text{Ph-Cl} + \text{Pt}(PCy_3)_2 \xrightarrow[20^\circ C]{14\ \text{days}} \underline{\text{trans}}\text{-PtClPh}(PCy_3)_2$$

Our attempts to prepare $Ni(C_2O_4)L_2$ complexes invariably led to formation of insoluble $Ni(C_2O_4)$. This may be attributed to the inability of the weak field oxalate ligand to stabilize square planar Ni(II). We thought that a strong field or soft version of the oxalate ligand might be useful. It seemed that the dithiooxalate ($S_2C_2O_2^{2-}$) ligand would exhibit photochemistry analogous to chelating oxalate. Therefore the series of dithiooxalate complexes $M(S_2C_2O_2)L_2$ have been prepared (53) where L = PMe_3 or, L_2 = diphos = $Ph_2PCH_2\text{-}CH_2PPh_2$ and depe = $Et_2PCH_2\text{-}CH_2PEt_2$, and M = Ni, Pd, and Pt. The IR stretch of the C=O group (1680-1750 cm^{-1}) proves sulfur coordination for the $S_2C_2O_2^{2-}$ ligand. Irradiation of the diphos derivatives in CH_2Cl_2 produced free SCO and MCl_2(diphos). Therefore it appears that the dithiooxalate ligand can also be reductively eliminated by photolysis. The chemistry of these systems is complicated by secondary reactions with SCO and is under investigation.

Synthesis and Photoreactivity of Surface-Bound Platinum Oxalates

The photochemically produced $Pt(PEt_3)_2$ fragment, stabilized as the cis- and trans-$PtH_2(PEt_3)_2$ complexes, has proved (54,55,56) to be an efficient and long lived homogeneous catalyst for H_2/D_2 exchange (Equation 17), deuteration of acetone or acetontrile (Equations 18 and 19), decomposition of formic acid (Equation 20), and hydrolysis of acetonitrile (Equation 21). Because of the catalytic promise

$$H_2 + D_2 \longrightarrow 2\ HD \tag{17}$$

$$3\ D_2 + CH_3C(O)CH_3 \longrightarrow 3\ H_2 + CD_3C(O)CD_3 \tag{18}$$

$$1\tfrac{1}{2}\ D_2 + CH_3CN \longrightarrow 1\tfrac{1}{2}\ H_2 + CD_3CN \tag{19}$$

$$HCOOH \longrightarrow H_2 + CO_2 \tag{20}$$

$$H_3C\text{-}C{\equiv}N + H_2O \longrightarrow H_3C\text{-}C(=O)\text{-}NH_2 \tag{21}$$

of these systems we decided to synthesize surface attached oxalate derivatives.

For certain catalyst applications (e.g., ease of separation) it is desirable to have heterogeneous rather than homogeneous catalysts. From a fundamental view there is interest in comparing the surface effect on catalytic rates and mechanisms of surface-attached homogeneous catalysts. Of the two commonly used supports (57), organic polymers or silica, we chose silica of high pore diameter (140Å) because of its rigidity and permeability in polar media.

Most previous (57) studies of phosphine supported transition metal complexes have employed arylphosphine ligands. This presents problems since arylphosphines often dissociate and catalyst leaching poses a problem. Small trialkylphosphines, by contrast, are among the most difficult ligands to displace from a metal center. Arylphosphines are bulky and hinder substrate access to the metal center. Cleavage of the P-C bond (i.e., degradation) as well as orthometallation occurs more readily with arylphosphine analogs (58). Several synthetic procedures were explored for synthesizing surface attached oxalate complexes. The best procedure (59) is outlined in Scheme III. In the synthesis of Scheme III we used Davison Silica (Grade 62, 140Å pore diameter, 340 m^2/g) and achieved a maximum surface coverage of 1 molecule/113$Å^2$, which amounts to 70% functionalization of the surface (with the assumption that 6 surface hydroxyls anchor one platinum complex). Key points of the synthesis include: 1) the volatility of reactants in step 1 and the high yield (97%) of the photochemical addition make it possible to prepare the L-PEt_2 ligand in greater than 99% purity; 2) the volatile SMe_2 ligand can be removed in step 2 and the derivatized platinum complex, which is an oil, can be isolated in high purity; 3) capping remaining surface hydroxyl groups with hexamethyldisilazane in step 4 prevents reactions of photogenerated Pt(0) with the support; 4) putting the complex on the support as a stable Pt(II) species protects the basic phosphine ligand from oxidation (60).

Besides the analytical data (Pt/P ratio = 1/2) that characterize the supported complex the IR spectrum exhibits stretches that are identical to those in the homogeneous analogue (49) $Pt(C_2O_4)(PEt_3)_2$. If the sample is irradiated (as a nujol mull) the oxalate stretches disappear and a new peak appears at 2330 cm^{-1}, attributed to CO_2. Thus, Equation 22 occurs on the surface. Recall (Scheme II) that

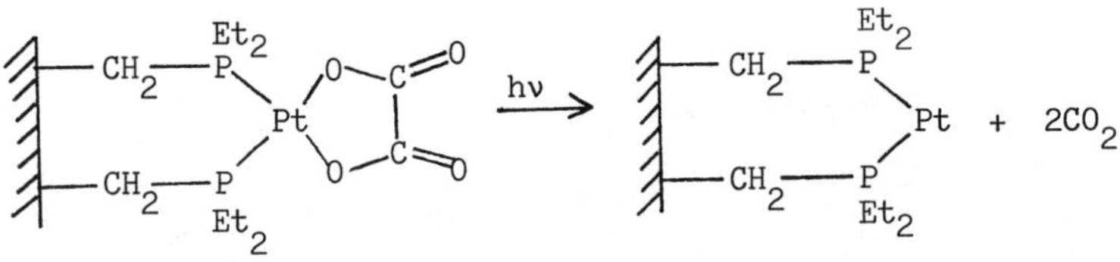

(22)

photogenerated $Pt(PEt_3)_2$ could be trapped with CO to form $Pt(PEt_3)_2(CO)_2$. Since the carbonyl stretches (1930 and 1973 cm^{-1}) are characteristic (47) of this complex we irradiated the surface supported complex under CO. Observation of the peaks at 1929 and 1965 cm^{-1}

Scheme III

(1) $MeO-Si-CH_2=CH_2$ + $HPEt_2$ $\xrightarrow{h\nu}$ $MeO-Si-(CH_2)_2-PEt_2$ = $L-PEt_2$
(with MeO groups on Si on both sides)

(2) $2L-PEt_2$ + $Pt(C_2O_4)(SMe_2)$ $\xrightarrow{C_6H_6}$ SMe_2 $\uparrow$ + $Pt(C_2O_4)(L-PEt_2)_2$

(3)

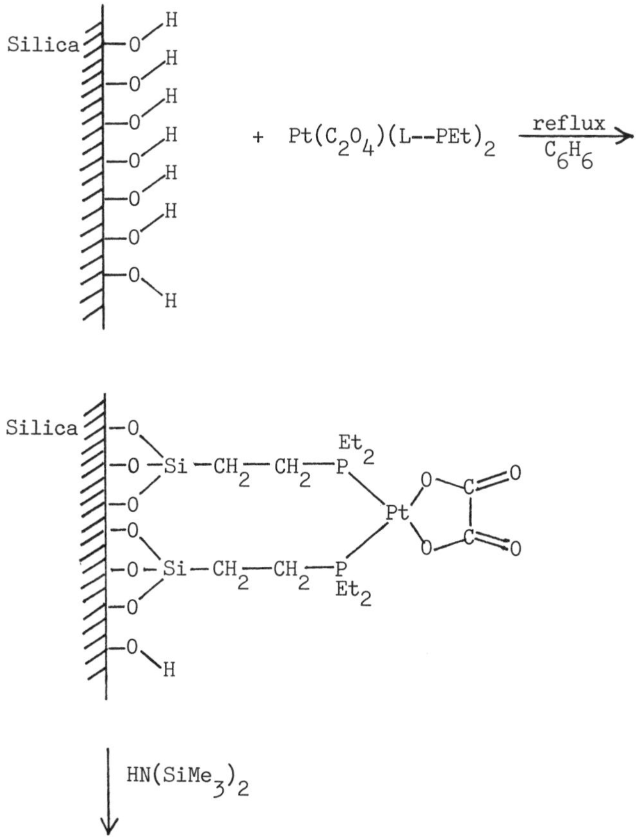

+ $Pt(C_2O_4)(L--PEt)_2$ $\xrightarrow[C_6H_6]{reflux}$

(4)

$\downarrow$ $HN(SiMe_3)_2$

caps off free surface OH with inert $SiMe_3$ group, and wash well.

suggests that the surface generated species of Equation 22 can be trapped, Equation 23.

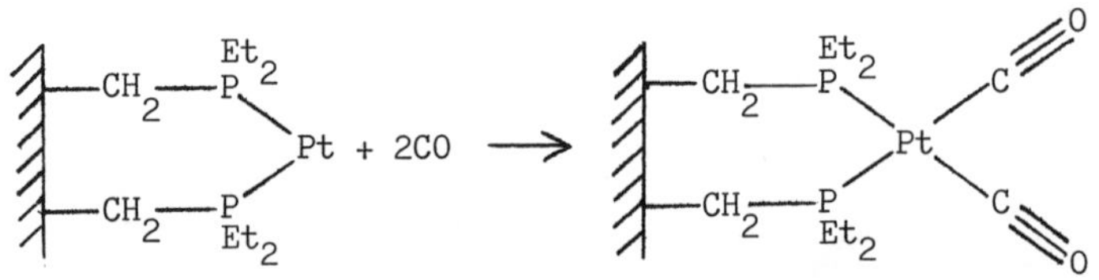

(23)

In contrast to homogeneous analogues (48,61) the silica bound PtL_2 fragment does not catalyze acetonitrile hydrolysis. Initial experiments showed hydrolysis activity < 1/1000 that of the homogeneous system. This puzzled us until we found that homogeneous catalyst systems where the phosphine ligands are constrained to be cis [e.g., $Pt(C_2O_4)$(diphos)] show similar low activity. Molecular modeling studies (CPK models) of the surface attached reagent of Scheme III suggest that the platinum center cannot adopt a trans configuration necessary for effective catalysis.

Previous work (48) with homogeneous analogues showed that Si-H oxidative additions yield cis products. A cis geometry of hydride and silyl may be allowed in catalytic hydrosilation. Because the industral homogeneous hydrosilation catalyst (62) is H_2PtCl_6 we tested the activity of our surface generated reagent for the reaction of Equation 24. A suspension of the catalyst was irradiated in 1-

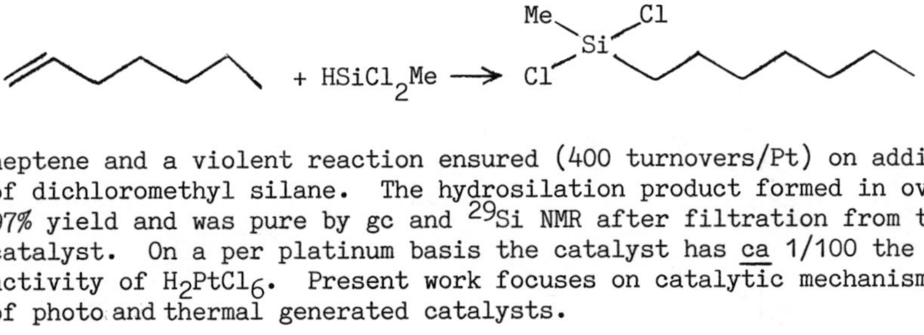

(24)

heptene and a violent reaction ensured (400 turnovers/Pt) on addition of dichloromethyl silane. The hydrosilation product formed in over 97% yield and was pure by gc and ^{29}Si NMR after filtration from the catalyst. On a per platinum basis the catalyst has ca 1/100 the activity of H_2PtCl_6. Present work focuses on catalytic mechanisms of photo and thermal generated catalysts.

Conclusions

Photochemical reactions of transition metal complexes that contain unsaturated chelates fall into three categories: 1) fragmentation of the ligand to yield two reactive functionalities; 2) elimination of the ligand to generate two reactive sites at the metal; 3) chelate localized excited states can function as photoreceptors to promote photodissociation of other metal-ligand bonds in the complex. These processes can be used as an entry to new reactive intermediates and catalysts.

Acknowledgments

I thank the students (C.E. Johnson, M.E. Gross, R.S. Paonessa, A.L. Prignano, D. Pourreau, R.L. Cowan), and postdoctorals (C.E. Jensen, M.J. Maroney) who contributed to the research program described. Financial support of our research by the Air Force Office of

Scientific Research, Army Research Office, and National Science Foundation is appreciated.

Literature Cited

1. Strohmeier, W. Angew. Chem. 1964, 76, 873.
2. Mok, C.Y.; Endicott, J.F. J. Am. Chem. Soc. 1977, 99, 1276.
3. Wrighton, M.S.; Ginley, D.S. J. Am. Chem. Soc. 1975, 97, 2065; Reinking, M.K.; Kullberg, M.L.; Cutler, A.R.; Kubiak, C.P. J. Am. Chem. Soc. 1985, 107, 3517.
4. Brand, J.C.; Snedder, W. Trans. Faraday Soc. 1957, 53, 894.
5. Green, M.L.H. Pure Appl. Chem. 1978, 50, 27.
6. Geoffroy, G.L. Prog. Inorg. Chem. 1980, 27, 123.
7. Bock, C.R.; von Gustorf, E.A.K. Adv. Photochem. 1977, 10, 222.
8. Geoffroy, G.L.; Wrighton, M.S. "Organometallic Photochemistry"; Academic Press: New York, 1979.
9. Trogler, W.C.; Ibers, J.A. Organometallics 1982, 1, 536.
10. Gross, M.E.; Trogler, W.C. J. Organomet. Chem. 1981, 209, 407.
11. Gross, M.E.; Johnson, C.E.; Maroney, M.J.; Trogler, W.C. Inorg. Chem. 1984, 23, 2968.
12. Gross, M.E.; Trogler, W.C.; Ibers, J.A. J. Am. Chem. Soc. 1981, 103, 192; Gross, M.E.; Trogler, W.C.; Ibers, J.A. Organometallics 1982, 1, 732.
13. Maroney, M.J.; Trogler, W.C. J. Am. Chem. Soc. 1984, 106, 4144.
14. Thorn, D.L.; Hoffmann, R. Nouv. J. Chim. 1979, 3, 39.
15. Gross, M.E.; Ibers, J.A.; Trogler, W.C. Organometallics 1982, 1, 530.
16. March, J. "Advanced Organic Chemistry", 2nd ed.; McGraw-Hill: New York, 1977.
17. Elian, M.; Chen, M.M.L.; Mingos, D.M.P.; Hoffmann, R. Inorg. Chem. 1976, 5, 1148.
18. Trogler, W.C.; Johnson, C.E.; Ellis, D.E. Inorg. Chem. 1981, 20, 980.
19. Shi, Q.-Z.; Richmond, T.G.; Trogler, W.C.; Basolo, F. Organometallics 1982, 1, 1033.
20. Kokkes, M.W.; Stufkens, D.J.; Oskam, A. J. Chem. Soc., Dalton Trans. 1983, 439.
21. Johnson, C.E.; Trogler, W.C. J. Am. Chem. Soc. 1981, 103, 6352.
22. Chang, C.-Y.; Johnson, C.E.; Richmond, T.G.; Chen, Y.-T.; Trogler, W.C.; Basolo, F. Inorg. Chem. 1981, 20, 3167.
23. Kokkes, M.W.; Stufkens, D.J.; Oskam, A. J. Chem. Soc., Dalton Trans. 1984, 1005.
24. Church, S.P.; Grevels, F.-W.; Hermann, H.; Schaffner, K. Inorg. Chem. 1985, 24, 418-422.
25. Wrighton, M.S.; Graff, J.L.; Reichel, C.L.; Sanner, R.D. Ann. N.Y. Acad. Sci. 1980, 333, 188.
26. Johnson, C.E., Ph.D. Thesis, Northwestern University, 1981.
27. Balk, R.W.; Stufkens, D.J.; Oskam, A. J. Chem. Soc., Dalton Trans. 1982, 275.
28. Wrighton, M.; Gray, H.B.; Hammond, G.S. Mol. Photochem. 1973, 5, 164.
29. Zink, J.I. Mol. Photochem. 1973, 5, 151.

30. Vanquickenborne, L.G.; Ceulemans, A. Coord. Chem. Rev. 1983, 48, 157.
31. Ford, P.C. Coord. Chem. Rev. 1982, 44, 61.
32. Adamson, A.W. Coord. Chem. Rev. 1968, 3, 169.
33. Robinson, G.W.; Frosch, R.P. J. Chem. Phys. 1962, 37, 1962; 1963, 38, 1187.
34. Chock, D.P.; Jortner, J.; Rice, S.A. J. Chem. Phys. 1968, 49, 610.
35. Jortner, J.; Rice, S.A.; Hochstrasser, R.M. Adv. Photochem. 1969, 7, 149.
36. Langford, C.H. Acc. Chem. Res. 1984, 17, 96.
37. Ouderkirk, A.J.; Wermer, P.; Schultz, N.L.; Weitz, E. J. Am. Chem. Soc. 1983, 105, 3354.
38. Seder, T.A.; Church, S.P.; Ouderkirk, A.J.; Weitz, E. J. Am. Chem. Soc. 1985, 107, 1432.
39. Tumas, W.; Gitlin, B.; Rosan, A.M.; Yardley, J.T. J. Am. Chem. Soc. 1982, 104, 55.
40. Poliakoff, M.; Weitz, E. Adv. Organomet. Chem. 1985, in press.
41. von Gustorf, E.A.K.; Leenders, L.H.G.; Fischler, I.; Perutz, R. Adv. Inorg. Chem. Radiochem. 1976, 19, 65.
42. Balzani, V.; Carassiti, V. "Photochemistry of Coordination Compounds"; Academic Press: New York, 1970.
43. Blake, D.M.; Nyman, C.J. J. Am. Chem. Soc. 1970, 92, 5359.
44. Addison, A.W.; Gillard, R.S.; Sheridan, P.S.; Tipping, L.R.H. J. Chem. Soc., Dalton Trans. 1974, 709.
45. Otsuka, S. J. Organomet. Chem. 1980, 200, 191.
46. Shaw, B.L. ACS Symp. Ser. 1982, 196, 101.
47. Paonessa, R.S.; Trogler, W.C. Organometallics 1982, 1, 768.
48. Paonessa, R.S.; Prignano, A.L.; Trogler, W.C. Organometallics 1985, 4, 647.
49. Trost, B.M. Acc. Chem. Res. 1980, 13, 385.
50. Stone, F.G.A. Angew. Chem., Intl. Ed. Engl. 1984, 23, 89.
51. Gerlach, D.H.; Kane, A.R.; Parshall, G.W.; Jesson, J.P.; Muetterties, E.L. J. Am. Chem. Soc. 1971, 93, 3543.
52. Fornies, J.; Green, M.; Spencer, J.L.; Stone, F.G.A. J. Chem. Soc., Dalton Trans. 1977, 1006.
53. Cowan, R.L.; Pourreau, D.; Trogler, W.C., to be published.
54. Paonessa, R.S.; Trogler, W.C. J. Am. Chem. Soc. 1982, 104, 1138.
55. Paonessa, R.S.; Trogler, W.C. J. Am. Chem. Soc. 1982, 104, 3529.
56. Paonessa, R.S.; Trogler, W.C. Inorg. Chem. 1983, 22, 1038.
57. Bailey, D.C.; Langer, S.H. Chem. Rev. 1981, 81, 109.
58. Parshall, G.W. Acc. Chem. Res. 1970, 3, 139.
59. Prignano, A.L.; Trogler, W.C., to be published.
60. Bemi, L.; Clark, H.C.; Davies, J.A.; Fyfe, C.A.; Wasylishen, R.E. J. Am. Chem. Soc. 1982, 104, 438.
61. Jensen, C.M.; Trogler, W.C., submitted.
62. Parshall, G.W. "Homogeneous Catalysis"; Wiley: New York, 1980.

RECEIVED November 8, 1985

Chemistry of Rhodium and Iridium Phosphine Complexes

Flash Photolysis Investigations of Reactive Intermediates

David Wink and Peter C. Ford

Department of Chemistry, University of California, Santa Barbara, CA 93106

Flash photolysis of the rhodium(I) and iridium(I) complexes $MCl(CO)(PPh_3)_2$ in benzene leads to formation of the unsaturated species $MCl(PPh_3)_2$, the reaction kinetics of which have been investigated. Reactions with CO to reform $MCl(CO)(PPh_3)_2$ occur with second order rate constants of 7×10^7 and 2.7×10^8 $M^{-1}s^{-1}$ for M = Rh and Ir, respectively. The $RhCl(PPh_3)_2$ species also undergoes fast reactions with PPh_3 ($k = 2.6 \times 10^6$ $M^{-1}s^{-1}$) and with ethylene ($> 2 \times 10^7$ $M^{-1}s^{-1}$) to form $RhCl(PPh_3)_2$ and $RhCl(H_2C{=}CH_2)(PPh_3)_2$, respectively; however, reaction with H_2 to form the dihydride is much slower, (1×10^5 $M^{-1}s^{-1}$). Also described are flash photolysis studies of the dinitrogen species $IrCl(N_2)(PPh_3)_2$ and the dihydride $H_2IrCl(CO)(PPh_3)_2$. In both cases, the transient $IrCl(PPh_3)_2$ is formed. These results indicate that CO labilization from the Ir(III) dihydride is a facile photochemical pathway and the photo-reductive elimination of H_2 is a more complicated mechanism than previously inferred.

Phosphine complexes of low valent metal complexes have a long history in the chemistry of homogeneous catalytic activation of small molecules(1,2). For example, the catalysis chemistry of rhodium(I) phosphine complexes continues to hold much interest several decades since the description of such reactions by Wilkinson(3). However, despite considerable quantitative scrutiny(3-10), the mechanistic details of key catalytic steps for even the original Wilkinson's catalyst $RhCl(PPh_3)_3$ are not fully resolved(7-10). The reason lies within the very nature of catalytic processes, namely that the activation of substrates often involves reactions of unstable transient species, the properties of which can only be inferred from kinetic rate laws or from spectral studies under conditions considerably different from those of an operating catalyst. In some cases it may be possible to use flash photolysis to generate significant concentrations of such a transient and to investigate the reactions of that species more directly. Here we describe some investigations

0097-6156/86/0307-0197$06.00/0

using flash photolysis techniques to probe the reaction dynamics of reactive intermediates in the chemistry of rhodium(I) and iridium(I) phosphine complexes.

Rhodium(I) Complexes

A key intermediate in proposed mechanisms for Wilkinson's catalyst is the tricoordinate species $RhCl(PPh_3)_2$ or its solvated analog (7-10). Thus, it would be particularly desirable to prepare this species and to interrogate its reactivity under catalytically relevant conditions. Given the commonly observed photolability of carbon monoxide complexes(11), a logical photochemical precursor to $RhCl(PPh_3)_2$ would be the carbonyl $RhCl(CO)(PPh_3)_2$. Although earlier investigators found that under continuous photolysis the latter did not display net photochemistry(12), it appeared likely that reversible ligand labilization would be the result of flash excitation.

When trans-$RhCl(CO)(PPh_3)_2$ in degassed benzene was subjected to flash photolysis ($\lambda_{irr} > 315$ nm, pulse duration about 20 μsec), transient absorption was observed with the spectral characteristics illustrated in Figure 1. This transient (when monitored at λ_{mon} 410 nm) decayed via second order kinetics over a period of several ms (Figure 2). When the solution was flashed under CO (1.0 atm, 0.006 M in benzene,(13)), no transient having a lifetime longer than the flash was detected; however, a long-lived transient with the same spectrum as $RhCl(PPh_3)_3$ was seen when the flash photolysis was carried out in the presence of excess PPh_3 (0.05 M, see below). Thus CO, not phosphine, photolabilization appears to be the major primary photoreaction (Equation 1).

$$RhCl(CO)(PPh_3)_2 \underset{k_{-1}}{\overset{h\nu}{\rightleftharpoons}} RhCl(PPh_3)_2 + CO \qquad (1)$$

When the flash photolysis of $RhCl(CO)(PPh_3)_2$ was carried out in the absence of other reactants but with λ_{mon} 450 nm, it was noted that the transient absorption decayed in two stages (Figure 3). The relatively rapid second order decay noted at 410 nm was followed by a slower first order decay back to $RhCl(CO)(PPh_3)_2$ with a k_{obs} of 1.8 s^{-1} (298 K). The spectrum of the longer lived transient (Figure 1) is very close to that of the dimer $[RhCl(PPh_3)_2]_2$, described previously(14) and discussed in the mechanistic schemes for Wilkinson's catalyst(7-10). Presumably, $[RhCl(PPh_3)_2]_2$ is formed via the dimerization of $RhCl(PPh_3)_2$:

$$2\ RhCl(PPh_3)_2 \underset{k_{-2}}{\overset{k_2}{\rightleftharpoons}} [RhCl(PPh_3)_2]_2 \qquad (2)$$

and decays by a unimolecular rate-limiting step, presumably dissociation to monomers, i.e., the k_{-2} step. (The stopped-flow kinetics of the reaction between $[RhCl(PPh_3)_2]_2$, prepared thermally,(14) and CO (P_{co} 0.1 - 1.0 atm) in benzene to give $RhCl(CO)(PPh_3)_2$ also gave first order rates with the nearly identical k_{obs}, 1.7 s^{-1} at 298 K).

Calculation of k_{-1} and k_2 from the flash photolysis data requires the extinction coefficient of $RhCl(PPh_3)_2$ at λ_{mon} in order

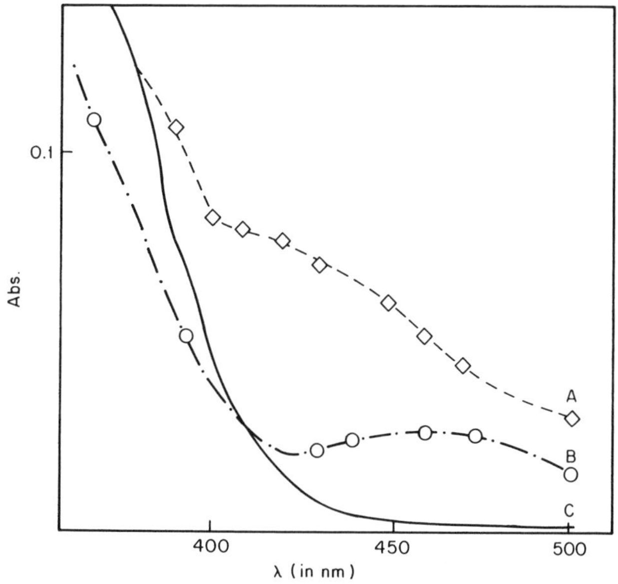

Figure 1. Transient spectra resulting from the flash photolysis of $RhCl(CO)(PPh_3)_2$ in benzene solution. A) Spectrum observed 100 μs after flash (λ_{irr} > 315 nm). Points indicated represent actual experimental observations; curve is drawn for illustrative purposes. B) Spectrum observed 20 ms after flash. C) Spectrum of $RhCl(CO)(PPh_3)_2$.

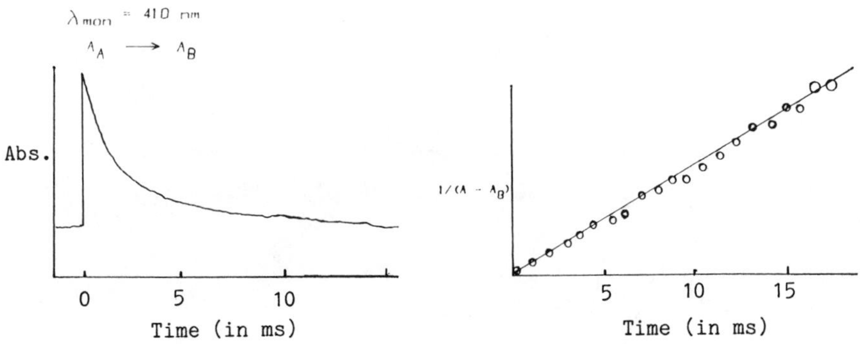

Figure 2. Left: Absorbance changes resulting from the flash photolysis (λ_{irr} > 415 nm) of $RhCl(CO)(PPh_3)_2$ in benzene solution under Ar at 25°. The monitoring wavelength was 410 nm. Right: A linear second-order plot $[(A - A_B)^{-1}$ vs t] for the data in the above curve.

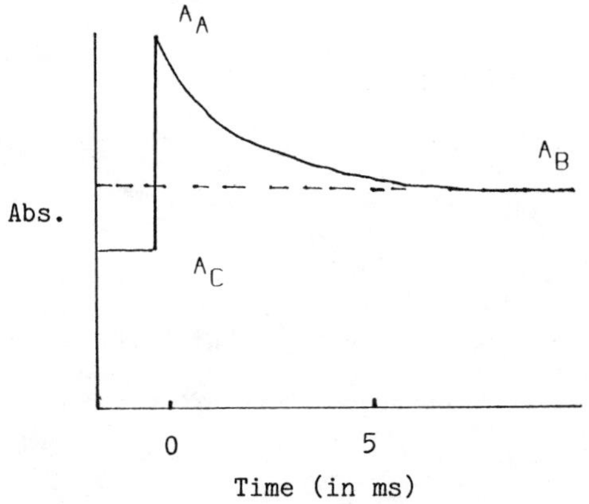

Figure 3. Decay curve for the flash photolysis of $RhCl(CO)(PPh_3)_2$ in 25° benzene under Ar showing the formation of another intermediate species (B) as a product of the second-order decay of the initial transient (A).

to determine the concentrations of this species. This was estimated by assuming that the reaction with excess PPh_3 (see above) trapped all $RhCl(PPh_3)_2$ as $RhCl(PPh_3)_3$. From the known spectrum of the latter species(14), the initial concentration, thus the extinction coefficient (8×10^2 M^{-1} cm^{-1} at 410 nm, the isosbestic point for the dimer intermediate and the starting material), of $RhCl(PPh_3)_2$ could be calculated. With this extinction coefficient, the second order rate constant for the disappearance of $RhCl(PPh_3)_2$ was determined to be 1×10^8 $M^{-1}s^{-1}$. The amount of dimer produced by the flash photolysis in the absence of added reactants (calculated from the spectrum of the long-lived intermediate) indicated that under these conditions about 40% of $RhCl(PPh_3)_2$ dimerized in competition with the back reaction with the photoliberated CO to give $RhCl(CO)(PPh_3)_2$. Thus k_{-1} and k_2 were estimated as 6×10^7 $M^{-1}s^{-1}$ and 4×10^7 $M^{-1}s^{-1}$, respectively.

The above k_{-1} value was confirmed by carrying out the flash photolysis experiments in the presence of excess CO, $(0.3 - 1.2) \times 10^{-4}$ M. Under these conditions, no dimer was observed and the transient decay was first-order with k_{obs} values linearly dependent on [CO]. The second order rate constant (k_{-1}) obtained from the plot of k_{obs} vs [CO] was $(6.9 \pm 0.2) \times 10^7$ $M^{-1}s^{-1}$.

The reaction of the transient $RhCl(PPh_3)_2$ with excess triphenyl phosphine tc give $RhCl(PPh_3)_2$ displayed first order rates dependent on the concentration of PPh_3. From these data, the second order rate constant for the reaction depicted in Equation 3 was calculated as 2.8×10^6 $M^{-1}s^{-1}$. Given the rate constant of 0.71 s^{-1} determined (4) for the dissociation of PPh_3 from $RhCl(PPh_3)_3$ in benzene (k_{-3}), the equilibrium constant for dissociation (k_{-3}/k_3) is calculated to be 0.25×10^{-6} M, consistent with the previous estimate of $< 10^{-5}$ M(4).

$$RhCl(PPh_3)_2 + PPh_3 \underset{k_{-3}}{\overset{k_3}{\rightleftharpoons}} RhCl(PPh_3)_3 \qquad (3)$$

When the flash photolysis of $RhCl(CO)(PPh_3)_2$ was carried out as above but under dihydrogen (1.0 atm, 0.0028 M)(5), the intermediate $RhCl(PPh_3)_2$ underwent first order reaction ($k_{obs} = 2.8 \times 10^2$ s^{-1}) to give a new transient spectrum having an even smaller absorbance than that of the carbonyl complex over the spectral range 360-450 nm (Figure 4). This new transient decayed over a period of seconds to give $RhCl(CO)(PPh_3)_2$ again. We interpret these observations in terms of the reaction of $RhCl(PPh_3)_2$ with H_2 to give the dihydride (Equation 4) followed by reaction of the latter with CO to regenerate $RhCl(CO)(PPh_3)_2$. If a second order rate law for Equation 4 is assumed, then the calculated value of k_4 is 1.0×10^5 $M^{-1}s^{-1}$, very much in agreement with Halpern's estimate of k_4 ($> 7 \times 10^4$ $M^{-1}s^{-1}$) drawn from kinetics analysis of the Wilkinson's catalysis (4). Flash photolysis under D_2 (1.0 atm) gave identical spectral changes and a calculated k_3 of 0.7×10^5 $M^{-1}s^{-1}$, i.e., a kinetic isotope effect k_4^h/k_4^d of about 1.4.

$$RhCl(PPh_3)_2 + H_2 \xrightarrow{k_4} H_2RhCl(PPh_3)_2 \qquad (4)$$

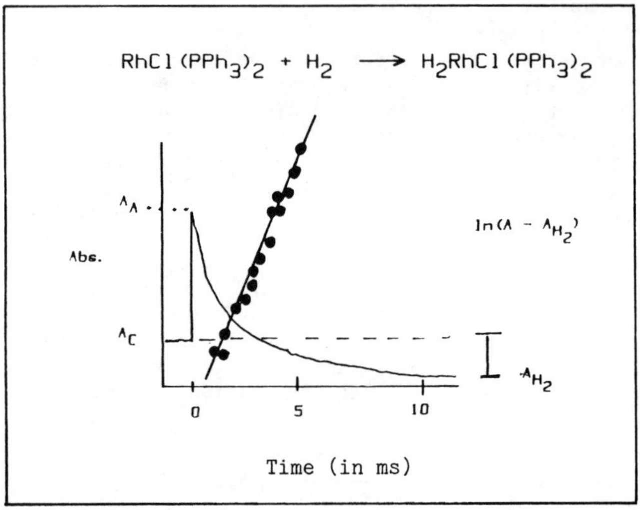

Figure 4. Flash photolysis of $RhCl(CO)(PPh_3)_2$ in 25° benzene under H_2 (1.0 atm) showing decay of the transient $RhCl(PPh_3)_2$ by reaction with excess hydrogen to form a new intermediate species (presumably $H_2RhCl(PPh_3)_2$) having an absorbance (A_H) less than that of the starting material A_C. The curve represents the temporal absorbance changes (scale to the left); the line represents a pseudo-first-order plot of this decay (scale to the right).

Flash photolysis of RhCl(CO(PPh$_3$)$_2$ under ethylene (0.01 atm, 0.0011 M) led to immediate spectral changes consistent with the formation of the ethylene complex(3) RhCl(H$_2$C=CH$_2$)(PPh$_3$)$_2$ within the duration of the flash. This observation provides a lower limit of 2×10^7 $M^{-1}s^{-1}$ for the second order rate constant for the reaction of RhCl(PPh$_3$)$_2$ with ethylene. The back reaction of the ethylene adduct with CO to reform RhCl(CO)(PPh$_3$)$_2$ was also rather rapid and occurred within a period of a few milliseconds.

In summary, the flash photolysis of RhCl(CO)(PPh$_3$)$_2$ in benzene leads principally to formation of the coordinatively unsaturated Wilkinson's catalyst intermediate RhCl(PPh$_3$)$_2$ (or its solvated analog). The reactions of this species are summarized in Scheme I. In the absence of other reactants, this intermediate recombines with CO or dimerizes via very rapid reactions, but in the presence of added PPh$_3$, ethylene or dihydrogen, adducts of RhCl(PPh$_3$)$_2$ are formed. Rate constants for the second order reactions of RhCl(PPh$_3$)$_2$ with various substrates are summarized in Table I. The rates for addition of the two electron donors CO, C_2H_4 and PPh$_3$ are significantly higher than for the oxidative addition of dihydrogen. In all cases, however, the initial adducts react eventually with the photolabilized CO to reform the more stable starting complex RhCl(CO)(PPh$_3$)$_2$. These observations are consistent with the continuous photolysis studies which reported no net photochemistry of this species in the absence of oxygen (12).

Table I. Second Order Rate Constants for the Reactions of Various Substrates with RhCl(PPh$_3$)$_2$ in 25° Benzene Solution.

Substrate	k (in $M^{-1}s^{-1}$)
CO	$(6.9 \pm 0.2) \times 10^7$
C_2H_4	$> 2 \times 10^7$
RhCl(PPh$_3$)$_2$	$(4 \pm 1) \times 10^7$
PPh$_3$	$(2.8 \pm 0.4) \times 10^6$
H_2	$(9.8 \pm 0.5) \times 10^4$
D_2	$(7.0 \pm 0.5) \times 10^4$

Iridium Complexes

Vaska's complex trans-IrCl(CO)(PPh$_3$)$_2$ has served as an important model for mechanistic investigation of catalytically relevant reactions such as the oxidative addition and reductive elimination of small molecules(15). The latter processes have also been the subject of some photochemical investigation. For example, the reductive elimination of H_2 depicted in Equation 5, which is a relatively slow thermally activated process ($k_1 = 3.8 \times 10^{-5}$ s^{-1} in 25° benzene solution (15)), has been shown to occur readily when the dihydride complex was subjected to continuous photolysis with 366 nm light(16). However, Vaska's compound itself was reported to be

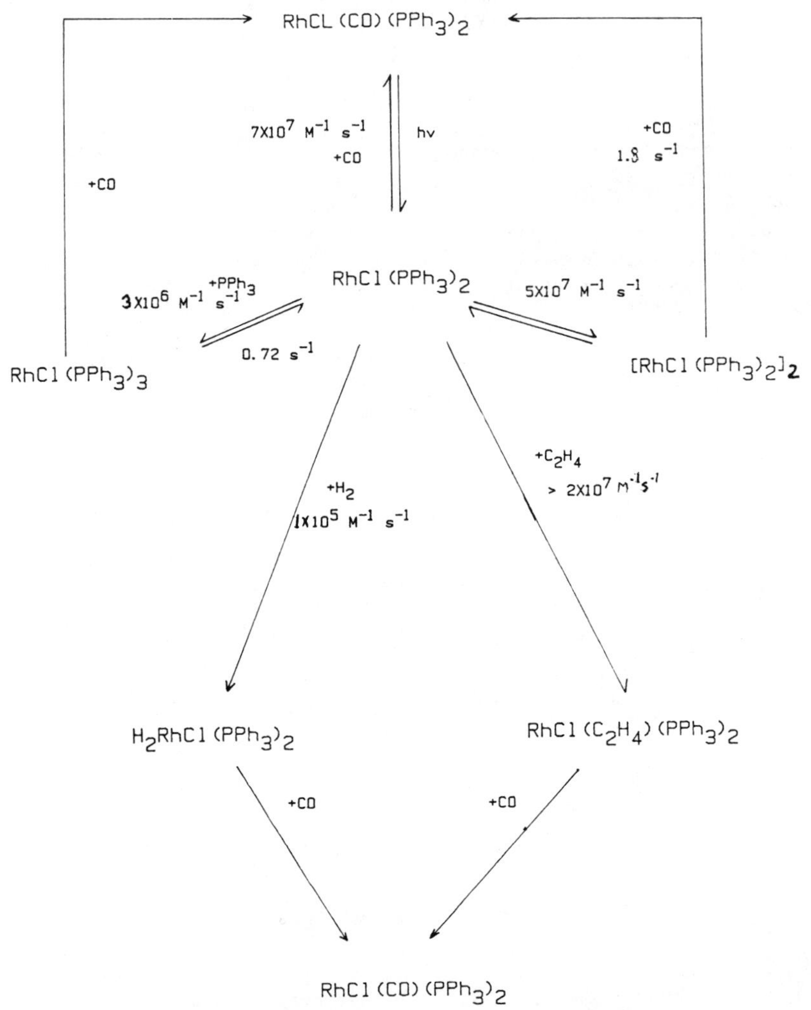

Scheme I. A summary of the reactions of $RhCl(PPH_3)_2$ as studied by the kinetic flash photolysis of $RhCl(CO)(PPh_3)_2$.

photoinert under continuous photolysis(16). In these contexts, we were interested in establishing whether Vaska's compound would show the same type of transient behavior upon flash photolysis as did the rhodium analog (above) and whether transients could be detected in the photoreactions of the related Ir(III) oxidative adducts.

$$H_2IrCl(CO)(PPh_3)_2 \rightarrow trans\text{-}IrCl(CO)(PPh_3)_2 + H_2 \quad (5)$$

Flash photolysis of trans-$IrCl(CO)(PPh_3)_2$ in stringently deaerated benzene solution under an argon atmosphere (λ_{irr} > 254 nm) resulted in the formation of a transient with strong absorption in the spectral region 390-550 nm ($\lambda_{max} \simeq$ 430 nm) which decayed to the initial baseline via cleanly second order kinetics (Figure 5). The return to the initial spectrum is consistent with the earlier report (16) that continuous photolysis of trans-$IrCl(CO)(PPh_3)_2$ gave no net photoreaction. When similar flash experiments were carried out under various pressures of CO, the transient decay kinetics were first order with the observed rate constants k_{obs} linearly dependent on P_{co}. Thus, we conclude that the intermediate species observed is the product of CO photodissociation (Equation 6) and that the decay process is the recombination to the starting complex (Equation 7). The second order rate constant $k_7 = (2.7 \pm 0.7) \times 10^8\ M^{-1}s^{-1}$ was determined from the plot of k_{obs} vs [CO]. These results are very similar to the chemistry induced by the flash photolysis of the rhodium(I) analogue above, although k_7 is about a factor of four faster for the Ir(I) transient.

$$trans\text{-}IrCl(CO)(PPh_3)_2 \xrightarrow{h\nu} IrCl(PPh_3)_2 + CO \quad (6)$$

$$IrCl(PPh_3)_2 + CO \xrightarrow{k_7} trans\text{-}IrCl(CO)(PPh_3)_2 \quad (7)$$

Flash photolysis of the analogous dinitrogen complex trans-IrCl$(N_2)(PPh_3)_2$ (17) demonstrates that flash photolysis leads in both cases to immediate appearance of a transient spectrum the same, within experimental uncertainty, as that attributed above to $IrCl(PPh_3)_2$. Continuous photolysis of $IrCl(N_2)(PPh_3)_2$ in C_6D_6 under otherwise analogous conditions results in the disappearance of the infrared band attributed to the coordinated N_2 (Figure 6), and the product solution displays a proton nmr resonance at -22.5 ppm indicating the formation of an iridium hydride (18). Thus, photolabilization of N_2 to give $IrCl(PPh_3)_2$ is irreversible (Equation 8), and, in the absence of the other reactants, this reactive intermediate apparently undergoes internal orthometallation of a triphenyl phosphine to give an iridium(III) hydride product(s). In the flash photolysis experiment, the latter process is evidenced by slow absorbance changes in the 340 to 550 nm range with isosbestic points at 460 and 334 nm consistent with formation of Ir(III) products.

$$IrCl(N_2)(PPh_3)_2 \xrightarrow{h\nu} IrCl(PPh_3)_2 + N_2 \quad (8)$$

$$IrCl(PPh_3)_2 \longrightarrow Ir(III)\ hydride \quad (9)$$

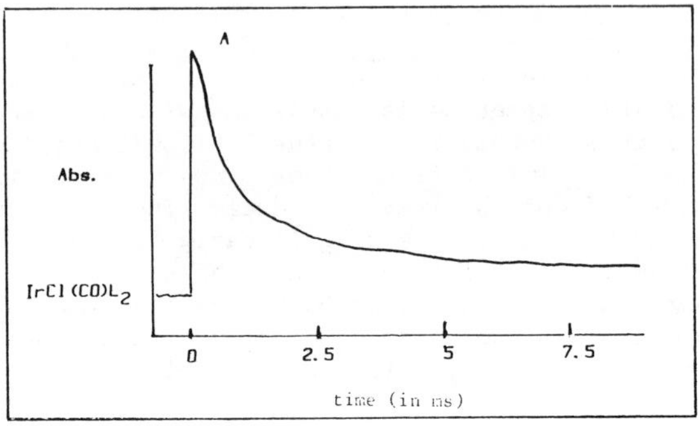

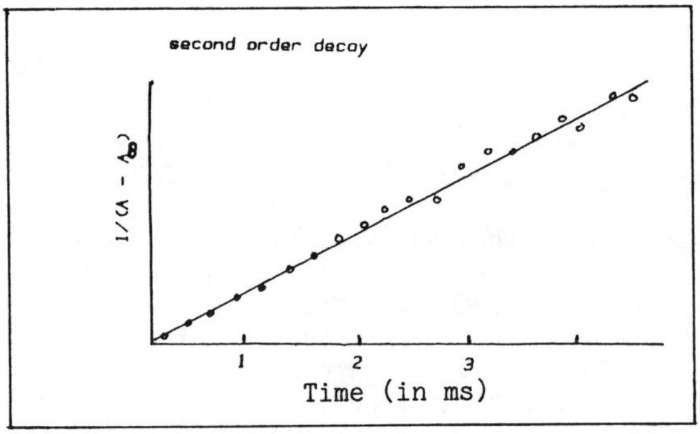

Figure 5. Top: Absorbance changes resulting from the flash photolysis of $IrCl(CO)(PPh_3)_2$ in 25° benzene under Ar. The monitoring wavelength was 420 nm. Bottom: A linear second-order plot for the data in the above curve.

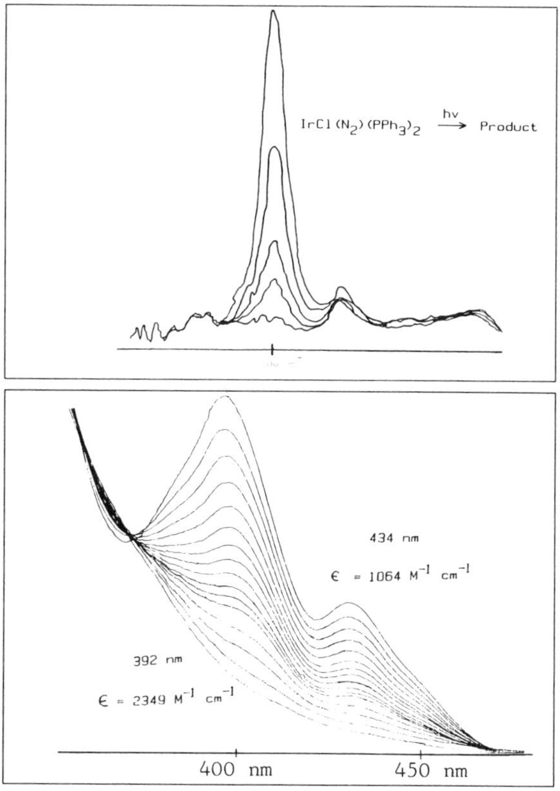

Figure 6. Top: IR absorbance changes resulting from the CW photolysis (λ_{irr} = 405 nm) or $IrCl(N_2)(PPh_3)_2$ in 25 benzene under Ar. Bottom: Electronic spectral changes resulting from a similar photolysis of $IrCl(N_2)(PPh_3)_2$. The top curve in each case represents the spectrum of the starting material.

In agreement with the earlier report (16), we found that $H_2IrCl(CO)(PPh_3)_2$ undergoes facile photoelimination of H_2 (Equation 5) when irradiated. The quantum yield was 0.56 moles/einstein for continuous photolysis at 313 nm, a value close to that reported for the similar complex $H_2IrCl(PPh_3)_3$ (0.55). However, a particularly striking observation was that flash photolysis of $H_2IrCl(CO)(PPh_3)_2$ in benzene under 1 atm. H_2 (λ_{irr} > 254 nm) resulted in transient absorbance in the spectral region 400-550 nm experimentally indistinguishable from that seen for the flash photolyses of trans-$IrCl(CO)(PPh_3)_2$ and of trans-$IrCl(N_2)(PPh_3)_2$. This transient decayed via second order kinetics to a product having the spectrum of Vaska's compound. Over a period of 10 minutes, the latter underwent subsequent reaction with H_2 to reform the starting complex according to Equation 10 for which the rate constant 1.2 $M^{-1}s^{-1}$ has previously been reported (15).

$$\text{trans-}IrCl(CO)(PPh_3)_2 + H_2 \longrightarrow H_2IrCl(CO)(PPh_3)_2 \quad (10)$$

The flash photolysis of $D_2IrCl(CO)(PPh_3)_2$ under 1 atm D_2 demonstrated no isotope effect on the relaxation of the first transient to give trans-$IrCl(CO)(PPh_3)_2$. However, flash photolysis of $H_2IrCl(CO)(PPh_3)_2$ under H_2/CO mixtures gave decay rates linearly dependent on P_{co}. Plots of k_{obs} vs [CO] as above gave the second order rate constant (2.6 ± 0.7) x 10^8 $M^{-1}s^{-1}$ within experimental uncertainty of the k_7 value reported above.

These results indicate that $H_2IrCl(CO)(PPh_3)_2$ first undergoes photodissociation of CO (Equation 11) followed by elimination of H_2 from the resulting pentacoordinated intermediate (Equation 12) to give the $IrCl(PPh_3)_2$ transient formed directly via flash photolysis of trans-$IrCl(CO)(PPh_3)_2$.

$$H_2IrCl(CO)(PPh_3)_2 \xrightarrow{h\nu} H_2IrCl(PPh_3)_2 + CO \quad (11)$$

$$H_2IrCl(PPh_3)_2 \underset{}{\overset{k_{12}}{\rightleftharpoons}} IrCl(PPh_3)_2 + H_2 \quad (12)$$

This view contrasts to the proposal that the dihydride photoelimination occurs by a single concerted step but is consistent with theoretical arguments(19) and experimental observations(20,21) that reductive elimination from d^6 complexes often occurs much more readily after ligand dissociation from the original hexacoordinate species to give a pentacoordinate intermediate. Given that formation of $IrCl(PPh_3)_2$ was complete within the lifetime of the flash (20 μs), a lower limit for k_{12} can be estimated as 5 x 10^4 s^{-1}. Thus, we conclude that dissociation of CO accelerates dihydrogen elmination by nine or more orders of magnitude. Notably, this rate acceleration occurs depsite the dissociation of the π-acid CO which would be expected to favor the lower oxidation state of the metal center. (*The following observation argues against the possibility of a sequential two-photon process involving initial H_2 photolabilization to generate $IrCl(CO)(PPh_3)_2$ followed by secondary photolysis of this product to give "$IrCl(PPh_3)_2$". The relative pulse intensity required to generate the same concentration of the latter transient was five times larger when the initial substrate was trans-IrCl(CO) $PPh_3)_2$ (under argon) than when $H_2IrCl(CO)(PPh_3)_2$ (under H_2) was the initial substrate. However, an alternative mechanism should also be considered, namely, the possibility that both CO and H_2 loss occur simultaneously or sequentially from the excited state of $H_2IrCl(CO)(PPh_3)_2$. At present such a mechanism, although unprecedented, can not be differentiated from the stepwise pathway proposed in Equations 11 and 12.)

Preliminary investigations of the Ir(III) species $HIrCl(CO)(PPh_2C_6H_4)(PPh_3)_2$, the orthometallated isomer of Vaska's compound(22), are closely analogous. Continous photolysis in this case leads to the formation of $IrCl(CO)(PPh_3)_2$. However, flash photolysis leads to a transient spectrum qualitatively the same as that attributed to $IrCl(PPh_3)_2$ and this transient decays by a second order pathway ([CO] dependent, i.e. Equation 7) to form trans-$IrCl(CO)(PPh_3)_2$ as the final product. Thus, again it appears that the starting complex has undergone CO photodissociation followed by H/aryl elimination to form $IrCl(PPh_3)_2$ within the 20 μs lifetime of the flash. This leads to a $5 \times 10^4\ s^{-1}$ lower limit for the rate constant of reductive elimination from the pentacoordinate intermediate, a value at least eight orders of magnitude faster than the rate of about $3 \times 10^{-4}\ s^{-1}$ we have measured for the thermal reaction of $HIrCl(CO)(PC_6H_4Ph)(PPh_3)_2$ to give trans-$IrCl(CO)(PPh_3)_2$ in 70° benzene.

In summary, we have observed that a common transient is produced in the flash photolysis of four species, trans-$IrCl(CO)(PPh_3)_2$, trans-$IrCl(N_2)(PPh_3)_2$, $HIrCl(CO)(PPh_2C_6H_4)(PPh_3)$ and $H_2IrCl(CO)(PPh_3)_2$. These resutls suggest that the photo-induced version of Equation 5 is a stepwise mechanism involving initial CO dissociation as the primary photoreaction of $H_2IrCl(CO)(PPh_3)_2$. The resulting pentacoordinated Ir(III) intermediate appears to be dramatically activated toward H_2 reductive elimination as predicted in theoretical treatments. These transformations are illustrated in Scheme II.

Experimental Section

The flash photolysis apparatus was that described previously(23) modified by the use of Biomation 805 transient digitizer interfaced to a Hewlett Packard 86 computer for data collection, analysis and plotting. Wavelength selection was accomplished by use of an aqueous $NaNO_3$ solution as a UV and IR filter. The benzene used in these studies was scrupulously deaerated by freeze/pump/thaw cycles and dried by distillation from a Na/K amalgam. Solutions were prepared by vacuum manifold techniques. Stopped-flow kinetics were carried out using a Gibson-Durram D110 spectrophotometer equipped for the handling of solutions under deaerated conditions(24). The data station described above was used for collection, analysis and plotting of digital data.

Acknowledgments

This research was sponsored by the National Science Foundation. The rhodium and iridium used in these studies was provided on loan by Johnson Matthey, Inc.

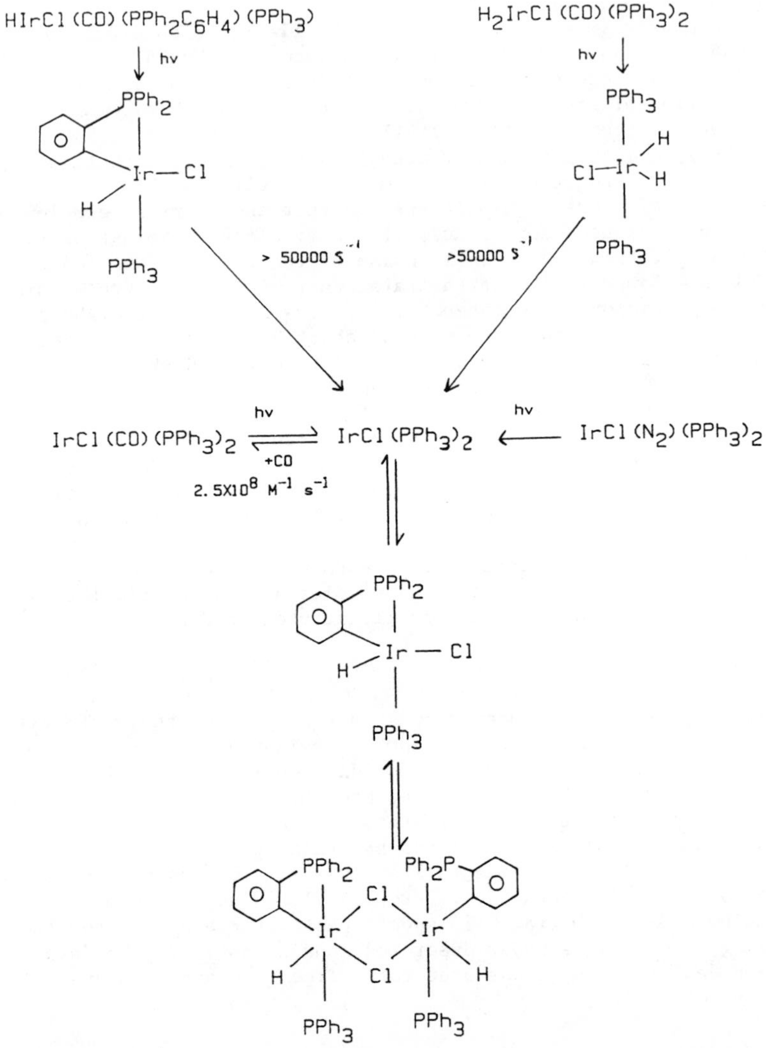

Scheme II. A summary of the photoreactions leading to the common intermediate $IrCl(PPh_3)_2$.

Literature Cited

1. Pignolet, L.H., Ed. "Homogeneous Catalysis with Metal Phosphine Complexes"; Plenum Press: New York, N.Y., 1983.
2. Halpern, J. Acc. Chem. Res. 1970, 3, 386-392.
3. Osborn, J.A.; Jardine, F.H.; Young, J.F.; Wilkinson, G. J. Chem. Soc.(A) 1966, 1711-1732.
4. Halpern, J.; Wong, C.S. J.C.S. Chem. Commun. 1973, 629-630.
5. Tolman, C.A.; Meakin, P.Z.; Lindner, D.L.; Jesson, J.P. J. Amer. Chem. Soc. 1974, 96, 2762-2774.
6. Halpern, J.; Okamoto, T.; Zakhariev, A. J. Mol. Catal. 1976, 2, 65-69.
7. Tolman, C.A.; Faller, J.W. Chapt. 2 in ref. 1.
8. Halpern, J. Trans. Amer. Cryst. Assoc. 1978, 14, 59-70.
9. Jardine, F.H. Prog. Inorg. Chem. 1982, 28, 63-201.
10. James, B.R. "Homogeneous Hydrogenation"; Wiley-Interscience: New York, N.Y. 1973, pp. 204-250.
11. Geoffroy, G.L.; Wrighton, M.S. "Organometallic Photochemistry"; Academic Press, New York, N.Y. 1979.
12. Geoffroy, G.L.; Denton, D.A.; Keeney, M.E.; Bucks, R.R. Inorg. Chem. 1976, 15, 238202385.
13. Braker, W.; Mossman, A.L. "The Matheson Unabridged Gas Databook: A Compilation of Physical and Thermodynamic Properties of Gases"; Matheson Gas Products: East Rutherford, NJ, 1974, Vol. I, p. 11.
14. Geoffroy, G.L.; Keeney, M.E. Inorg. Chem. 1977, 16, 205-207.
15. Vaska, L. Acc. Chem. Res. 1968, 1, 335.
16. Geoffroy, G.L.; Hammond, G.S.; Gray, H.B. J. Amer. Chem. Soc. 1975, 97 3933-3936.
17. Collman, J.P.; Kubota, M.; Vastine, F.D.; Sun, J.Y.; Kang, J.W. J. Amer. Chem. Soc. 1968, 90, 5430-5437.
18. Bennett, M.A.; Milner, D.L. J. Amer. Chem. Soc. 1969, 91, 6983-6994.
19. Tatsumi, K.; Hoffman, R.; Yamamoto, A.; Stille, J.K. Bull. Chem. Soc. Japan 1981, 51, 1857-1867.
20. Basato, M.; Morandini, F.; Longato, B.; Bresadola, S. Inorg. Chem. 1984, 23, 649-653.
21. Milstein, D.; Calabrese, J.C. J. Amer. Chem. Soc. 1982, 104, 3773-3774; 5227-5228.
22. Valentine, J. J.C.S. Chem. Commun. 1973, 857-858. This paper reports that the orthometallated product is a mixture of two geometric isomers.
23. Durante, V.A.; Ford, P.C. Inorg. Chem. 1979, 18, 588-593.
24. Trautman, R.J.; Gross, D.C.; Ford, P.C. J. Amer. Chem. Soc. 1985, 107, 2355.

RECEIVED November 8, 1985

15

Intrazeolite Organometallics

Chemical and Spectroscopic Probes of Internal Versus External Confinement of Metal Guests

Geoffrey A. Ozin and John Godber

Lash Miller Chemical Laboratories, University of Toronto, Toronto, ON M5S 1A1 Canada

The process of loading zeolites with organometallic complexes always brings to the forefront the question of internal versus external confinement of the metal guest. In this paper we present some experiments based on size exclusion, metal loading and intrazeolite chemistry which in conjunction with FT-FAR-IR, EPR and UV-visible reflectance spectroscopy, critically probe the question of internal versus external location for the case of five representative organometallics, $(\eta^6\text{-}C_6H_6)_2V$, $(\eta^5\text{-}C_5H_5)_2Cr$, $(\eta^5\text{-}C_5H_5)_2Fe$, $(C_6H_5CH_3)_2Co$, and $Co_2(CO)_8$.

It is now widely accepted that certain organometallics can be anchored to the surfaces of supports such as, oxides, carbons, polymers (1). However, the placement of metal guests into the internal voids of zeolite hosts using organometallic precursors, is by comparison a rather new field of investigation (2-4). These intrazeolitic guests may be the original organometallic itself, or some agglomerated and/or oxidized form, the latter being contingent upon the choice of zeolite and pre- and post-treatment conditions. Whatever the chemical fate of the metal, an intimate knowledge of the physical location, geometry, distortions and support-interactions of the metal guest in the zeolite is a prerequisite of any subsequent application of the metal-zeolite composition. In the one-to-one replacement of intrazeolite metal cations by ion-exchange loading procedures, the distribution of cationic guests at internal lattice sites is well established (5). However, the intrusion of uncharged, organometallic molecules to the internal lattice of a zeolite is not so clear cut and special procedures are required to probe the phenomenom. An elegant recent example of this is Schwartz's $Rh(\eta^3\text{-}C_3H_5)_3$/HY system, in which the presence of supercage $ZO\text{-}Rh(\eta^3\text{-}C_3H_5)H$ species was demonstrated by size selective olefin hydrogenation and phosphine poisoning experiments; only those olefins that were small enough to pass through the 8Å twelve ring window and enter the supercage were hydrogenated (6,7); moreover, only those phosphines that were sterically accessible to the supercage were able to poison the active

0097-6156/86/0307-0212$07.00/0

rhodium site (6,7). Some other approaches which are equally as effective for establishing the outcome of a zeolitic organometallic impregnation are described in this paper.

A. Metal Loading Studies: Bis(toluene)cobalt(0)

Bis(toluene)cobalt(0), (decomposition temperature -50°C) was synthesized by depositing cobalt vapour into liquid toluene held at around -90°C in a metal vapour rotary reactor (8). The solution containing $(C_6H_5CH_3)_2Co$ was transferred at -80°C to a Schlenk tube held at the same temperature. Equal volume aliquots of $(C_6H_5CH_3)_2Co$ were then transferred to equal weights of dehydrated sodium and acid forms of four different faujasite samples having Si/Al ratios of 1.25, 2.5, 3.8, 6.3 (acid only) and 18.0 (sodium only). Following a 36 hour impregnation, each sample was anaerobically filtered at -80°C, washed with -80°C toluene, brought to room temperature slowly under a dynamic vacuum, and flushed with dry Ar until free of residual toluene (GC analysis). The cobalt loading of each sample was determined by NAA (neutron activation analysis) and the results are depicted graphically in Figure 1.

The key features of these data are that (i) the Co loading of the sodium faujasite NaFAU is essentially invariant to the Si/Al ratio, (ii) the Co loading of the acid faujasite HFAU analogues are consistently higher than their sodium counterparts and monotonically increase with the Si/Al ratio.

An explanation of these observations takes the following tack. The molecular dimensions estimated for $(C_6H_5CH_3)_2Co$, 5.85x7.08Å (9) are such that there is no steric barrier to passage through the 12-ring window of faujasite, having a kinetic diameter of 8.1 Å (10). The results for sodium and acid faujasite impregnated with $(C_6H_5CH_3)_2Co$ are fascinating. Both show the presence of cobalt but the trends are quite distinct. Let us first consider the sodium forms. Here we know from ^{13}C nmr, and neutron diffraction (11-12) studies that aromatics reside over Na^+ site II cations in the supercage (C_{3v} axial interactions) of Y zeolite. We are also informed from crystallographic and FT-FAR-IR studies (13), that the number of accessible Na^+ site II cations diminishes at the expense of inaccessible Na^+ site I cations, as the faujasites become more silicious. Concomitantly, the basicity of the lattice decreases while the hydrophobicity increases. Thus one can argue that diminishing interactions of the Na^+ site II cations with the electron density of the toluene ligands of $(C_6H_5CH_3)_2Co$, (or decreasing Van der Waal interactions with the lattice oxygens) are approximately counterbalanced by the increasing hydrophobicity of the faujasites with increasing Si/Al ratios; hence a rationale for the essentially invariant cobalt loading with increasing silicon content in sodium faujasites impregnated with bis(toluene)cobalt(0). Thus the Bronsted acidity and hydrophobicity of the H-faujasites both increase with the Si/Al ratio, providing solvent compatibility, favourable interactions between the supercage acidic protons and the charge density of the coordinated toluene, as well as oxidation of the $(C_6H_5CH_3)_2Co(0)$ to strongly bound supercage located site II and III Co^{2+} cations (optical reflectance and FT-FAR-IR spectroscopy, see later). These chemical potentials work together to enhance the cobalt loading of the acid faujasites with increasing silicon content.

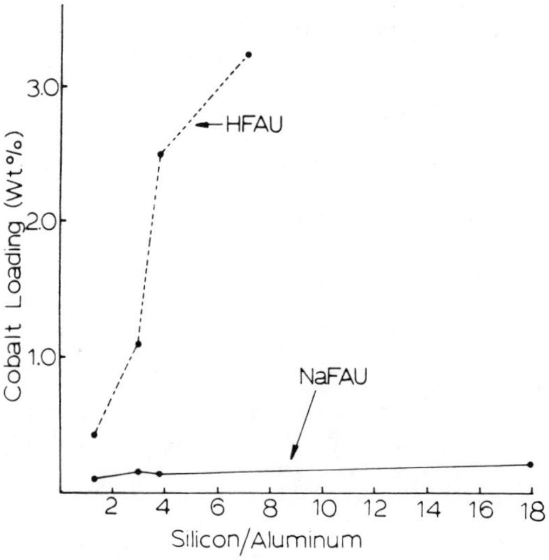

Figure 1. $(C_6H_5CH_3)_2Co$ impregnation of HFAU and NaFAU. Cobalt loading as a function of the Si/Al ratio (see text).

Metal loading studies of this type serve to enlighten the factors which control the impregnation of organometallics into zeolites, as well as providing an indirect probe of internal versus external location of the metal guests. In the following, a direct spectroscopic probe of this kind of process is presented.

B. Direct Probe FT-FAR-IR Spectroscopy: Dicobalt Octacarbonyl

In a number of recent publications (14-19) we demonstrated the great utility of the 30-350 cm^{-1} region of the far-IR spectrum of faujasite and A-type zeolites for identifying metal cations and establishing their site distributions in the lattice. The impressive power of the method stems from the realization that absorptions between 30-250 cm^{-1} are essentially metal localized modes, whose number, frequencies and intensities are straightforward signatures of metal cation type, site symmetry, site location and population (20). For details of this work the reader is referred to the original papers. Thus in Figures 2A and 2B, the four and three crystallographically located Na^+ ion sites of NaY (II, I, I', III,) and NaA (A, E, H) are easily identified, as indicated by the respective letter designations in the Figures. Absorptions to higher energy than 250 cm^{-1} are assigned to pore opening modes (21) and will not be further discussed here. Bands marked with an asterisk denote the IR active symmetric counterpart Na^+ ion modes that accompany the more intense asymmetric modes indicated in Figures 2A, 2B. Before describing the $Co_2(CO)_8$ impregnation experiments some preliminary remarks are necessary concerning the assignment of residual Na^+ ion modes in dehydrated sodium and acid faujasites and A-zeolites with increasing Si/Al ratio. A summary of the far-IR data for such samples is shown in Figures 3 and 4. From these traces, it is possible to assess cation site occupancies and estimate relative site populations as the sodium, Bronsted acid and silicon content is altered within a faujasite and an A-zeolite series. The far-IR spectra of the Co^{2+} ion-exchanged versions of the above samples are equally informative concerning Co^{2+} site locations. Some of this data is summarized in Figure 5 and 6 which show the effect of increasing Co^{2+} loading and Si/Al ratio, on the distributions of Co^{2+} ion sites in a series of sodium faujasites. An accumulation of and evaluation of data of this kind allows one to (i) pinpoint the characteristic far-IR signatures of Co^{2+} ions in sites II, I, III'' and I', (ii) identify residual Na^+ ion sites, (iii) establish any overlaps between the absorptions of Co^{2+} and Na^+ site II cations (iv) evaluate site preferences and distributions for Co^{2+} and Na^+ ions and (v) probe crystal-field effects and metal-support interactions for Co^{2+} sites. The reader is referred to the original papers for details of the above (13-20).

With all of these preliminaries at hand, let us now move to the question of probing internal versus external metal guests, following for example, sublimation of $Co_2(CO)_8$ into faujasite and A-zeolites using FT-FAR-IR spectroscopy. The molecular dimensions of $Co_2(CO)_8$ are estimated to be 9.76 x 6.29 x 5.45 Å for the non-bridged form and 8.72 x 5.95 x 5.92 Å for the bridged isomer (31); hence this molecule should be excluded from entering A zeolites but should be able to enter the pores of faujasite zeolites. Consider first, the far-IR spectrum of dehydrated $Na_9H_{47}Y$ depicted in Figure 7. Because

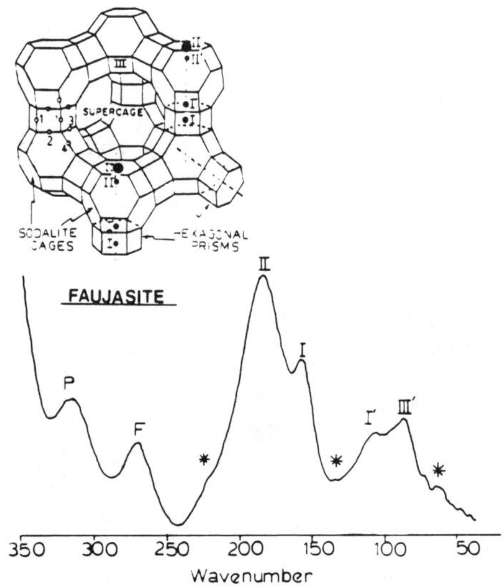

Figure 2A. In situ FT-FAR-IR spectrum of vacuum thermally dehydrated $Na_{56}Y$. Insert shows the faujasite unit cell, framework oxygen numbering and cation site designations. (Asterisks refer to symmetric metal cation modes - see text).

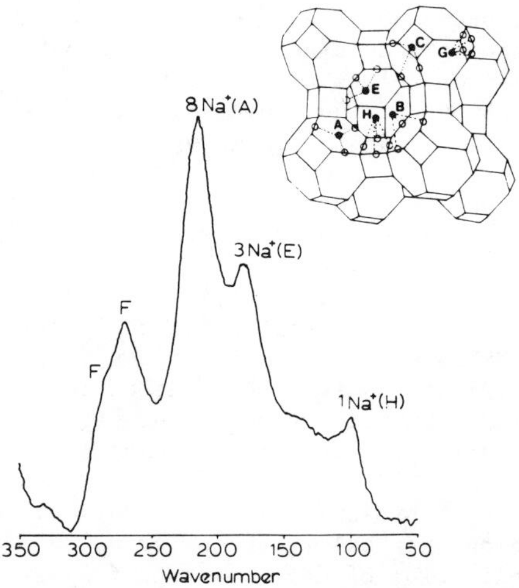

Figure 2B. In situ FT-FAR-IR spectrum of vacuum thermally dehydrated $Na_{12}A$. Insert shows the A-zeolite pseudo unit cell and cation site designations.

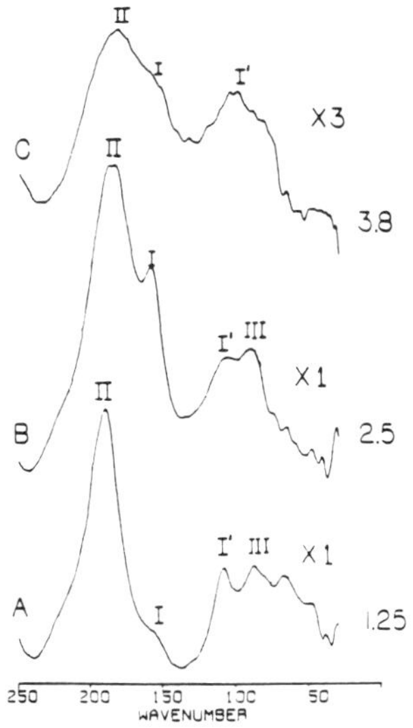

Figure 3. In situ FT-FAR-IR spectrum of vacuum thermally dehydrated NaFAU for Si/Al = 1.25, 2.5 and 3.8. Sodium cation sites are designated, II, I, III' and I'.

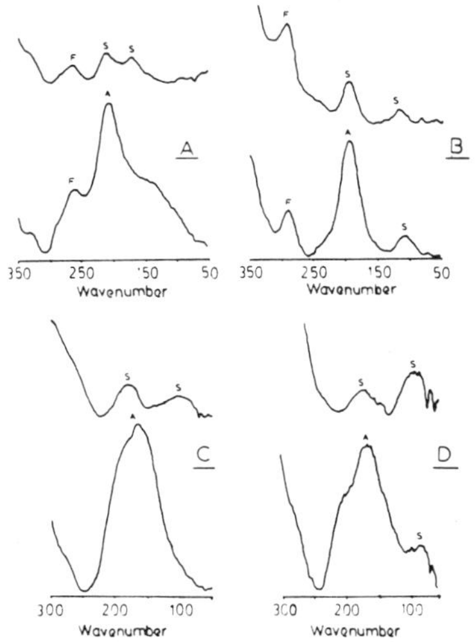

Figure 4. In situ FT-FAR-IR spectra of NH_4NaZ (lower) and HNaZ (upper) for (A) zeolite A (B-D) faujasite, Si/Al=1.25, 2.5, 3.8. (F represents a framework mode, A is a supercage NH_4^+ cation mode and S are residual sodium cation modes).

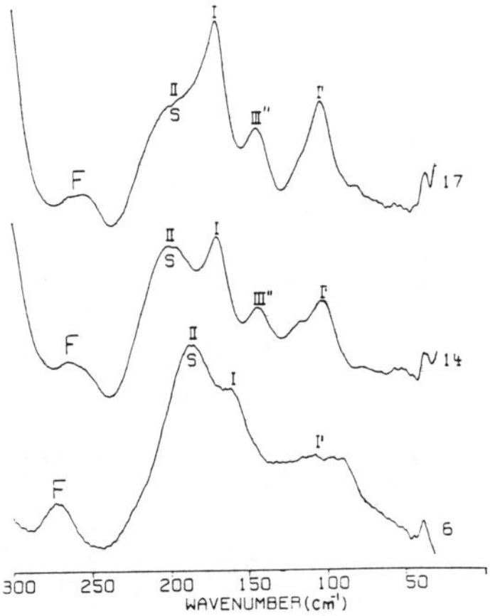

Figure 5. In situ FT-FAR-IR spectra of Co^{2+} ion-exchanged $Na_{56}Y$ containing 6, 14 and 17 Co^{2+} cations per unit cell. (S represents a residual Na^{+} site II cation mode, I, III'' and I' designate the respective Co^{2+} cation site modes. Higher loading studies show that site II Co^{2+} lies in the same region as site II Na^{+}).

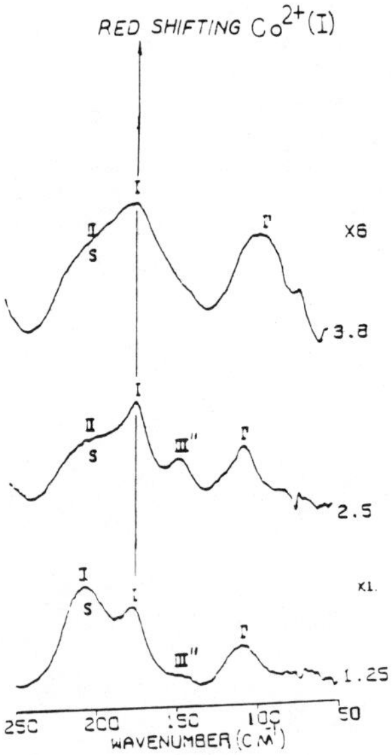

Figure 6. In situ FT-FAR-IR spectra of Co^{2+} ion-exchanged NaFAU (Si/Al=1.25, 2.5, 3.8). Note the red-shifting of the Co^{2+} site I and I' modes with increasing silicon content (see text).

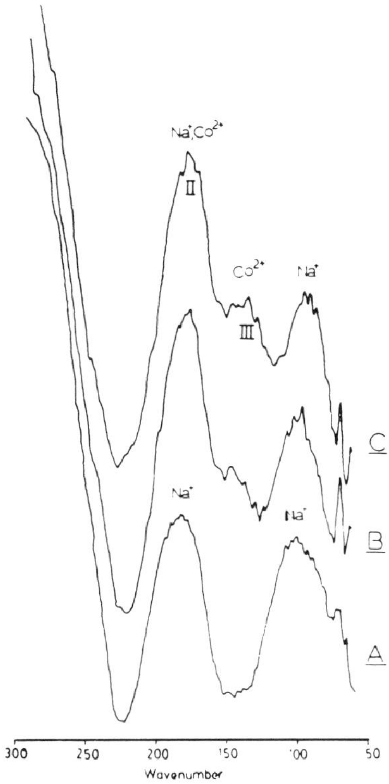

Figure 7. In situ FT-FAR-IR spectra of (A) vacuum thermally deamminated/dehydrated $Na_9H_{47}Y$ showing residual sodium sites, and (B-C) the outcome of exposure to $Co_2(CO)_8$ vapour for 5 and 15 minutes respectively, showing the formation of accessible Co^{2+} site II and III'' cations (see text).

of the low Na^+ ion content of this sample, the spectrum is displayed on a x5 ordinate scale expansion. The presence of residual Na^+ ions in site II and I' is quite apparent, although the band widths are such that they do not unequivocally exclude the contribution of some Na^+ ions in sites I and III'.

Following a controlled in situ sublimation of $Co_2(CO)_8$ into the $Na_9H_{47}Y$ zeolite wafer, the far-IR spectrum for increasing loadings of $Co_2(CO)_8$ clearly depict the growth of two new absorptions at 180 and 145 cm^{-1}, flagging the generation of Co^{2+} cations occupying sites II and III'' in the supercage of the faujasite. (Note that in separate experiments, $Co_2(CO)_8$ sublimed onto polyethylene and ALPO-5 shows no absorptions of any intensity in the 350-30 cm^{-1} region of the far-IR). The sample changes from white to blue following the impregnation of the $Co_2(CO)_8$ into $Na_9H_{47}Y$ indicating the presence of Co^{2+} ions within the zeolite. The observation of $Co(CO)_4^-$ in the corresponding mid-IR experiment suggests the following intrazeolitic redox chemistry: $Co_2(CO)_8 + ZOH \longrightarrow ZOCo^{2+} + 4CO + 1/2\ H_2 + Co(CO)_4^-$. Similar experiments performed with $Na_{56}Y$ and $Co_{17}Na_{22}Y$ display only $Co_2(CO)_8$ adsorption-induced cation vibrational shifts (Figures 8A,B), comparable to those reported for the adsorption of pyridine on the same faujasite samples (13-22). In contrast, the sublimation of $Co_2(CO)_8$ onto sodium and acid A-zeolites (the former having a kinetic diameter of 3.9Å (10)) has essentially no effect on the far-IR spectra.

The above observations are significant for a number of reasons. Firstly, they unequivocally demonstrate that $Co_2(CO)_8$ enters the supercages of $Na_9H_{47}Y$, facilitating intrazeolitic oxidation by the acid sites, to Co^{2+} ions located "exclusively" in accessible sites II and III''. This is to be sharply contrasted with Co^{2+} ion-exchanged and $[Co(NH_3)_6]^{3+}$ ion-exchanged/deamminated/autoreduced faujasite samples (the latter shown in Figure 9) which clearly illustrate the preference of Co^{2+} ions for both accessible (II, III'') and inaccessible (I, I') sites, located throughout the zeolite lattice. Thus the $Co_2(CO)_8/Na_9H_{47}Y$ preparation yields a NaCoHY zeolite in which all of the Co^{2+} sites are accessible to reagents, whereas the $[Co(NH_3)_6]^{3+}$/NaY route yields a NaCoHY zeolite with some of the Co^{2+} sites located in lattice regions inaccessible to most chemical reagents, (namely sites I, I'). In catalytic applications for example, such differences can often be exploited to advantage. A second point that emerges from this study, is that intrazeolitic oxidation of $Co_2(CO)_8$ is enjoyed only by $Na_9H_{47}Y$ and not by $Na_{56}Y$, $Co_{17}NaY$, NaA or HA. For the $Na_{56}Y$ sample intrusion of the $Co_2(CO)_8$ guest into the supercage of the faujasite is flagged by the observed far-IR adsorption induced vibrational shifts of the Na^+ ions. By contrast, the exclusion of $Co_2(CO)_8$ from the intrazeolitic voids of NaA and HA is seen by invariance of the far-IR cation spectra of these samples, when exposed to the vapour of $Co_2(CO)_8$.

Thus by using $Co_2(CO)_8$ as a representative organometallic guest, and a combination of size exclusion experiments and intrazeolite oxidation chemistry, as probed through the far-IR cation spectra of sodium and acid faujasites and A-zeolites, one is able to distinguish between internal and external confinement of the metal guest(s) in the zeolite.

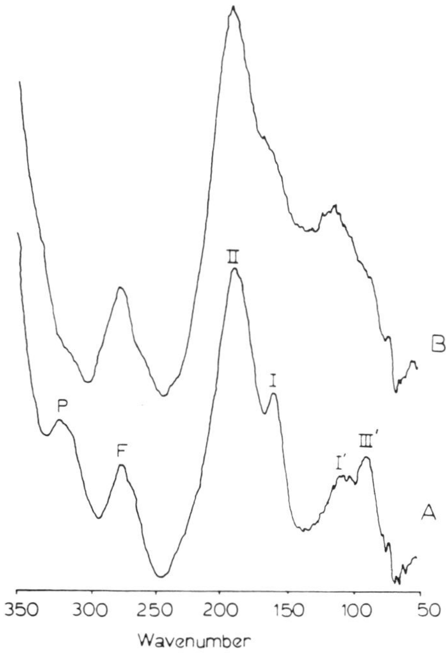

Figure 8A. In situ FT-FAR-IR spectra of A) vacuum thermally dehydrated $Na_{56}Y$ and B) the outcome of exposure to the vapour of $Co_2(CO)_8$ for 15 minutes (see text).

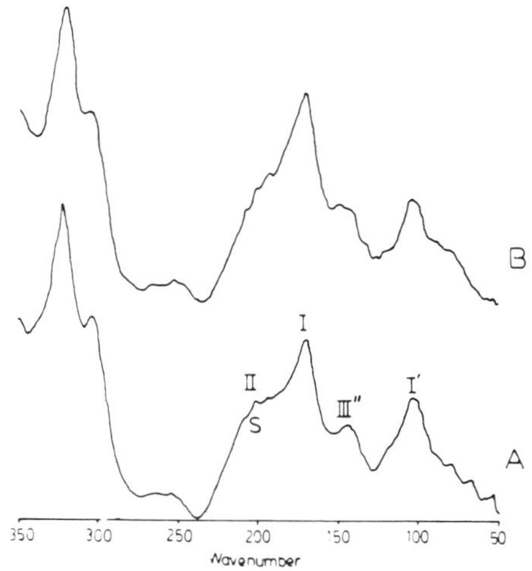

Figure 8B. In situ FT-FAR-IR spectra of A) vacuum thermally dehydrated $Co_{17}Na_{22}Y$ and B) the outcome of exposure to the vapour of $Co_2(CO)_8$ for 15 minutes (see text).

C. Organometallic EPR Spin Probes

I. Bis(benzene)vanadium(0): Size exclusion and diffusion studies. The molecular dimensions of $(\eta^6\text{-}C_6H_6)_2V$ are estimated to be 3.88Å x 5.24Å, (23), hence it should be essentially excluded from entering the supercages of A-zeolite but not Y-zeolite. The choice of $(\eta^6\text{-}C_6H_6)_2V$ for these zeolite impregnation studies was also predicated on its ideal epr properties, that is, a non-degenerate,totally symmetrical $^2A_{1g}$ electronic ground state and a highly diagnostic pattern of ^{51}V (I = 7/2)hyperfine and ^1H (I = 1/2) superhyperfine lines, whose coupling constants and line shapes are quite revealing as to its interactions with its immediate surroundings, as well as its ring and whole-molecule dynamical motions. For details of this work the reader is referred to the original papers (24). As a brief preamble to the epr zeolite work that follows, one finds that $(\eta^6\text{-}C_6H_6)_2V$ in weakly interacting solvents like pentane, shows at room temperature an isotropic 8-line V hyperfine spectrum with partially resolved superimposed proton superhyperfine components (Figure 10). In the rigid limit below 123K this epr spectrum dramatically transforms to one displaying axial symmetry, with full resolution of a 13-line isotropic proton superhyperfine pattern on each of the 8 perpendicular V-hyperfine lines (Figure 10). The parallel V-hyperfine components in this case have been proven to have a much smaller splitting and give rise to the small central feature in the epr spectrum (24). The magnetogyric g, hyperfine a^V, and superhyperfine a^H parameters which emerge from a line shape simulation of the rigid limit epr spectrum are listed below:

$$g_{xx} = g_{yy} = g_{\perp} = 1.9810G$$

$$g_{zz} = g_{\parallel} = 1.9857G$$

$$a^V_{\perp} = 91.4955G$$

$$a^V_{\parallel} = 6.733G$$

$$a^H_{\perp} = 4.264G$$

$$a^H_{\parallel} = 4.264G$$

Thus the room temperature spectrum in pentane characterizes a freely tumbling $(\eta^6\text{-}C_6H_6)_2V$ molecule, with coordinated benzene ring rotation, but experiencing some electron spin exchange line broadening due to frequent molecular collisions. In the rigid limit below 123K the axial nature of $(\eta^6\text{-}C_6H_6)_2V$ is revealed in the powder spectrum of the frozen solid; however, the molecule is trapped in a void of solid pentane of sufficiently spacious dimensions that ring rotation is still permitted, as seen by the magnetic equivalency of the 12 protons of the coordinated benzenes. The intermediate temperature regime 300-123K has been investigated in detail in pentane (Figure 10) as well as more sticky solvents like aromatics and ethers (24). In pentane, the effect of lowering the temperature (synonymous with increasing viscosity) is seen first as an enhancement of the resolution of the proton superhyperfine (molecular tumbling with diminished electron spin exchange broadening) and then as the appearance of

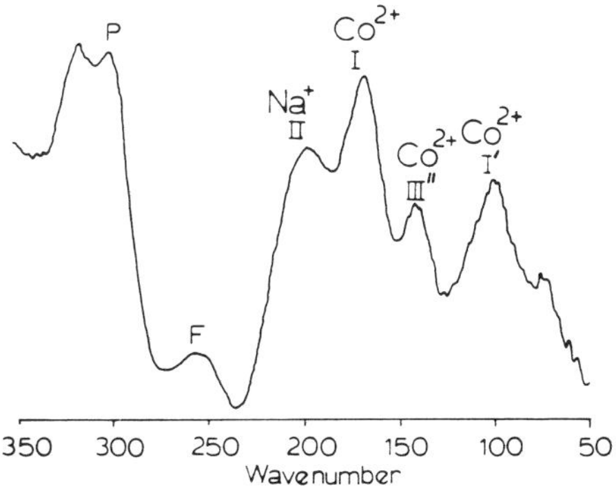

Figure 9. In situ FT-FAR-IR spectrum of vacuum thermally dehydrated/deamminated/autoreduced $[Co^{III}(NH_3)_6]_{8.6}Na_{30.2}Y$ giving $Co^{II}_{8.6}\ H_{8.6}\ Na_{30.2}Y$, (see text).

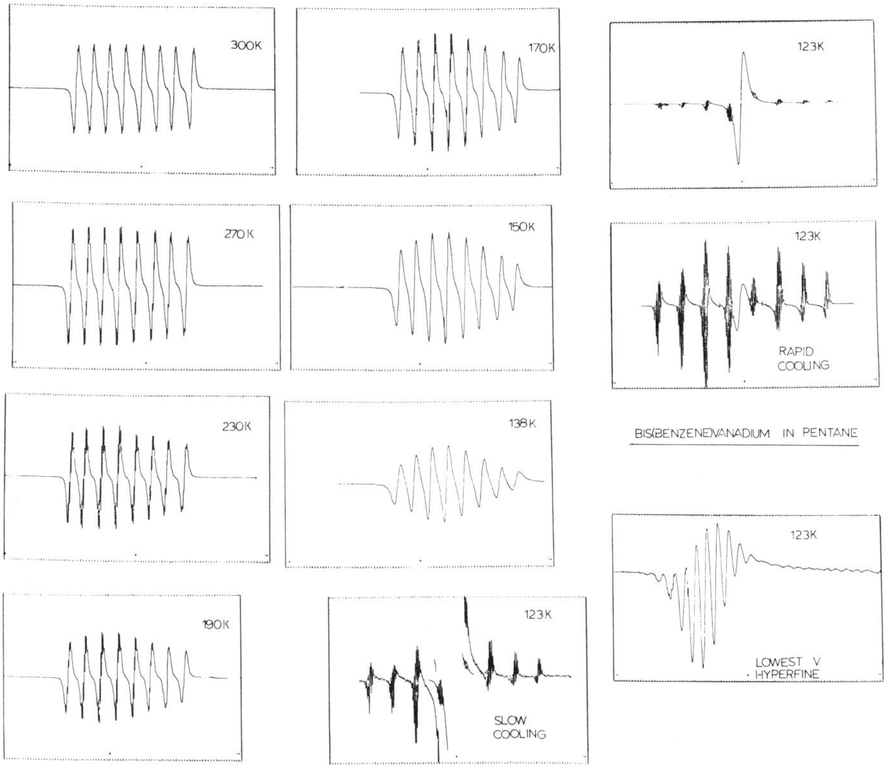

Figure 10. Temperature dependent epr spectra of $(\eta^6\text{-}C_6H_6)_2V$ in pentane over the range 300-123K (see reference 24 for details).

anisotropy (different line widths) on the V-hyperfine lines (viscous drag on the tumbling dynamics). At around 138K the coalescence regime is met, below which the axial rigid limit spectrum persists (Figure 10). Aggregation of $(\eta^6\text{-}C_6H_6)_2V$ in solid pentane is diagnosed by total collapse of the eight vanadium hyperfine components (dipole-dipole broadening) to a single line centred around g = 2 (more pronounced on slow cooling than rapid cooling - see Figure 10). The effect of more strongly interacting solvents is manifest as partial or complete loss of proton superhyperfine splitting, the observation of unequal line widths for the V-hyperfine lines and the attainment of the rigid limit axial spectrum at higher temperatures than those observed for pentane under identical concentration conditions. These effects are traceable to greater solute-solvent interactions in aromatic and ether media, and the concomitant perturbations of the ring and molecule dynamics.

With this background at hand, we are now in a position to assess the outcome of impregnating $(\eta^6\text{-}C_6H_6)_2V$ into dehydrated NaY and NaA. For these materials we will employ the epr spin probe intensities as a diagnostic of the extent of impregnation of $(\eta^6\text{-}C_6H_6)_2V$ into the supercages and the line shapes as an indicator of the dynamical properties (and possible location) of supercage encapsulated $(\eta^6\text{-}C_6H_6)_2V$.

Consider first the room temperature pentane impregnation of $(\eta^6\text{-}C_6H_6)_2V$ into NaY followed by a thorough pentane wash. The epr spectrum was recorded in the range 300-123K and showed little change in line shape. Features of interest to the present study include (i) the observation of a rigid limit spectrum at room temperature (Figure 11A) which is unaffected on cooling to 123K (Figure 11B), (ii) the loss of proton superhyperfine coupling, (iii) adsorption of oxygen or water (38) irreversibly changes the spectrum to one characteristic of the vanadyl ion VO^{2+}, (Figure 11C), (the low intensity and bowed baseline in this figure is an inherent property of the O_2/H_2O oxidation product (38)), (iv) adsorption of weakly interacting solvents like toluene or pentane to the sample in Figure 11A have no effect on the spectrum, (Figure 11D). Taken together with the observation that a similar room temperature treatment of NaA with $(\eta^6\text{-}C_6H_6)_2V$/pentane yields a very much less intense spectrum (∿2% of NaY) suggestive of residual external surface confinement of $(\eta^6\text{-}C_6H_6)_2V$, we can draw the following conclusions: (a) $(\eta^6\text{-}C_6H_6)_2V$ enters the supercages of NaY but not NaA; (b) $(\eta^6\text{-}C_6H_6)_2V$ is rigidly locked in the supercage of NaY either by way of Na^+ site II cations or lattice oxygen interactions with the coordinated benzenes or the vanadium atom itself. Interestingly, the Na^+ cation-arene ring interaction model is roughly in line with (i) the distance between the centres of benzene adsorbed onto site II Na^+ cations in Y zeolite (25) and the vanadium-benzene ring distances (1.66Å) thereby able to fix $(\eta^6\text{-}C_6H_6)_2V$ in the supercage (ii) the loss of proton superhyperfine splitting, through the Na^+ -benzene(or lattice oxygen) interactions restricting the free rotation of the coordinated benzenes, thereby creating magnetically inequivalent ring protons (3 sets of 4 equivalent protons; ENDOR experiments are planned to test this idea) with concomitant line broadening and loss of resolution and (iii) the observation of an axial limit epr spectrum with a slightly increased $a_{\perp}^{V}$ hyperfine coupling compared to $(\eta^6\text{-}C_6H_6)_2V$ in pentane.

The observation (Figure 11D) that weakly interacting solvents like toluene or pentane do not perturb the spectrum shown in Figure 11A is in line with the hypothesis of a strong interaction between the zeolite and $(\eta^6\text{-}C_6H_6)_2V$. It might be expected that adsorption of toluene or pentane would serve to solvate the $(\eta^6\text{-}C_6H_6)_2V$ and release it from interaction with the walls of the zeolite.

In a series of separate experiments the effect of impregnating equal weights of dehydrated NaY with $(\eta^6\text{-}C_6H_6)_2V$/pentane solution having increasing concentrations of $(\eta^6\text{-}C_6H_6)_2V$ was investigated using the above type of spin-probe epr method; one of the most well defined V-hyperfine component lines (unchanging band width) was used as a measure of the spin concentration of supercage immobilized $(\eta^6\text{-}C_6H_6)_2V$. The results of these experiments are graphed in Figure 12 for the anaerobic procedure performed at room temperature, with a 2 hour impregnation time, followed by a thorough pentane wash. From these data one can deduce that there exists a proportionate relationship between the concentration of $(\eta^6\text{-}C_6H_6)_2V$ in the pentane phase and the amount of intrazeolitic $(\eta^6\text{-}C_6H_6)_2V$. Clearly under these impregnation conditions there appears to exist no appreciable diffusional barrier to the entry of $(\eta^6\text{-}C_6H_6)_2V$ into the supercages of NaY.

Thus by using organometallic epr spin probes, size exclusion and diffusion experiments, one is able to distinguish in a fairly convincing way between internal and external confinement of the metal guest in the zeolite.

II. Bis(cyclopentadienyl)chromium(II); Size Exclusion and Intrazeolite Chemistry

The choice of $(\eta^5\text{-}C_5H_5)_2Cr$ as an epr spin probe in these applications, offers numerous opportunities for investigating the ultimate form and location of the metal guest. Its molecular dimensions are estimated to be 4.34Å x 4.44Å (26) and so it should be a useful candidate for size exclusion experiments in faujasite and A-zeolites. The chromium is formally in oxidation state +2 with an electronic ground state term of $^3E_{2g}$ (27). In this form the molecule has a scarlet hue and has a characteristic electronic spectrum showing UV-visible ligand-field and UV charge-transfer excitations and a D_{5d} centrosymmetric sandwich structure as shown in Scheme I.

In frozen pentane glass, down to 123K, $(\eta^5\text{-}C_5H_5)_2Cr$ displays a very broad almost isotropic epr line ($g \cong 3.89, \Delta H \cong 3400$ G) (28) which at room temperature broadens even further, almost to disappearance. This typifies the epr behaviour of a first transition series complex having a high spin, orbitally degenerate electronic ground state, and a resulting short T_1 spin lattice relaxation time. Under these conditions the Jahn-Teller instability of $(\eta^5\text{-}C_5H_5)_2Cr$ could be a dynamic type. However, when entrapped in the supercage of a zeolite one can anticipate a quite different state of affairs for the epr properties of chromocene.

Consider first the impregnation of rigorously dehydrated/calci-

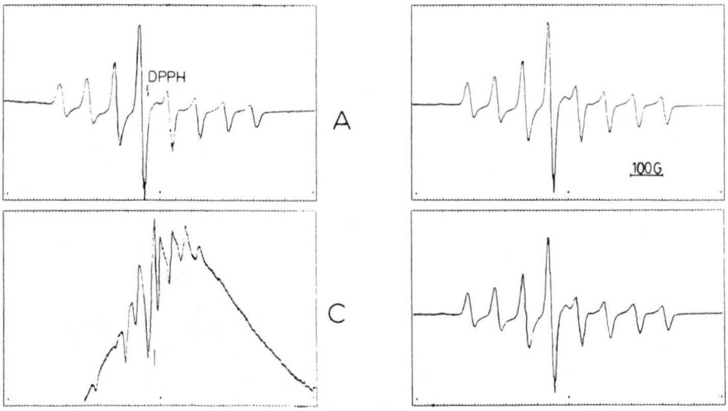

Figure 11. EPR spectra of NaY encapsulated $(\eta^6\text{-}C_6H_6)_2V$, (pentane impregnation), recorded (A) at room temperature, Scale: 2X, (B) at 123K, Scale: 1X, (C) after addition of H_2O vapour at room temperature and recorded at 123K, Scale: 10X, (D) sample in (A) wet with pentane, recorded at room temperature, Scale: 2X.

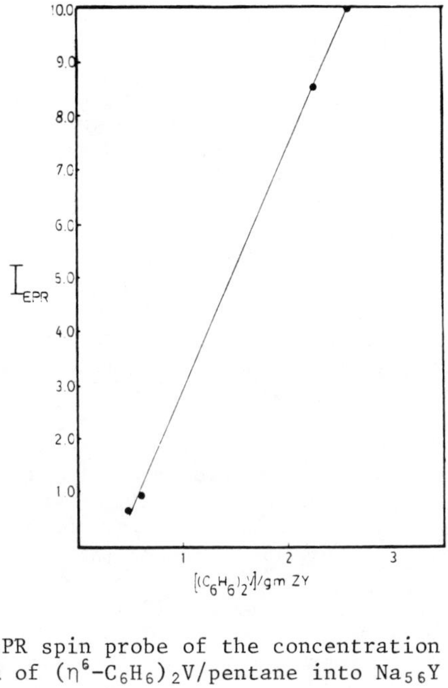

Figure 12. EPR spin probe of the concentration dependence of the diffusion of $(\eta^6\text{-}C_6H_6)_2V$/pentane into $Na_{56}Y$ (see text).

ned/defect removed NaY with a pentane solution of Cp_2Cr. After a thorough pentane wash an epr spectrum attributable to Cp_2Cr can be observed (Figure 13C,D). At room temperature the spectrum consists of a single absorption at g = 2.084 which is almost isotropic (Figure 13D). On cooling to -150°C, the sample exhibits growth of a resonance at g = 4.079, and the high field signal shows an axial disto tion, the g value changing slightly to g = 2.065 (Figure 13C). When the same experiment is repeated but with an equal weight of acid zeolite $Na_9H_{47}Y$, epr signals corresponding to orthorhombic Cp_2Cr^+ are observed (Figure 13B, all spectra are on the same scale) (30). At -150°C the epr spectrum has g_{xx} = 4.13, g_{yy} = 3.86 and g_{zz} = 1.98; when warmed to room temperature the spectrum remains essentially the same except for a decrease in intensity, with resonances g_{xx}= 4.12, g_{yy} = 3.89, g_{zz} = 1.98. The magnetogyric (g) tensor and temperature behaviour of this sample is similar to that observed for Cp_2Cr^+ in other diamagnetic hosts, except that zeolite encapsulated Cp_2Cr^+ exhibits an orthorhombic distortion (27-30). In sharp contrast, when Cp_2Cr in pentane is contacted with NaA zeolite, only very low intensity high and low field resonances are observed, (Figure 13A, run at -150°C). This we attribute to residual, surface confined chromocene or chromicenium.

Finally we wish to point out that when Cp_2Cr is reacted with SiO_2, vacuum thermally pretreated at 400°C, roughly one cyclopentadiene ligand is released with formation of the proposed surface anchored organometallic CpCr-OSIL (37). While it is not impossible that such a half sandwich Cr^{II} species could be responsible for the esr spectrum observed from impregnating Cp_2Cr into HY, we find that the optical reflectance spectrum of the latter support the proposed Cp_2Cr^+ZY formulation (see also Cp_2Fe^+ZY in the next section). Nevertheless, studies are underway to further clarify this intriguing detail. From these observations we can draw the following conclusions:

(i) $(\eta^5-C_5H_5)_2Cr$ can enter the supercages of NaY and HY but is excluded from entering NaA.

(ii) The observation of a temperature dependent epr spectrum for Cp_2Cr/NaY, and its distinct difference to Cp_2Cr/HY suggests that Cp_2Cr has penetrated into the pores of NaY and is locked in place in the supercage in a distorted configuration not previously attained for Cp_2Cr, (see Scheme II); the latter effect could be responsible for its observable epr spectrum. For such a low symmetry triplet state Cp_2Cr species one can expect besides short T_1 relaxation times, that zero field splittings, dynamic Jahn- Teller effects and $\Delta M_S=\pm2$ transitions can provide additional factors responsible for the unusual epr spectrum and its temperature sensitivity. The analysis of these fascinating Cp_2Cr epr spectra are under continuing study in our laboratory.

(iii) When Cp_2Cr is impregnated into acid faujasite, it is oxidized to epr detectable Cp_2Cr^+ which has undergone an orthorhombic distortion in its supercage location. (Note that $Cp_2Cr \rightleftharpoons Cp_2Cr^+ + e^-$ has $E_{\frac{1}{2}}$ = 0.55 V vs SCE (27)).

D. Optical Reflectance Probe: Bis(cyclopentadienyl)iron(II); Size Exclusion and Intrazeolite Chemistry

The optical properties of the architypical metallocene Cp_2Fe, the

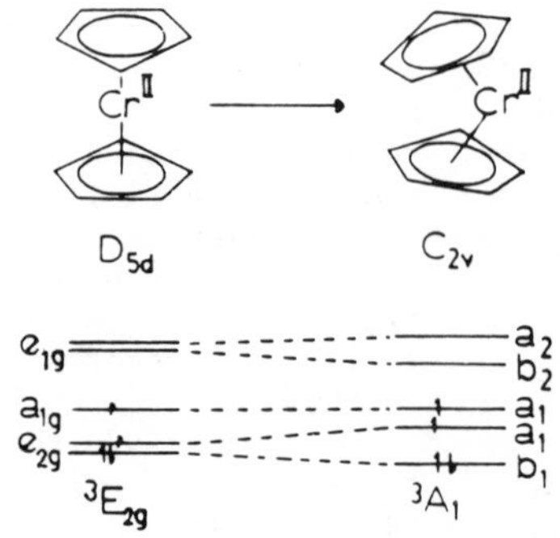

Scheme I

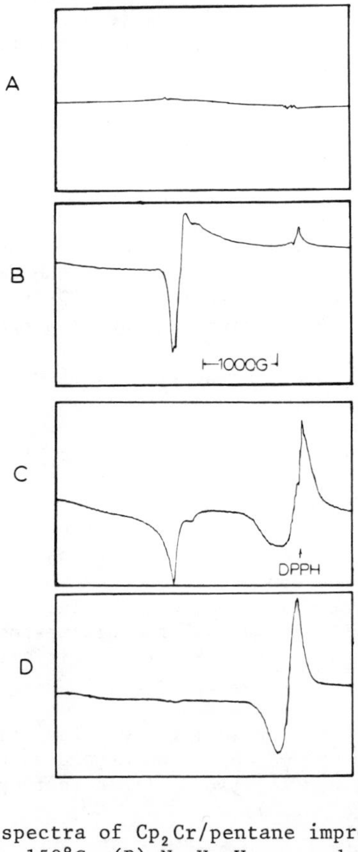

Figure 13. EPR spectra of Cp_2Cr/pentane impregnated into(A) NaA, recorded at -150°C, (B) $Na_9H_{47}Y$, recorded at -150°C, (C-D) $Na_{56}Y$ recorded at -150°C and RT respectively (see text).

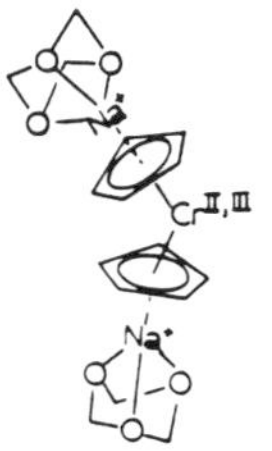

Cation Trapping

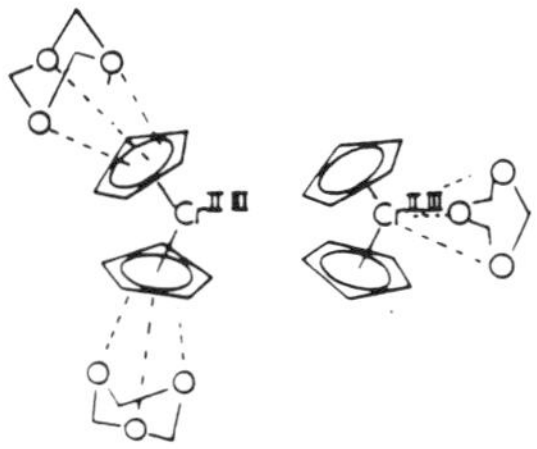

Framework Oxygen Trapping

Scheme II

cation Cp_2Fe^+ and complexes containing substituted rings have been extensively studied (29). The molecular dimensions of ferrocene, 4.13 x 4.20Å (26) are sufficient for unhindered diffusion into faujasite zeolites, but are too large to penetrate A zeolites. Ferrocene has an electronic ground state term of $^1A_{1g}$ and in pentane has characteristic absorptions at 440 and 325 nm due to d-d transitions, and charge transfer absorptions in the UV at 265, 240 and 200 nm (Figure 14A). The ferricenium cation has an orbitally degenerate ground state term which has been shown by esr and magnetic susceptibility measurements to be $^2E_{2g}$ (32,33). In aqueous solution the cation exhibits d-d transitions at 565, 524, 467 and 380 nm, and charge transfer absorptions at 617, 283 and 251 nm (Figure 14G). The intense charge transfer band at 617 nm is especially diagnostic for the presence of Cp_2Fe^+ as Cp_2Fe has no absorptions in this region.

The optical reflectance spectra of the products obtained from a room temperature impregnation of Cp_2Fe/pentane into dehydrated/calcined/defect removed NaY, $Na_9H_{47}Y$ and NaA are displayed in Figures 14 B-F, H. Ferrocene is excluded from NaA as evidenced by the lack of any bands in the spectrum which could be assigned to either Cp_2Fe or Cp_2Fe^+. The UV bands that are observed are due to zeolite $O \rightarrow T$ (where T = Al, Si) charge-transfer transitions. This result also strengthens our contention that Cp_2Fe is contained within the pores of NaY rather than confined to the external surface. Diffusion of Cp_2Fe into rigorously pretreated NaY proceeds with slight partial oxidation to the ferricenium cation, possibly via Lewis acid sites (the samples were carefully defect removed by washing with NaCl solution and meticulously dehydrated and calcined). This interesting observation is currently under study. Even with the small amount of oxidation as evidenced by the weak Cp_2Fe^+ CT band at 617 nm, the important feature to be noted from the reflectance spectrum is the existence of an essentially unperturbed ferrocene within the confines of the zeolite; no distortion of the molecule is apparent. If ferrocene had been bent one might have expected a lowering in energy of the $^1A_{1g} \rightarrow {}^1E_{1g}$ transition as observed in $(CpCH_2CH_2Cp)Fe$, in which the molecule is 'tethered' into a bent configuration (34). Of particular interest is the effect that air exposure has on the Cp_2Fe/NaY sample, as illustrated in Figures 14C, D; the ferrocene is slowly oxidized (whether it is oxygen or water or both is still under investigation) to the ferricenium cation as reflected by the drop in intensity of the 440 nm ferrocene band and the concomitant growth of the characteristic 617 nm charge transfer absorption of Cp_2Fe^+.

The fate of ferrocene after impregnation into $Na_9H_{47}Y$ is easily discernible from the reflectance spectrum (Figure 14E) as it shows essentially complete absence of ferrocene absorptions and only ferricenium bands. To confirm the existence of $[Cp_2Fe]NaHY$ we independently prepared $Cp_2Fe^+BF_4^-$ (33) and ion-exchanged it into $Na_{56}Y$ in aqueous 0.01M $HClO_4$. The resulting blue-green zeolite, with a composition of $[Cp_2Fe]NaHY$, was washed until free of perchlorate and dried at 100°C. The reflectance spectrum of this sample (Figure 14F) indicates that Cp_2Fe^+ has ion-exchanged into the zeolite lattice presumably to occupy a supercage cation site.

Regarding the reaction of Cp_2Fe with an acid faujasite, it is possible to exclude a simple acid-base reaction, yielding protonated

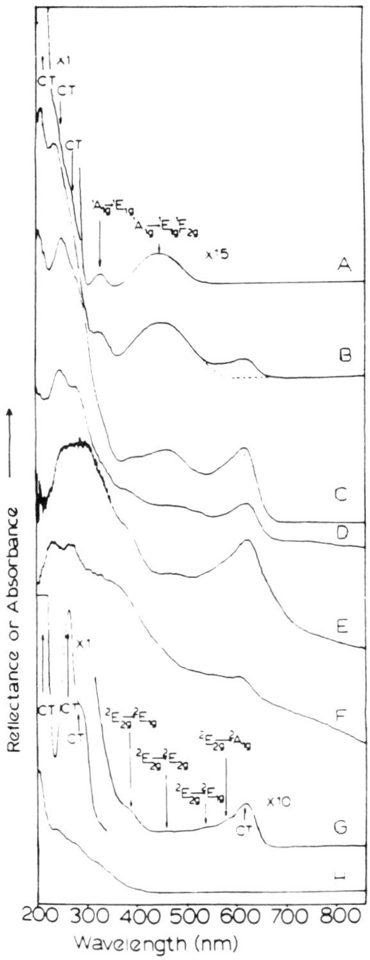

Figure 14. Optical spectra of (A) Cp_2Fe/pentane (ABS), (B) Cp_2Fe/NaY (REFL), (C-D) the same as (B) but after 6 and 24 hours exposure to air (REFL), (E) $Cp_2Fe/H_{47}Na_9Y$ (REFL), (F) $[Cp_2Fe]BF_4$/0.01M $HClO_4$ ion-exchanged into $Na_{56}Y$ (REFL), (G) $[Cp_2Fe]BF_4$/0.01M $HClO_4$ (ABS), (H) Cp_2Fe/NaA (REFL).

ferrocene $[Cp_2FeH]^+$ as a candidate for the observed product based on the reflectance spectrum. Thus $[Cp_2FeH]^+$ exhibits an optical spectrum quite similar to ferrocene, except that the $^1A_{1g} \rightarrow {}^1E_{1g}$ d-d transition at 325 nm is broadened (35). Note also that while "tethered" $(CpCH_2CH_2Cp)Fe$, with the rings fixed in a tilted position can be protonated by less than 0.1% H_2SO_4 in ethanol, Cp_2Fe by contrast is unaffected by the presence of for example 10% H_2SO_4 in ethanol (36). In the light of this information we postulate that the intrazeolitic oxidation of ferrocene in acid faujasite involves an initial bending back of the rings followed by protonation and oxidation. From a combination of intrazeolite oxidation chemistry of Cp_2Fe/HY, Cp_2Fe^+ ion-exchanged NaY and pentane impregnation of Cp_2Fe into NaY and NaA, one can deduce that (i) Cp_2Fe enters the α-cage of faujasite but is excluded from the internal voids of NaA; (ii) the acid strength of $Na_9H_{47}Y$ is sufficient to oxidize Cp_2Fe to Cp_2Fe^+ ($Cp_2Fe^+ + e^- \rightleftharpoons Cp_2Fe, E_{\frac{1}{2}} = 0.41$ V vs SCE (27)), confirming that it enters the α-cage; (iii) Cp_2Fe^+ can be effectively ion-exchanged into NaY; (iv) Cp_2Fe impregnated into rigorously dehydrated NaY when exposed to air gradually oxidizes to Cp_2Fe^+.

Overall it is quite clear, that a combination of epr spin-probes, UV-vis reflectance spectroscopy, size exclusion and intrazeolite oxidation experiments are able to effectively differentiate those organometallic-zeolite impregnations which place metal guests within the internal voids of the zeolite compared to those on the external surface of the zeolite lattice.

Conclusions

This paper describes some new zeolite organometallic impregnation experiments, in which a combination of metal loading, size exclusion, intrazeolite chemistry and diffusion considerations in conjunction with epr, far-IR and UV-visible reflectance spectroscopic probes, serve to distinguish metal guests located in the intracrystalline voids of the zeolite from those located on the external surface.

Representative organometallics $(C_6H_5CH_3)_2Co$, $Co_2(CO)_8$, $(\eta^6\text{-}C_6H_6)_2V$, $(\eta^5\text{-}C_5H_5)_2Cr$ and $(\eta^5\text{-}C_5H_5)_2Fe$ impregnated into sodium and acid faujasites and A-zeolites were selected for this initial investigation. The key points to emerge from this study are listed below:

a) All of the above organometallics pass freely into the supercages of sodium and acid faujasites, whereas they are all effectively size-excluded from sodium A-zeolites.

b) The loading of $(C_6H_5CH_3)_2Co$ into sodium faujasites is essentially insensitive to the Si/Al ratio whereas in the corresponding acid forms it follows the Si/Al ratio. Supercage Na^+ cations (and/or lattice oxygen charge density) and hydrophobicity work in opposition in the sodium faujasites, while the Bronsted acidity, cobalt oxidation and hydrophobicity reinforce the impregnation/diffusion for the acid faujasites.

c) $Co_2(CO)_8$ adsorbs in the molecular form on accessible supercage Na^+ cations of NaY but oxidizes to supercage located Co^{2+} site II, III'' cations in the HY form.

d) $(\eta^6\text{-}C_6H_6)_2V$ freely enters the supercages of NaY and is locked in place by specific interactions with either Na^+ supercage cation or lattice oxygens.

e) $(\eta^5\text{-}C_5H_5)_2Cr$ freely enters the supercages of NaY and is locked in place, probably in a Jahn-Teller distorted form, whereas on entering the supercages of HY it is oxidized to an orthorhombic distorted form of intrazeolite Cp_2Cr^+.

f) $(\eta^5\text{-}C_5H_5)_2Fe$ freely passes into the supercage of NaY whereas on impregnation into HY it suffers intrazeolite oxidation to the ferricenium cation.

Acknowledgments

The financial assistance of the Natural Sciences and Engineering Research Council of Canada's Major Equipment, Operating and Strategic Grants programmes and the Connaught Foundation of the University of Toronto is gratefully acknowledged. We wish also to acknowledge the donation of high purity and high crystallinity zeolite samples from Drs. Edith Flanigen (Union Carbide), Nicolas Spencer (W.R. Grace) and Paul Kasai (IBM), as well as helpful technical discussions. The assistance of Ms. Caroline Gil in obtaining the far-IR spectra, and Mr. Ted Huber in the synthesis of some of the materials is also greatly appreciated.

Literature Cited

1. Yermakov, Yu. I.; Kuznetsov, B.N.; Zakharov, Y.A., "Studies in Surface Science and Catalysis Vol. 8, Catalysis by Supported Complexes"; Elsevier; New York, 1981.
2. Bein, T.; Jacobs, P.A., J. Chem. Soc. Faraday Trans. I. 1983, 79, 1819.
3. Bein, T.; Jacobs, P.A., J. Chem. Soc. Faraday Trans. I., 1984, 80, 1391.
4. Schneider, R.L.; Howe, R.F.; Watters, K.L., Inorg. Chem., 1984, 23, 4600-4607.
5. Mortier, W.J., "Compilation of Extra Framework Sites in Zeolites"; Butterworth, 1982.
6. Schwartz, J.; Huana, T., J. Am. Chem. Soc., 1982, 104, 5244.
7. Corbin, D.R.; Seidel, W.C.; Abrams, L. Herron, N.; Stucky, G.D.; Tolman, C.A., Inorg. Chem., 1985, 24, 1800-1803.
8. Nazar, L.F.; Ozin, G.A.; Hugues, F.; Godber, J.; Rancourt, D., J. Mol. Cat., 1983, 21, 313.
9. These dimensions are based on the assumption that (i) the Co-ring distance will not be significantly different than that found in $[\eta^6\text{-}C_6(CH_3)_6]\ Co^+PF_6^-$. This is true for the crystallographically defined $(\eta^6\text{-}C_6H_6)_2Cr$, (Cr-ring = 1.61Å) and $(\eta^6\text{-}C_6H_6)_2Cr^+I^-$ (Cr-ring = 1.60Å); Muetterties, E.L.; Blecke, J. R.; Wucherer, E.J.; Albright, T.A.; Chem. Rev., 1982, 82,499; and (ii) that the C-C and C-H bond distances and angles are not significantly different from those of $[\eta^6\text{-}C_6(CH_3)_6]_2Co^+PF_6^-$, Thompson, M.R.; Day, V.W.; Minks, R.F.; Muetterties, E.L.; J. Am. Chem. Soc., 1980, 102, 2979.
10. The crystallographic diameter is 7.4Å but the kinetic diameter of 8.1Å is the more applicable in this situation, Breck, D.W. "Zeolite Molecular Sieves"; Wiley Interscience; New York, 1974.
11. Fitch, A.N.; Jobic, H.; Renouprez, A., J. Chem. Soc. Chem. Comm. 1985, p.284.

12. Lechert, H.; Wittern, K.P.; Ber. Bunsenges Phys. Chem.,1978, 82, 1054.
13. Ozin, G.A.; Baker, M.D.; Godber, J. in "Heterogeneous Catalysis"; Shapiro, B. Ed.; Texas A and M University Press: College Station, 1984. See also reference 5.
14. Ozin, G.A.; Baker, M.D.; Godber, J., J. Phys. Chem., 1984, 88, 4902.
15. Ozin, G.A.; Baker, M.D.; Godber, J., J. Phys. Chem., 1985, 89, 305.
16. Ozin, G.A.; Baker, M.D.; Godber, J., J. Phys. Chem., 1985, 89, 2299.
17. Ozin, G.A.; Baker, M.D.; Godber, J.; Shiuhua, W., J. Am. Chem. Soc., 1985, 107, 1995.
18. Ozin, G.A.; Baker, M.D.; Helwig, K.; Godber, J., J. Phys. Chem., 1985, 89, 1846.
19. Ozin, G.A.; Baker, M.D.; Godber, J., Catal. Rev. -Sci. Eng., 1985, 27, 591.
20. Ozin, G.A.; Baker, M.D.; Godber, J., J. Am. Chem. Soc., 1985, 107, 3033.
21. Flanigen, E.M., in "Zeolite Chemistry and Catalysis"; Rabo, J.A., Ed.; ACS Monograph No.171, American Chemical Society: Washington, D.C., 1976.
22. Butler, W.M.; Angell, C.L.; McAllister, W.; Risen, W.M., J. Phys. Chem., 1977, 81, 2061.
23. Fischer, E.O.; Fritz, H.P.; Manchot, J.; Driebe, E.; Schneider, R., Chem. Ber. , 1963, 96, 1418.
24. Andrews, M.P.; Mattar, S.; Ozin, G.A., J. Phys. Chem., (in press).
25. Lechert, H.; Wittern, K.P.; Schweitzer, W. Acta. Phys. Chem., 1978, 24, 201.
26. Haaland, A., Topics Curr. Chem., 1975, 53, 1.
27. Robbins, J.L.; Edelstein, N.; Spencer, B.; Smart, J.C., J. Am. Chem. Soc., 1982, 104, 1882.
28. Warren, K.D., Struct and Bonding 1976, 27, 45.
29. Sohn, V.S.; Hendrickson, D.N.; Gray, H.B., J. Am. Chem. Soc., 1971, 93, 3603.
30. Ammeter, J.H., J. Magn. Res., 1978, 30, 299.
31. Somner, G.G.; Klug, H.P.; Alexander, L.E. Acta. Cryst., 1964, 17, 732.
32. Prins, R. Mol. Phys.,1970, 19, 603.
33. Hendrickson, D.N.; Sohn, V.S.; Gray, H.B., Inorg. Chem., 1971, 10, 1559.
34. Barr, T.H.; Watts, W.E., J. Organometal. Chem., 1968, 15, 177.
35. Rosenblum, M.; Santer, J.O.; Howells, W.G., J. Am. Chem. Soc., 1963, 85, 1450.
36. Lentzner, H.L.; Watts, W.E., J. Chem. Soc. Chem. Comm., 1970, p.26.
37. Karol, F.G.; Kavapinka, G.L.; Wu, C.; Dow, H.W.; Johnson, R.N.; Carrick, W.L., J. Polym. Sci. A-1, 1972, 10, 2621.
38. We find that low concentration pentane impregnation of $(\eta^6\text{-}C_6H_6)_2V$ into a wide range of rigourously pretreated zeolites (sodium and acid faujasities, Si/Al=1.25/1 to 3.8/1, and ALPO-5) yielded samples which displayed well defined epr spectra (level baselines, cf Figure 11C) of vanadyl, VO^{2+} the temperature and solvent dependence of which indicate the VO^{2+} moiety to be

strongly bound to the zeolite. However, the vanadyl species generated in this way, although similar to that produced by intrazeolite O_2 or H_2O oxidation of $(\eta^6-C_6H_6)_2V$, is nevertheless sufficiently different to suggest that the oxidant in the former involves framework oxygen or defect sites, rather than trace O_2/H_2O contaminants.
39. Using the SKM theory for the quantitative analysis of reflectance spectra as detailed by Klier (ref. J. Opt. Soc. Am., 62, 882 (1972), our preliminary estimates of the metallocene loadings in the samples of the present study, fall in the range of 0.5-2.5 molecules per unit cell.

RECEIVED December 20, 1985

16

Spectroscopic Studies of Active Sites

Blue Copper and Electronic Structural Analogs

Edward I. Solomon, Andrew A. Gewirth, and Susan L. Cohen

Department of Chemistry, Stanford University, Stanford, CA 94305

An understanding of the electronic structure of metal ion active sites is essential in understanding their high reactivity. An important example of the contributions of spectroscopic, crystallographic and theoretical studies in elucidating electronic structure is in investigations of the blue copper active site in plastocyanin. These studies underscore the large role of covalent delocalization in determining the electronic and spectroscopic properties of the site. Further confirmation of the role of covalent delocalization comes from photoemission studies of small molecule spectral analogs.

Inorganic spectroscopy has evolved to the point where a great deal of insight into electronic structure and its contribution to reactivity can be obtained from detailed studies on high symmetry transition metal complexes. Attention can now be directed toward some rather unusual inorganic complexes which are active sites involved in catalysis. These include metalloproteins involved in enzymatic catalysis and metal ions on surfaces involved in heterogenous catalysis. These active sites often exhibit unique spectral features compared to high symmetry inorganic complexes. These unique features generally derive from unusual geometric and electronic structures imposed on the metal by the biopolymer or the surface. An understanding of these geometric and electronic structures should provide significant insight into the highly specific reactivity of these active sites.

A clear example of the contribution of inorganic spectroscopy in understanding the unique properties associated with an active site is the blue copper center (1-3) in plastocyanin.

The blue copper site exhibits unique spectral properties when compared with those of normal copper complexes. These spectral features include an unusually small copper hyperfine splitting of the EPR signal in the $g_{\parallel}$ region ($A_{\parallel} < 70 \times 10^{-4}$ cm^{-1} as compared to $A_{\parallel} \cong 150 \times 10^{-4}$ cm^{-1} for normal tetragonal copper) [Figure 1] and an extremely intense low energy absorption band (ν = 600 nm, $\varepsilon = 4000$ M^{-1} cm^{-1} compared to $\varepsilon = 100$ M^{-1} cm^{-1} for normal copper complexes). The original goal of spectroscopy on the blue copper site was to under-

0097-6156/86/0307-0236$08.75/0

stand these features and use them to generate a "spectroscopically effective" working model of the active site. In particular, infrared circular dichroism (IRCD) studies (4-5) [Figure 2] demonstrated that at least three d-d transitions existed in blue copper proteins to energies below 5000 cm^{-1}. A ligand field analysis (4-5) of these transitions then indicated that the site should have a geometry close to tetrahedral and that all d-d transitions occur at energies below 800 nm. Therefore, the intense 600 nm absorption band must involve a charge transfer (CT) transition which, based on other chemical and spectroscopic studies (6-8), probably derived from cysteine ligation at the site. In 1978, high resolution structures (9-10) appeared which confirmed the general tetrahedral geometry and cysteine ligation and demonstrated that the remaining ligands are two imidazoles of histidine and a thioether from methionine. While the length of the imidazole to copper bond is fairly normal when compared to model complexes, the Cu-S (thiolate) bond is found to be quite short (2.1 Å) and the Cu-S (thioether) bond is quite long (2.9 Å) [Figure 3].

With establishment of the crystal structure, three major features concerning the electronic structure of the blue copper site can be addressed. These features are 1) the nature of the thiolate and thioether bonds, 2) the nature of the ground state wavefunction and 3) the extent of covalency. We have also become strongly involved in using photoelectron spectroscopy as a powerful approach toward determining covalency in transition metal complexes. These will be discussed in turn.

Thiolate and Thioether Bonds

The first experiments to be discussed involved polarized single crystal optical (11) studies on the charge transfer region of plastocyanin. The plastocyanin crystal has four symmetry related molecules in the P $2_1 2_1 2_1$ (orthorhombic) unit cell. The crystal morphology combined with the optical properties of crystals allowed polarized spectra to be obtained parallel and perpendicular to the $\underline{a}$ axis of the (011) face. The spectra in Figure 4A are observed to be strongest in the parallel (to "a") polarization. As the Cu-S thioether bond is oriented approximately along the $\underline{c}$ axis, CT transitions associated with this ligand should appear dominately in the perpendicular (to "a") polarization. Thus, Cu-S thioether CT transitions contribute at most weakly to the absorption spectra, a feature which raised significant concern with respect to the nature of a copper thioether bond. The absence of a long bond between the copper and the thioether at the blue site has, in fact, been considered as a possiblity based on resonance Raman (12) and EXAFS (13) studies.

A combination of variable temperature absorption, CD and MCD spectroscopies (4-5) indicated that at least five transitions are present in the CT region of plastocyanin. A correlation of these with the polarized single crystal absorption spectra gave the possible band assignments shown in Figure 4B. Clearly, the imidazole contributes in this region. In addition, these results indicate that three thiolate to Cu CT transitions may be present. Originally, two were considered: a low energy doubly degenerate pi set and a higher energy, more intense sigma transition (bands 5 and 4, respectively). This pattern however considers only the bonding of a sulfur to a copper. If the C-S-Cu angle is significantly less than 180°, the

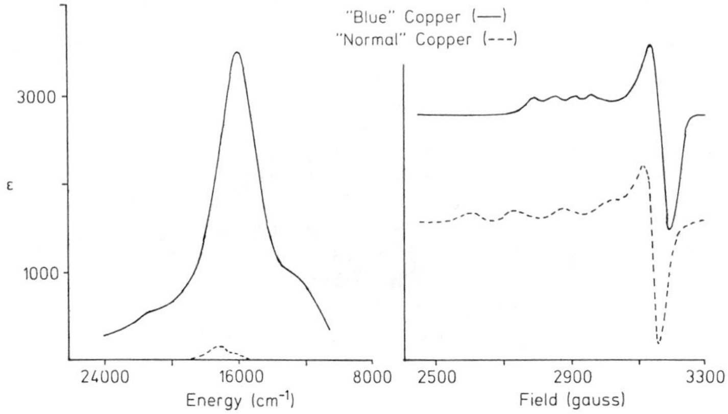

Figure 1. Optical (left) and EPR (right) spectra of a blue copper protein (solid line) and a tetragonal copper site (dashed line). Reproduced from Ref. 1. Copyright Wiley.

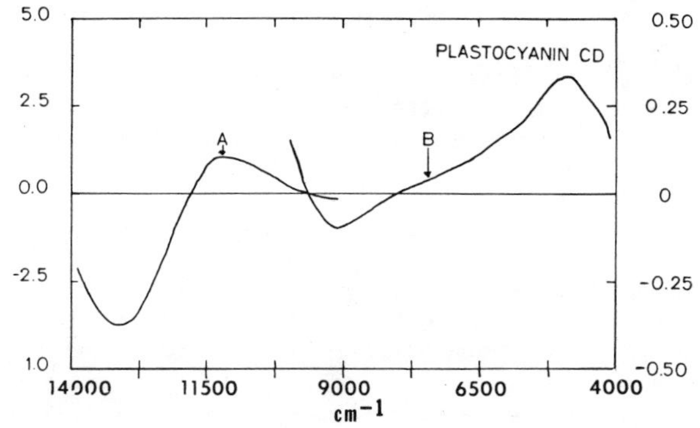

Figure 2. Near-infrared circular dicrhoism spectrum of plastocyanin in D 2 O at 290 K. Spectrum A corresponds to scale on left. Spectrum B corresponds to scale on right. Reproduced from Ref. 5. Copyright 1980, American Chemical Society.

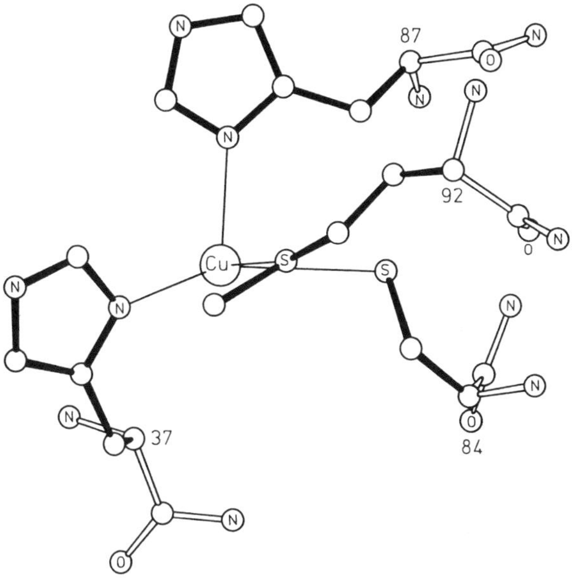

Figure 3. The blue copper site in plastocyanin as determined by X-ray crystallography. Ligands (and copper-ligand bond lengths) are histidine 37 (2.04 A), cysteine 84 (2.13 A), histidine 87 (2.10 A) and methionine 92 (2.90 A). Reproduced with permission from Ref. 9. Copyright 1983, Journal of Molecular Biology.

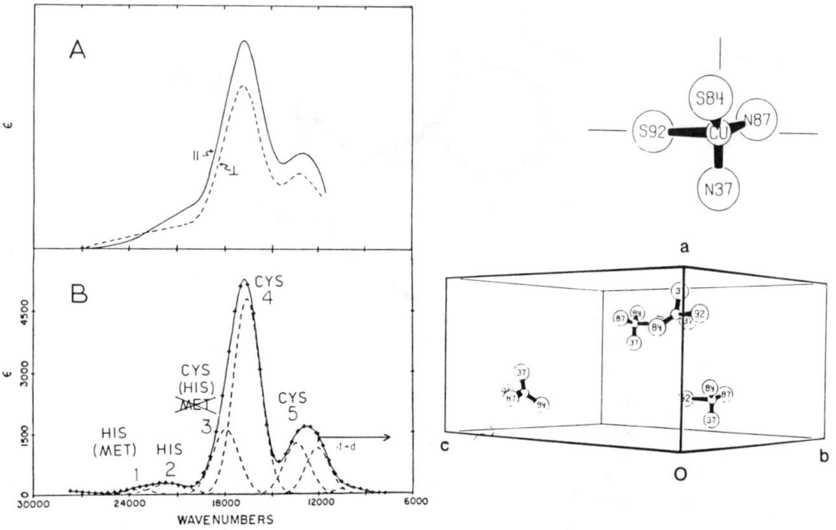

Figure 4. A) Room-temperature optical spectrum of a single crystal of plastocyanin obtained with light incident on the (0,1,1) face and polarized parallel (solid line) and perpendicular (dashed line) to a (from Ref. 11). B) Gaussian resolution of the 35 K visible absorption spectrum of a plastocyanin film with suggested assignments; the symbols (•) represent the experimental absorption spectrum. Right: plastocyanin unit cell projected on the (0,1,1) plane, showing the positions of the four symmetry-related Cu atoms at their first coordination shells.

strong interaction of the sulfur with the carbon of the cysteine residue would result in three transitions. The R-S-Cu angle in the plastocyanin structure is in fact 107°.

These results have led us to a more quantitative evalutation of the bonding in the blue copper site through a many electron SCF-Xα-SW calculation (14). The structures calculated include the free ligands and approximations to the site shown in Figure 5 with the Xα-SW parameters as indicated.

First considering the thiolate bond, the valence orbitals of the free thiolate ligand include the highest energy occupied doubly degenerate 2e level consisting of sulfur $p_{x,y}$ orbitals oriented perpendicular to the S-C bond, and to approximately 2 eV deeper binding energy, the $3a_1$ level which is the sulfur p_z orbital involved in sigma bonding with the carbon. Coordination of the thiolate ligand to the copper splits these valence orbitals into three roughly equally spaced levels, [Figure 6] each with significant bonding interactions with the orbitals on the copper (7a"-35%, 9a′-43%, 7a′-20%). This splitting results from one of the sulfur $p_{x,y}$ orbitals of the free thiolate being significantly stabilized due to bonding and mixed with the C-Sp_z orbital, which has the same symmetry. The contour diagrams associated with these three bonding levels are given in Figure 7. The highest energy occupied orbital (7a") is the sulfur p_x which is perpendicular to the C-S-Cu plane and involved in a strong pi bond with the $d_{x^2-y^2}$ orbital on the copper. The middle level involves the sulfur p_y, which is in plane and mixes with the sulfur p_z, forming a pseudo-sigma bond with the copper. Here, the electron density is no longer maximized along the S-Cu bond. The level to deepest binding energy involves a molecular orbital which is sigma bonding with copper but also significantly delocalized into the S-C bond.

The long (2.9 A) Cu-S(thioether) bond is next considered. The valence orbitals of the free ligand are $2b_2$ which is a p_y orbital of the sulfur, perpendicular to the C-S-C plane, and, to 2.1 eV deeper binding energy, the $4a_1$ level which is the p_z orbital of the sulfur involved in a sigma bonding interaction with the symmetric combination of methyl valence orbitals. Coordination of the thioether to the copper at the same distance and angle as in the protein leads to a stabilization of the $4a_1$ orbital by 0.4 eV relative to other valence orbitals due to bonding with about 36% delocalization of the wavefunction onto the copper [Figure 8]. The nature of the bonding interaction is shown in Figure 9. Here the sulfur p_z orbital is involved in a pseudo sigma type bond into the d_{z^2} orbital of the copper. As mentioned previously, a number of physical methods have raised the question as to whether there is a Cu-S (methionine) bond. The lack of CT intensity is now seen to be a consequence of the orientation of the S(thioether) p_z donor orbital which is orthogonal to the half-occupied $d_{x^2-y^2}$ acceptor (vida infra). The calculation does, however, indicate spectral features which are sensitive to thioether interaction with the copper. The effect of axial thioether coordination in the Xα calculations is presented in Figure 10. Here, the relative calculated energy of the copper d orbitals are given for the copper site in C_s symmetry with and without the thioether. Upon addition of the axial ligand, three of the four levels go down slightly in energy relative to the half-occupied level. The d_{z^2} level, however, goes up in energy due to antibonding interactions

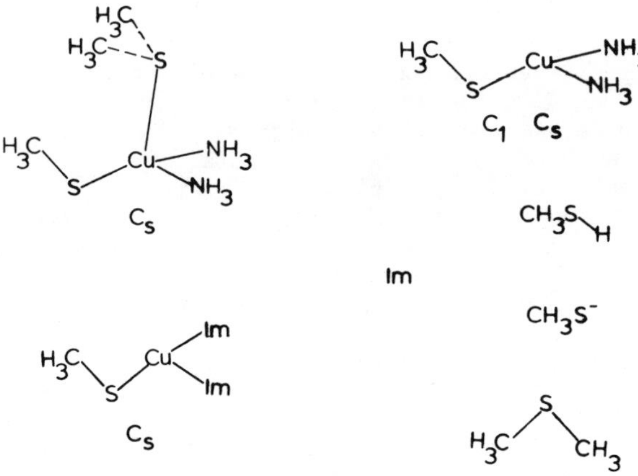

Xα PARAMETERS FOR $Cu(S(CH_3)_2)(SCH_3)(NH_3)_2$

	L_{MAX}
OUTER SPHERE	3
CU	2
S	1
N	1
C	1
H	0

OVERLAP	
CU - S(THIOLATE)	2%
CU - S(THIOETHER)	0%
CU - N	0%

CS SYMMETRY

CONVERGED AFTER 29 ITERATIONS

(EPS=0.009; -2T/V=0.9998)

Figure 5. Top: Approximations considered by SCF-Xα-SW calculation. Bottom: Xα parameters for $Cu(S(CH_3)_2)(SCH_3)(NH_3)_2$.

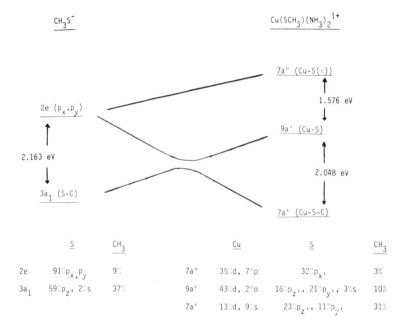

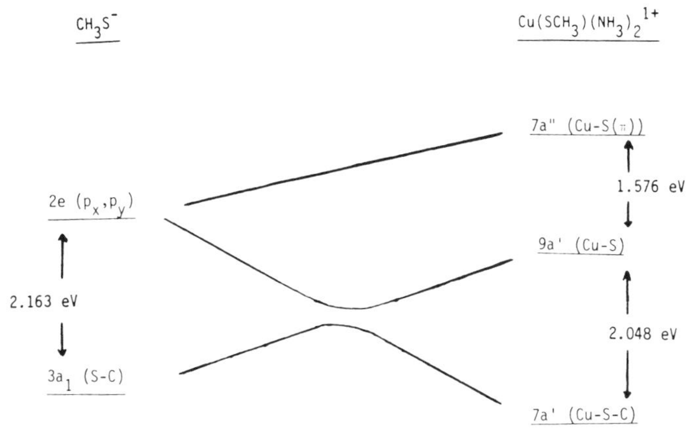

	S	CH_3
2e	91% p_x, p_y	9%
$3a_1$	59% p_z, 2% s	37%

	Cu	S	CH_3
7a"	35% d, 7% p	32% $p_{x'}$	3%
9a'	43% d, 2% p	16% $p_{z'}$, 21% $p_{y'}$, 3% s	10%
7a'	13% d, 9% s	23% $p_{z'}$, 11% $p_{y'}$	31%

Figure 6. Representation of the interaction of three highest energy orbitals of methylthiolate with a copper(II) ion. Top: shifts in energies relative to the sulfur 2s orbital. Bottom: character of the orbitals in terms of atomic orbitals. Primed coordinates of the copper-bound sulfur p orbitals indicate a ligand-based coordinate system. Reproduced from Ref. 14. Copyright 1985, American Chemical Society.

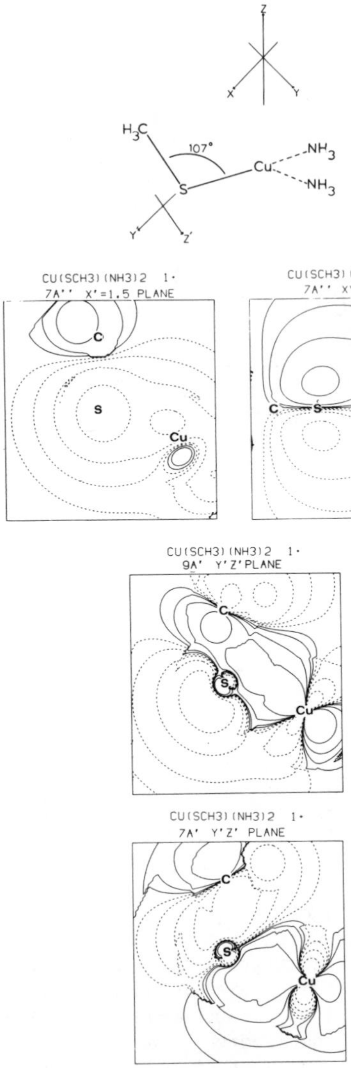

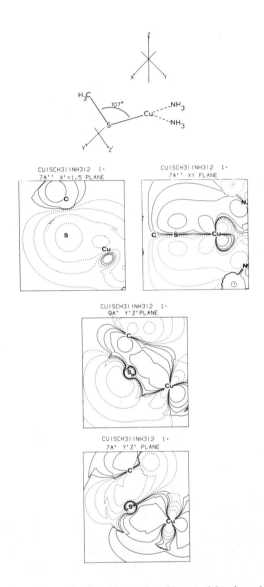

Figure 7. Contours of the three bonding orbitals with substantial thiolate sulfur 3p character and the geometry of $Cu(SCH_3)$-$(NH_3)_2^+$ with copper and thiolate based coordinate systems indicated. All nuclei indicated are in the planes of the figures except for the contour of the 7a" orbital in the xy plane. Only the copper nucleus is in the plane of this figure. Values of the contours are +0.003, +0.009, +0.027 and +0.081 Reproduced from Ref. 14. Copyright 1985, American Chemical Society.

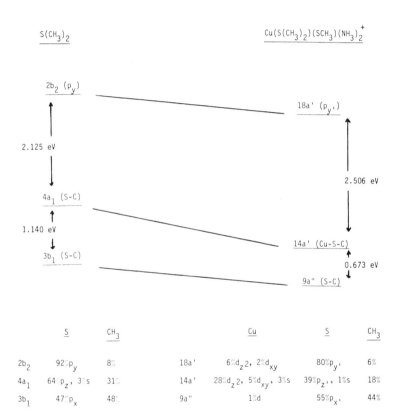

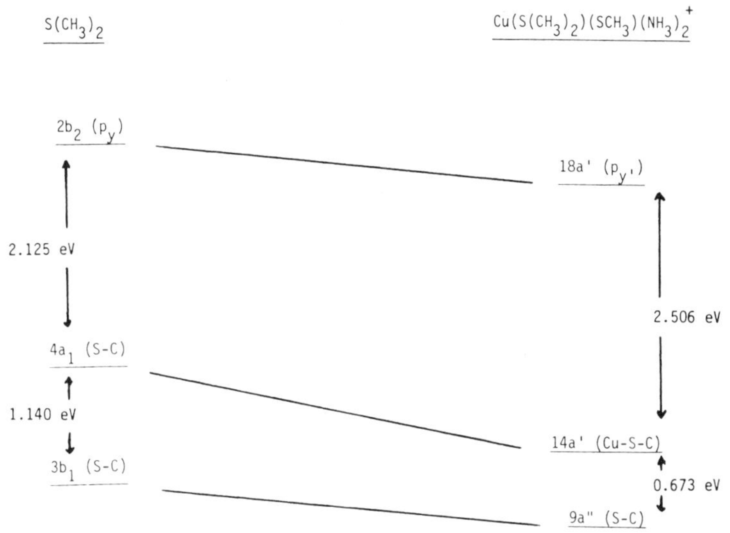

	S	CH_3		Cu	S	CH_3
$2b_2$	92% p_y	8%	18a'	6% d_{z^2}, 2% d_{xy}	80% $p_{y'}$	6%
$4a_1$	64% p_z, 3% s	31%	14a'	28% d_{z^2}, 5% d_{xy}, 3% s	39% $p_{z'}$, 1% s	18%
$3b_1$	47% p_x	48%	9a"	1% d	55% $p_{x'}$	44%

Figure 8. Interaction of the three highest energy occupied orbitals of dimethylsulfide with copper. Top: shifts in energies relative to the sulfur 2s orbital. Bottom: character of the orbitals in terms of atomic orbitals. Primed coordinates of the copper-bound sulfur p orbitals indicate a ligand-based coordinate system. Reproduced from Ref. 14. Copyright 1985, American Chemical Society.

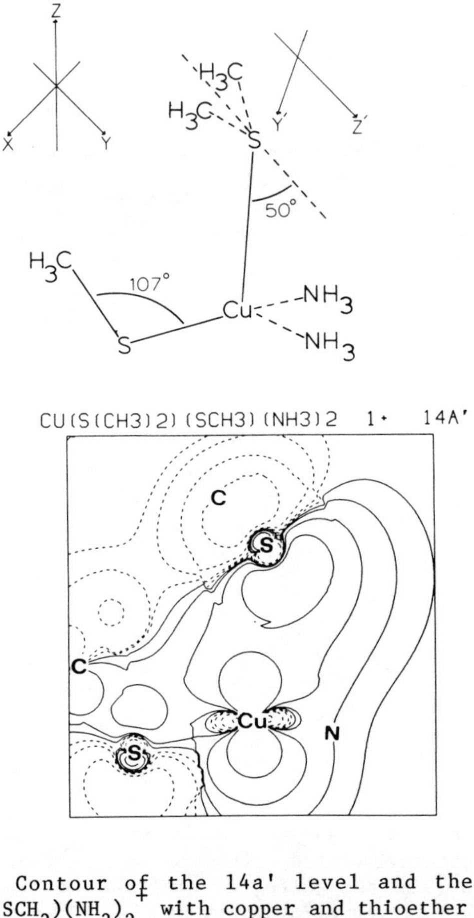

Figure 9. Contour of the 14a' level and the geometry of $Cu(S(CH_3)_2)(SCH_3)(NH_3)_2^+$ with copper and thioether based coordinate systems indicated. In the contour, all nuclei indicated are in the plane of the diagram except for those of the amine nitrogens and thioether carbons. Values of the contours are the same as in Figure 7. Reproduced from Ref. 14. Copyright 1985, American Chemical Society.

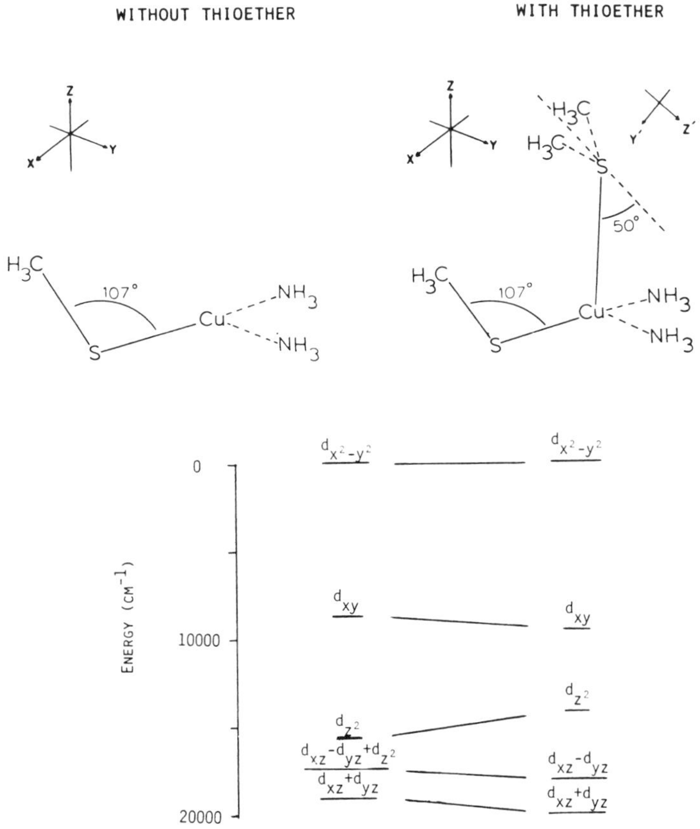

Figure 10. Calculated effects of axial thioether ligation upon copper d orbitals. The relative energies of the antibonding orbitals and their predominant copper d character are indicated for the two sites shown. The energies of the half-occupied level have been set to zero. Reproduced from Ref. 14. Copyright 1985, American Chemical Society.

with the sulfur p_z orbital of the thioether. Thus, the d-d transition associated with the d_{z^2} orbital should go down in energy. While removing the thioether from the blue copper site via site directed mutagenesis and at the same time retaining the remaining geometric features is unrealistic, copper chloride spectral analog studies do clearly demonstrate destabilization of the d_{z^2} orbital due to antibonding interactions with chloride ligands at approximately 3 Å distance as shown in Figure 11. At the top of the figure is the ligand field spectrum of bis(N-methylphenethylammonium) $CuCl_4$ (15), a square planar complex which contains no axial ligand. At the bottom is the spectrum of bis(ethylammonium) $CuCl_4$ (16) which does contain axial chlorides. The assigned transitions, based on polarized spectra for each complex are indicated at the top of the absorption bands in Figure 12. Upon addition of apical chlorides, (Figure 11 top to bottom), the transition from the d_{z^2} orbital decreases in energy from 16000 cm^{-1} to 11000 cm^{-1}, while the other transitions occur at approximately the same energies in both complexes. Thus, a distant apical ligand has significant effect upon the energy of the d_{z^2} orbital of the complex.

Ground State Wavefunction and Covalency

Single crystal EPR studies (11) of plastocyanin in conjunction with a ligand field calculation enabled a determination of the orientation of the g^2 tensor relative to the copper site. Figure 12 presents the EPR spectra of a single crystal of plastocyanin, oriented with the magnetic field (H) perpendicular to **a** and rotated about the **a** axis. An approximately $g_{\parallel}$ spectrum is observed when H is along **c** indicating that $g_{\parallel}$ is orientated in the general direction of the thioether-Cu bond. Simulation of four different rotations for the four molecules in the unit cell demonstrate that g_z and A_z are colinear and that g_z is 8° off **c** and 5° out of this plane. Thus, the $d_{x^2-y^2}$ orbital, which is perpendicular to g_z and contains the unpaired electron, is less than 15 degrees above the plane formed by the S(cys) and the two N (his) ligands. This orientation of the $d_{x^2-y^2}$ orbital is reproduced in Figure 13, along with the energy level diagram associated with the ligand field of plastocyanin. While this energy level diagram reflects a low symmetry, rhombically distorted site, it is of importance to consider the axial limits so as to evaluate the close to axial nature of the experimental X-band EPR spectrum, shown in Figure 1. Two axial subgroups of a tetrahedron are possible, D_{2d} and C_{3v}. If the rhombic splitting of the d orbitals is removed from that given in Figure 13 only the C_{3v} energy level ordering is reasonable. This energy ordering is consistent with an elongated C_{3v} structure, the long axis being along the Cu thioether bond, as shown in Figure 13.

This elongated C_{3v} effective symmetry of the blue copper site raises a significant problem with respect to the present interpretations (17-18) of the small copper hyperfine splitting observed in the EPR spectrum shown in Figure 1. The small splitting had been attributed to a D_{2d} mixing of Cu $4p_z$ into the $d_{x^2-y^2}$ ground state orbital. Electron spin in $4p_z$ would produce oppositely signed dipolar coupling with the nuclear spin than that arising from an electron in $d_{x^2-y^2}$. The two together would thus tend to cancel the anisotropic part of the hyperfine tensor. For D_{2d} $CuCl_4$, 12% $4p_z$ mixing was invoked (19) to explain the small copper hyperfine splitting

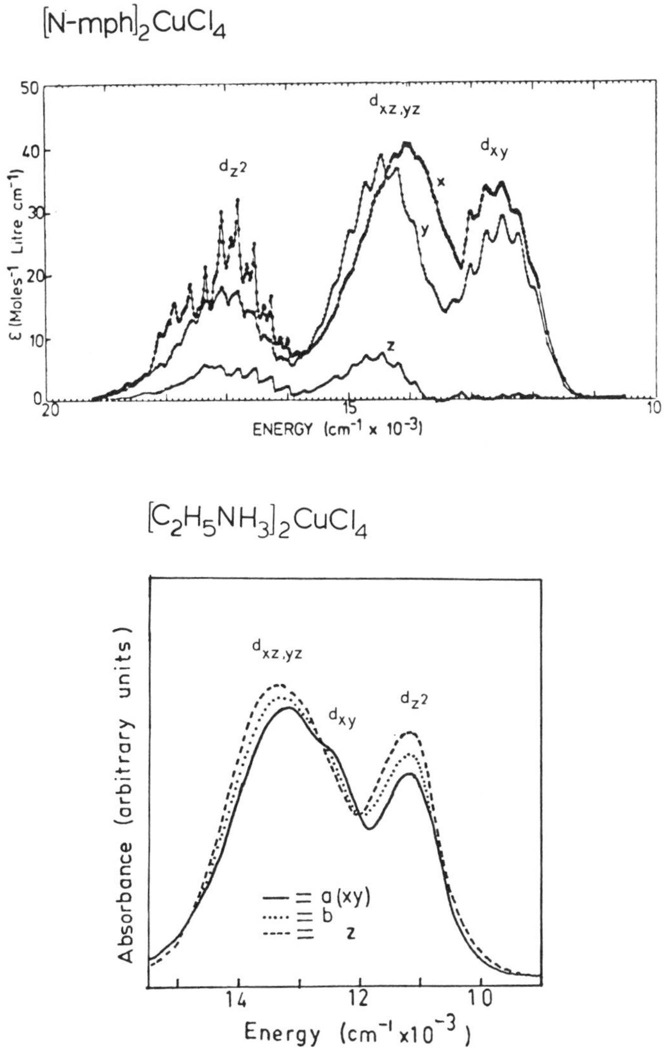

Figure 11. Top: molecular spectrum of the planar $CuCl_4^{=}$ ion in $(nmph)_2CuCl_4$. Adapted from Reference 16. Bottom: molecular spectrum of the tetragonal copper site in bis(ethylammonium) $CuCl_4$. Adapted from Reference 15. Polarization and assignments are indicated in each frame.

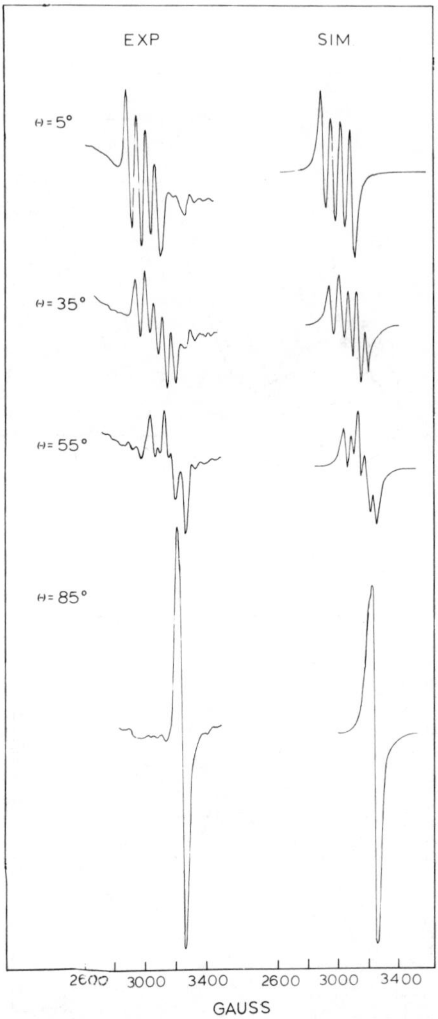

Figure 12. Experimental and simulated EPR spectra of a single crystal of plastocyanin with the crystal _a_ axis perpendicular to the plane of the magnetic field; θ is the angle between the crystal _c_ axis and the magnetic-field direction. The simulation was performed with four molecules in a unit cell, g_z 8 degrees off _c_ in the bc plane and 5 degrees out of the bc plane and g_z and A_z co-linear (from Ref. 11).

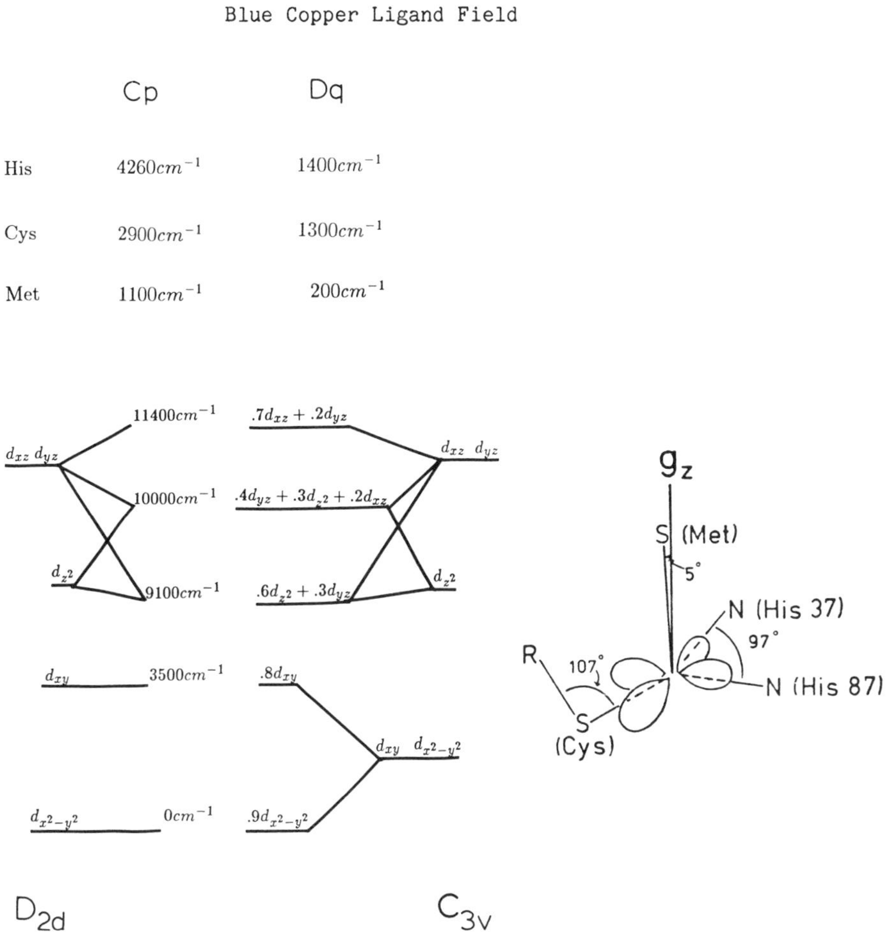

Figure 13. Left: ligand field energy-level diagram calculated for plastocyanin. Center contains energies and wavefunctions of the copper site. Energy levels determined after removing the rhombic distortions to give D_{2d} and C_{3v} symmetries are shown in the left and right columns, respectively (from Ref. 11). Right: electronic structural representation of the plastocyanin active site derived from ligand field calculations (from Ref. 11).

observed. Alternatively, a C_{3v} distortion can only mix p_x and p_y with the $d_{x^2-y^2}$ orbital. These are the same sign for dipolar coupling and thus would tend to increase the hyperfine splitting.

Thus, it was necessary to experimentally probe 4p mixing in $d_{x^2-y^2}$ directly. This was accomplished through analysis of the polarized single crystal spectral studies of the 1s to $3d_{x^2-y^2}$ transition in the x-ray absorption Cu K-edge at 9000 eV (in collaboration with Prof. Keith Hodgson at Stanford). We first calibrated this transition through polarized single crystal edge studies (20) of square planar (D_{4h}) and D_{2d} $CuCl_4^=$. The 90 degree period in modulation of intensity of the 8979 eV transition in the square planar complex (Figure 14) indicates that most of the intensity here is electric quadrupole in origin. This is consistant with the fact that p_z mixing with $d_{x^2-y^2}$ is forbidden for a complex with inversion symmetry. The intensity increases by a factor of four on going to the D_{2d} $CuCl_4^=$ structure demonstrating the effect of p_z mixing which is now symmetry allowed (Figure 15). The increase in intensity however can be well accounted for by the 2% p_z mixing calculated, not the 12% mixing that had been invoked earlier to explain the small hyperfine.

Extension of these edge studies (21) to the blue copper site in plastocyanin indicates that there is less than 1% p orbital mixing into $d_{x^2-y^2}$ and in particular that this involves the p_x and p_y orbitals. The 1s to $3d_{x^2-y^2}$ transition at 8979 eV occurs only when the polarization vector of the synchrotron radiation is in the xy plane. This can be seen from the 1s to 3d transition reproduced in Figure 15 which appears only with the electric vector perpendicular to the Cu-methionine bond. As emphasized above, this p_x, p_y mixing cannot account for the small copper hyperfine coupling constant of the blue copper site. In turn, this focuses attention on the possible role of covalent delocalization.

A significant feature of our Xα calculation of the blue copper active site is the large contribution of the sulfur 3p orbital, indicating that the ground state is very covalent. The magnitude of delocalization computed (only 33% Cu character) is quite large, making it important to consider the performance of the calculation with regard to delocalization. The ground state parameters most sensitive to delocalization are the EPR g and superhyperfine coupling values. In plastocyanin however, the delocalization occurs primarily over the sulfur p orbitals. As only ^{33}S (natural abundance 0.76%) has nonzero nuclear spin among naturally occuring sulfur isotopes, no superhyperfine from the sulfur in plastocyanin is observed. Any empirical fit of the ground state wavefunction must therefore be made to the g values.

Several methods exist for calculating g values. The use of crystal field wave functions and the standard second order perturbation expressions (22) gives g_z = 3.665, g_y = 2.220 and g_x = 2.116 in contrast to the experimental values (at X-band resolution) of g_z = 2.226 and $g_{x,y}$ = 2.053. One possible reason for the d screpancy is the use of perturbation theory where the lowest excited state is only 5000 cm^{-1} above the ground state and the spin-orbit coupling constant is -828 cm^{-1}. A complete calculation which simultaneously diagonalizes spin orbit and crystal field matrix elements corrects for this source of error, but still gives g_z = 3.473, g_y = 2.195 and g_x = 2.125. Clearly, covalent delocalization must also be taken into account.

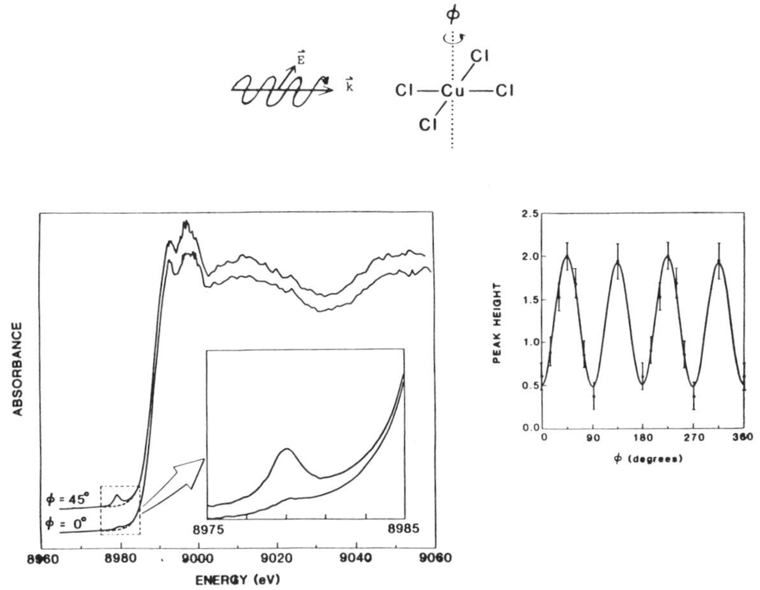

Figure 14. Top: representation of square planar $CuCl_4^{=}$. Left: edge spectra of square planar $CuCl_4^{=}$ showing variation in intensity of 8987 eV transition. Right: angular dependence of intensity for the 8979 eV peak.

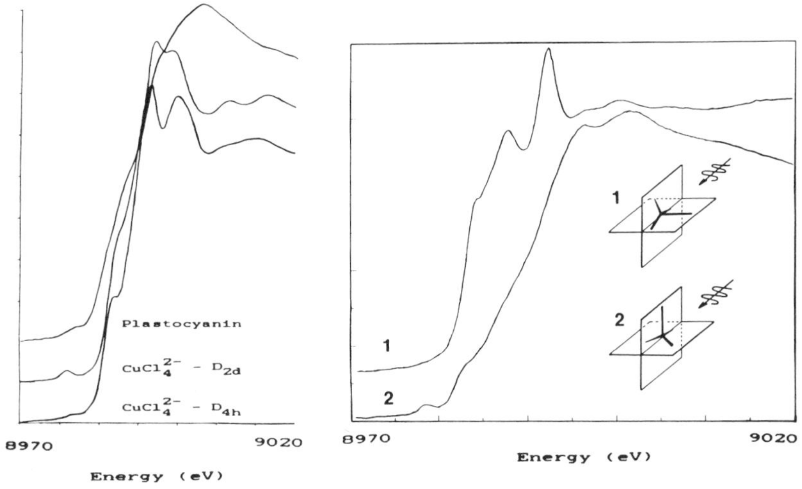

Figure 15. Left: edge spectra for plastocyanin, D_{2d} $CuCl_4^{=}$ and D_{4h} $CuCl_4^{=}$. Right: variation in intensity of the 8979 eV transition for two different orientations of plastocyanin.

A flow chart detailing the calculation (14) of g values from the Xα charge decomposition is shown in Figure 16. First, the partitioned charge decomposition is orthogonalized, producing a complete set of basis functions which therefore form a large matrix of states. Spin is included, doubling the size of the matrix, and spin orbit coupling is evaluated as off-diagonal elements, while the diagonal elements are the observed transition energies. The spin-orbit coupling constants used in calculating the off-diagonal elements were $\lambda_{Cu} = -828\ cm^{-1}$, $\lambda_S = -382\ cm^{-1}$ and $\lambda_N = -76\ cm^{-1}$. While there is evidence (23) to suggest that the spin-orbit coupling parameter of copper does not change significantly from the free ion value with increasing delocalization, such evidence is lacking for sulfur. However, varying this number by several tens of wavenumbers produced relatively minor changes in the calculated g values. Diagonalizing the matrix gave spin-orbit corrected energies and wavefunctions. The spin-orbit corrected ground state was then used to calculate a g^2 tensor. The g^2 tensor was diagonalized and the principal g values were obtained as the square root of the eigenvalues. The eigenvectors of this diagonalization represent the direction cosines of the diagonal g^2 tensor with respect to the orientation of the original basis set.

A test of this approach on D_{4h} $CuCl_4^{=}$ (14) gave $g_{\parallel} = 2.144$, $g_{\perp} = 2.034$, while the experimental values are $g_{\parallel} = 2.221$, $g_{\perp} = 2.040$. Since the calculated values are smaller than those obtained experimentally, it is apparent that the Xα calculation overestimates the degree of delocalization. To adjust the delocalization, and thereby to obtain an experimentally calibrated ground state wave function, a single adjustable parameter was included in the g value calculation which increased proportionally the amount of metal character in ligand field orbitals while decreasing proportionally the metal character in charge transfer levels. Application of this approach to D_{4h} $CuCl_4^{=}$ gives 64% metal character in the ground state with $g_{\parallel} = 2.221$ and $g_{\perp} = 2.047$.

For D_{4h} $CuCl_4^{=}$, we can obtain additional estimates for the delocalization using a variety of complementary experimental techniques. These results are presented in Table 1. A calculation of hyperfine parameters for D_{4h} $CuCl_4^{=}$ using the results of a crystal field calculation gives $A_{\parallel} = 269 \times 10^{-4}\ cm^{-1}$ and $A_{\perp} = 38 \times 10^{-4}\ cm^{-1}$, while the experimental (24) values are $A_{\parallel} = 164 \times 10^{-4}\ cm^{-1}$ and $A_{\perp} = 34 \times 10^{-4}\ cm^{-1}$. In contrast, a solution of simultaneous equations to fit the experimental values following the method of McGarvey (25,26) gives 67% metal character in the ground state wavefunction. Similarly, a consideration of superhyperfine parameters in the crystal field limit gives $A_{\parallel}^{Cl} = 0.28 \times 10^{-4}\ cm^{-1}$ and $A_{\perp}^{Cl} = 0.14 \times 10^{-4}\ cm^{-1}$, in contrast to the experimental (24) values of $18.5 \times 10^{-4}\ cm^{-1}$ and $5.0 \times 10^{-4}\ cm^{-1}$, respectively. A fit to these values, however, gives 64% delocalization. While both of these methods are necessarily approximate, their consistency in giving delocalizations in the 65% range lend support to the results obtained through an empirical fit of the Xα wavefunction. These results are further supported by the photoionization cross sections observed in the UPS spectra of $CuCl_4^{=}$ as discussed in the next section.

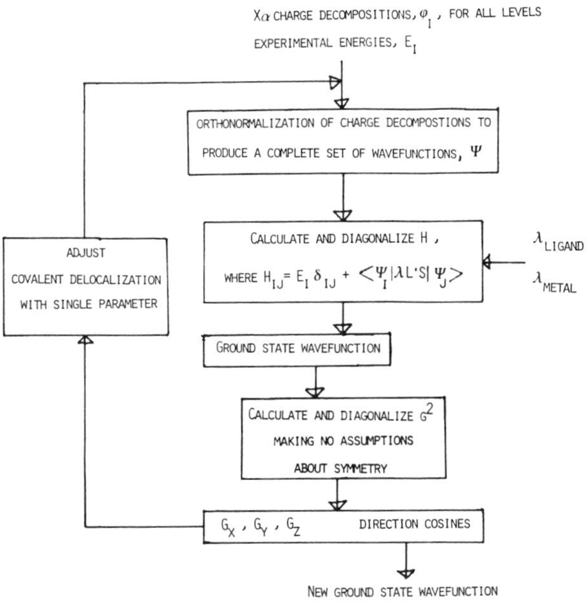

Figure 16. Flow chart showing method of calculating g values from Xα charge decompositions.

Table I. Determination of delocalization for D_{4h} $CuCl_4^{=}$ from a variety of complementary experimental techniques.

Spin Hamiltonian Parameter	Experimental Value	Values Given By Crystal Field Calculation	Value After Adjustment	% Metal Character After Adjustment
$g_{\parallel}$	2.221	2.743	2.222	64%
$g_{\perp}$	2.040	2.177	2.048	
$A_{\parallel}^{Cu}$ x 10^{-4} cm^{-1}	164	269	164	67%
$A_{\perp}^{Cu}$ x 10^{-4} cm^{-1}	34.5	38	34	
$A_{\parallel}^{Cl}$ x 10^{-4} cm^{-1}	18.5	0.28	18.5	64%
$A_{\perp}^{Cl}$ x 10^{-4} cm^{-1}	5.0	0.14	5.0	

In plastocyanin, the ground state wavefunction obtained directly from the Xα calculation consists of 31% copper character and gives $g_x = 2.046$, $g_y = 2.067$ and $g_z = 2.159$. Adjusting the metal character to 40% however raises the g values to $g_x = 2.059$, $g_y = 2.076$ and $g_z = 2.226$ and leads to a final spin-orbit corrected wavefunction:

$$\Psi_o = 40\%\ Cu\ [0.99(x^2-y^2) - 0.10(xz) - 0.10(yz)] + 0.8\%\ Cu[0.71(px) + 0.71(py)]$$
$$+36\%\ S_{cys}\ [P\pi] + 4.8\%\ N[0.99\ P\sigma] + 4.8\%\ N[0.99\ P\sigma] + 0.01\%\ S_{met}$$

A contour diagram of the ground state wavefunction which includes the orientation of the g^2 tensor calculated for the C_s(met) approximation to the plastocyanin site is shown in Figure 17.

It is evident that the unpaired electron is strongly delocalized over both the copper (40%) and sulfur (36%) centers. The contour diagram in Figure 17 shows two lobes of the $d_{x^2-y^2}$ orbital directed toward the nitrogen atoms, while the other two lobes are involved in a pi-antibonding interaction with a thiolate sulfur p orbital. The large amount of delocalization over the sulfur suggests that it is the antibonding interaction between the sulfur and the copper which controls the orientation of the $d_{x^2-y^2}$ orbital.

A small rhombic splitting in the g values is also calculated ($g_y - g_x = 0.017$). As a rhombic splitting was also indicated by our crystal field computations (11), a higher resolution EPR spectrum than the previously considered X-band data was obtained. Figure 18 shows the results of Q-band EPR on a frozen sample of spinach plastocyanin. A rhombic splitting of 0.017 is clearly discernable, in agreement with the adjusted Xα calculation.

The corrected ground state wavefunction from the Xα calculation can also be used to calculate hyperfine values. Over a manifold

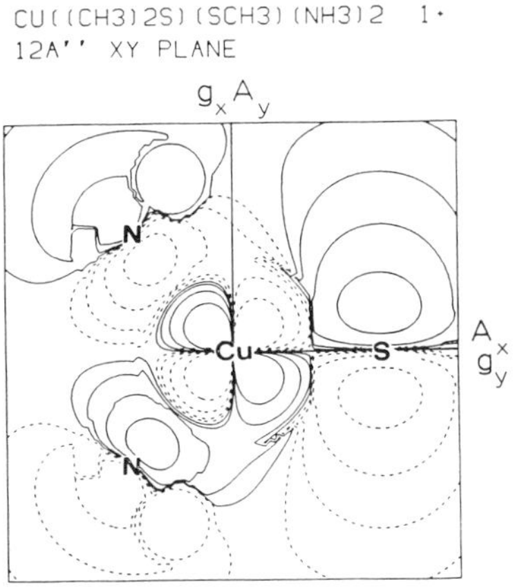

Figure 17. Orientation of the ground state wavefunction for the C_s (met) approximation to plastocyanin. The contour indicates the orientation of the g^2 and A^2 tensors in the plane containing both nitrogens and the copper. Reproduced from Ref. 14. Copyright 1985, American Chemical Society.

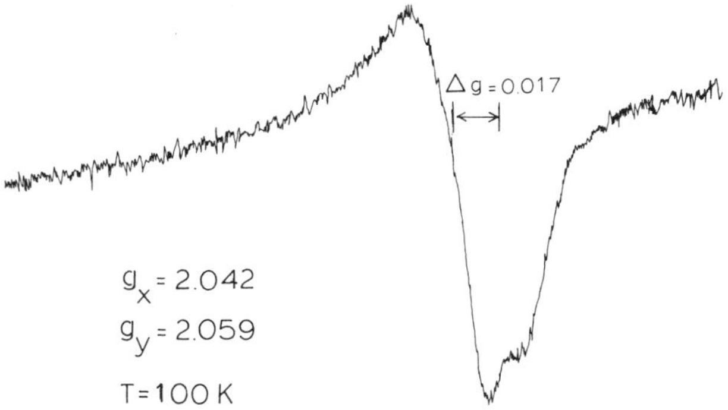

Figure 18. Q-band EPR spectrum of the g perpendicular region of spinach plastocyanin. Field sweep was between 11,275 and 12,275 gauss while the microwave frequency was 34.282 GHz. Reproduced from Ref. 14. Copyright 1985, American Chemical Society.

|LS>, Abragam and Pryce have obtained expressions for the hyperfine operator:

$$A_i = P_d\,[L_i + (\xi L(L+1) - \kappa)S_i - 3/2\,\xi\, L_i\,(L\cdot S) - 3/2\,\xi\,(L\cdot S)\,L_i\,]$$

where i=x,y,z, ξ =2/21, κ is the Fermi contact term and P_d is $g_e g_n \beta_e \beta_n / \langle r^3 \rangle$. By treating A_i in a manner analogous to the Zeeman operator considered above and operating only over the metal part of the spin-orbit corrected wave function, one can obtain the elements of the A^2 tensor. Diagonalization yields the square of the principal values of the hyperfine interaction as well as eigevectors which will transform the starting basis into one in which A^2 is diagonal.

The results of calculation of hyperfine parameters for the C_s(met) approximation to the plastocyanin site, along with those for D_{4h} $CuCl_4$, are presented in Table II. Two factors account for the difference between the "normal" hyperfine values of D_{4h} $CuCl_4$ and the "anomolous" small values of plastocyanin. The first factor is the reduction in dipolar coupling which is directly due to covalency. The second factor is the reduction in κ which again reflects delocalization (27). In addition, a small amount of 4s mixing can contribute to the reduction in values; experimentally on the order of 0.7% mixing is required, while the Xα calculation predict only 0.1%.

Photoelectron Spectroscopic Studies of Covalency

In order to further experimentally probe the covalent delocalization of metal and ligand valence orbitals in the $CuCl_4$ complex, variable energy photoelectron spectroscopy of the valence band of D_{4h} and D_{2d} $CuCl_4$ has been employed. Through knowledge of the atomic photoionization cross sections of the Cu 3d and Cl 3p orbitals as well as the resonant photoemission behavior of the valence band features at the Cu 3p→ 3d absorption edge at 73 eV, a consistent experimental evaluation of covalent mixing in these materials can be obtained.

The atomic photoionization cross section (σ) is governed by the electric dipole matrix element between the initial state atomic orbital radial wavefunction (Ψ_i) and the final state free electron wavefunction (Ψ_f) with kinetic energy E_k.(28,29) As the input photon energy increases the kinetic energy of the photoelectron increases thus decreasing its deBroglie wavelength, and changing the value of the electric dipole matrix element (Figure 19). The electric dipole matrix element is also strongly affected by the nature of the initial state orbital. The photoionization cross sections for Cl 3p and Cu 3d atomic orbitals are shown in Figure 20. The Cl 3p cross section has a maximum in intensity right above threshold while the Cu 3d cross section increases more slowly to a delayed maximum ~70 eV above threshold. This delayed maximum is due to a repulsive centrifugal potential in the radial Schrodinger equation and it depends on the l value of the ionized orbital (28,29). In addition, for a Cl 3p orbital which has a radial node, the positive and negative contributions to the electric dipole matrix element can cancel giving a minimum in the photoionization cross section. For Cl 3p this Cooper Minimum (30) occurs at ~50 eV as shown in Figure 20.

The variable energy photoelectron spectra of the valence bands of Cu(I) and Cu(II) chlorides (31-33) are shown in Figure 21. For

Table II. Results of calculation of hyperfine parameters for plastocyanin and D_{4h} $CuCl_4^{=}$ broken down by terms in the hamiltonian

Term in Hamiltonian:	$-P_d\kappa$	$P_d\xi L(L+1)S_i$	$-P_d\frac{3}{2}\xi[L_i(L\cdot S)+(L\cdot S)L_i]$	$P_d L_i$	Total	experimental
	"Fermi Contact"	"Dipolar"	Off-diagonal			
values (10^{-4} cm^{-1}):						
$CuCl_4$:						
$A_{x,y}$	-121	+72	-4	+19	-34	34.5
A_z	-121	-144	+7	+94	-164	164
C_s (met) plastocyanin:						
$A_{x,y}$	-79	+45	-5	+24	-15	<17
A_z	-79	-90	+9	+97	-65	63

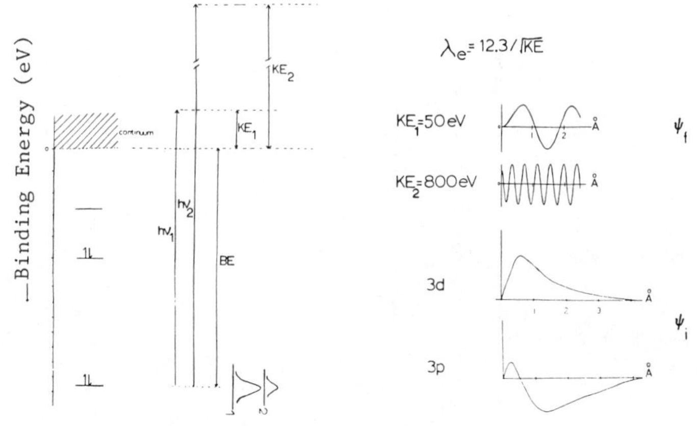

Figure 19. Photon energy dependence of the photoionization cross section: intensity of photoelectron peak varies depending upon input photon energy, due to the variation in the deBroglie wavelength of outgoing electron. These wavelengths are compared to the radial wavefunctions of Cu 3d and Cl 3p orbitals. Reproduced from Ref. 28. Copyright 1985, American Chemical Society.

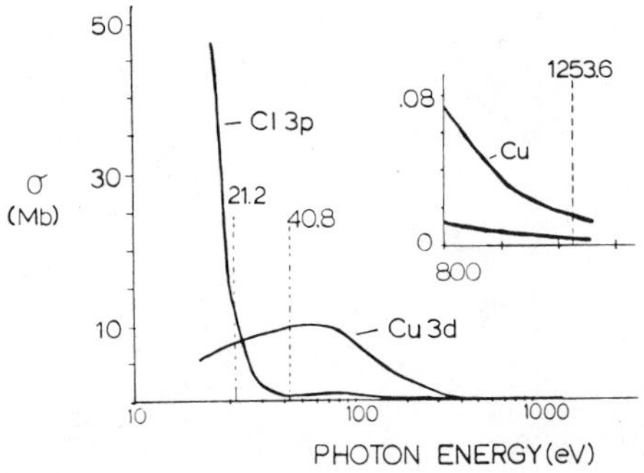

Figure 20. Energy dependence of atomic photoionization cross sections of Cu 3d and Cl 3p orbitals: dashed lines indicate He I, He II, and Mg K α sources. Insert gives high photon energy region at higher sensitivity. 1mb = 10^{-18} cm^{-2}
Reproduced from Ref. 28. Copyright 1985, American Chemical Society.

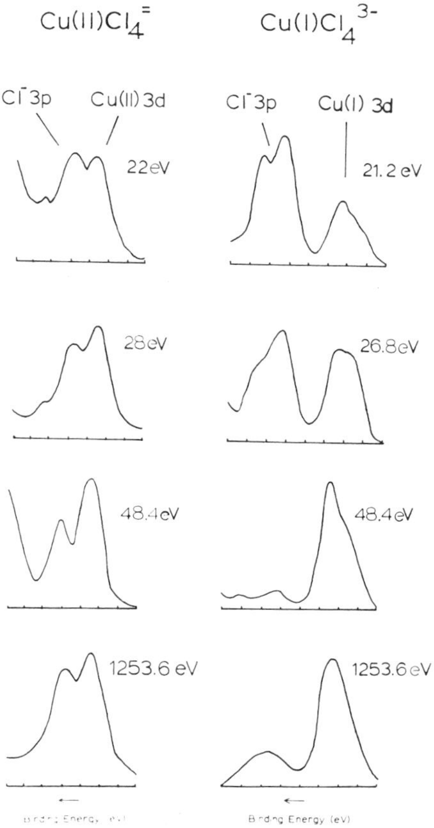

Figure 21. Variable photon energy photoelectron spectra of the valence band region of D_{2d}-$Cu(II)Cl_4^{=}$ and T_d-$Cu(I)Cl_4^{3-}$

Reproduced from Ref. 28. Copyright 1985, American Chemical Society.

$Cu(I)Cl_4^{3-}$ in CuCl, two bands are observed separated by 3.5 eV. The deeper binding energy peak is most intense in the 21.2 eV spectrum but its relative intensity drops dramatically as the photon energy increases, reaching a minimum at 48.4 eV and then increasing slowly with higher photon energy. This is the behavior predicted by the atomic photoionization cross sections for Cl 3p and Cu 3d orbitals in Figure 20, thus allowing general spectral assignment of the peaks as noted in Figure 21.

For D_{2d} $Cu(II)Cl_4^{=}$ in Cs_2CuCl_4, two bands again appear in the UPS spectra but now they are separated in energy by only 1.5 eV. As the photon energy increases from 22 to 1253.6 eV, the same qualitative changes in relative intensity of the two peaks are observed, however, the changes are quantitatively much less pronounced. The greater overlap of the Cu 3d and Cl 3p bands as well as the less atomic like behavior of the photoionization cross sections indicates that there is significant mixing of the Cu 3d and the Cl 3p orbitals into the molecular orbitals of the $CuCl_4^{=}$ molecule.

Further evidence for strong mixing of Cu 3d and Cl 3p orbitals is obtained from the resonant photoemission data shown in Figure 22 for D_{4h} $(CH_3NH_3)_2CuCl_4$ (31). In addition to the Cu 3d and Cl 3p photoemission features, a satellite is observed 8.5 eV below the d band (the features marked with asterisks are due to the $(CH_3NH_3)^+$ counterion). As the photon energy is tuned through the Cu(II) 3p$\rightarrow$ 3d absorption edge at 73 eV, dramatic changes in intensity of the valence band features are observed (see resonant profiles Figure 22). The lineshape of the absorption profiles is characteristic of a Fano interference (34) between a discrete and continuum state. For Cu(II) these competing processes can be described by:

$$3p^6 3d^9 \xrightarrow{\text{absorption}} 3p^5 3d^{10} \xrightarrow{\text{SCK decay}} 3p^6 3d^8$$

photoemission

where the Cu 3p$\rightarrow$ 3d absorption and subsequent super Coster-Kronig decay yields the same final state as in direct photoemission. The resonance of the Cl 3p band and the intense resonance of the satellite and the dip in the Cu 3d band are all characteristic of resonant photoemission from a molecule with strong mixing of metal 3d and ligand 3p orbitals (35).

A reasonable first estimate of the coefficients of mixing of Cu 3d and Cl 3p character in the valence bands of D_{4h} and D_{2d} $CuCl_4$ can be obtained using a simple initial state model. The relative intensities of the photoemission features in spectra taken at high photon energy are fit assuming that the $CuCl_4^{=}$ molecular orbital cross section can be approximated as the sum of atomic orbital cross sections (36-39). The D_{4h} and D_{2d} $CuCl_4$ spectra ($h\nu$= 200 eV) are compared in Figure 24 showing the gaussian deconvolution of the D_{4h} spectra into its components.(31,40) Using calculated photoionization cross sections(30), and measuring the relative intensity of the features at both 200 eV and 1253.6 eV, a consistent set of mixing coefficients has been determined (see Table III). SCF-Xα-SW calculations (also shown in Table III) are in reasonable agreement with the experiment, although they overestimate the covalency in the case

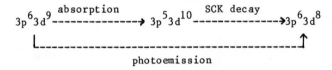

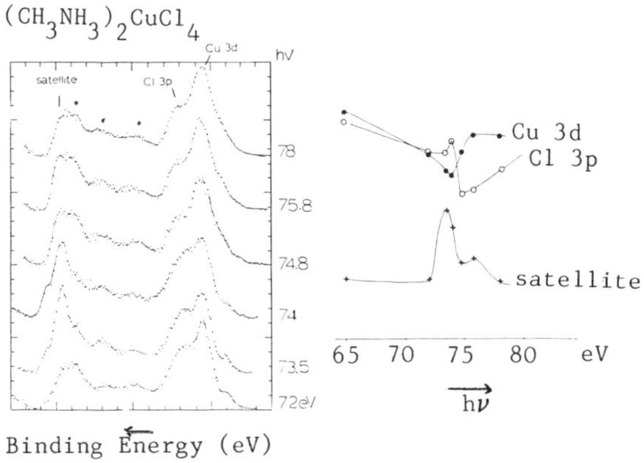

Figure 22. Resonant photoemission spectra of the valence band of D_{4h}-Cu(II)$Cl_4^=$ as the photon energy is tuned through the Cu 3p ---> 3d absorption edge (h ν = 73 eV). The areas of the peaks are plotted as a function of photon energy (on right) giving absorption profiles with a characteristic Fano lineshape (see text).

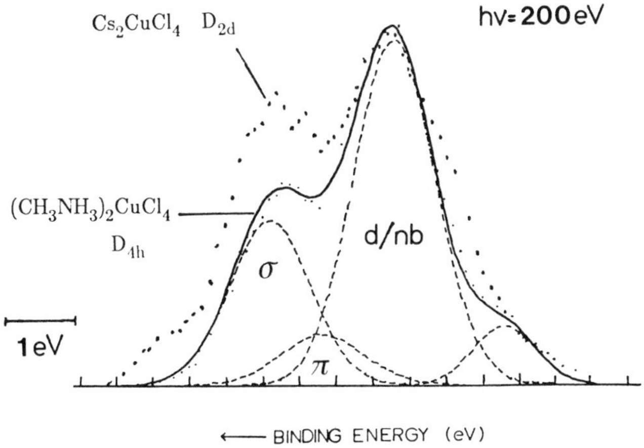

Figure 23. Valence band photoemission spectra of D_{4h} and D_{2d} Cu(II)$Cl_4^=$ (h ν = 200 eV). The relative intensity of the Cu 3d estimated. Reproduced from Ref. 13. Copyright 1982, American Chemical Society.

of the cupric chlorides. This analysis is also in good agreement with the EPR g value and hyperfine analysis discussed in the previous section.

Table III. Percent d character in d bands of $Cu(I)Cl_4^{3-}$ and $Cu(II)Cl_4^{=}$ as determined by variable energy photoelectron spectroscopy and SCF-Xα -SW calculations.

	%d Experimental	%d Xα
CuCl T_d	84 ± 3%	85% (avg.)
$(CH_3NH_3)_2CuCl_4$ D_{4h}	67 ± 5%	59% (avg.)
Cs_2CuCl_4 D_{2d}	62 ± 5%	50% (avg.)

In addition, the strong resonance of the satellite peak at the Cu 3p absorption edge indicates that changes in electron-electron repulsion induced upon ionization must also be considered. Thus a more complete calculation which includes final state configuration interaction has also been pursued thus providing estimates of both the initial and final state wavefunctions (31,35). However, from the initial state analysis presented above, it is clear that as the energy separation of the metal 3d and ligand 3p orbitals decreases going from $CuCl_4^{3-}$ to $CuCl_4^{=}$, the mixing between these levels increases as predicted from simple molecular orbital theory.

Summary

Over the past several years, a detailed picture of the electronic structure of the blue copper active site has been developed through a combination of spectroscopy, crystallography and theory. Remaining to be answered are questions relating this geometric and electronic structure to the reactivity of the site: the role of the orientation of the ground state wavefunction in determining electron transfer pathways, the contribution of the covalent electronic structure described above to the redox properties of the site and the contribution of Jahn-Teller forces connected with the ground state to the determination the geometric structure of the site.

Acknowledgment

The authors would like to acknowledge the National Science Foundation (CHE-82-04841) for support of this research.

Literature Cited

1. Gray, H.B.; Solomon, E.I. In "Copper Proteins"; T.G. Spiro, Ed.; Wiley; 1-39.
2. Solomon, E.I.; Penfield, K.W.; Wilcox, D.E. Struct. Bond. 1983, 53, 1-57.

3. Solomon, E.I. in Karlin, K. and Zubieta, J. "Copper Coordination Chemistry: Biochemical & Inorganic Perspectives"; Adenine Press: Guilderland, N.Y., 1982; 1-22.
4. Solomon, E.I.; Hare, J.W.; Gray, H.B. Proc. Natl. Acad. Sci. U.S.A. 1976, 73, 1389-1393.
5. Solomon, E.I.; Hare, J.W.; Dooley, D.M.; Dawson, J.H.; Stephens, P.J.; Gray, H.B. J. Am. Chem. Soc. 1980, 102, 168-178.
6. Katoh, S.; Takamiyama, A. J. Biochem. 1964 (Tokyo) 55, 378.
7. Solomon, E.I.; Clendening, P.J.; Gray, H.B.; Grunthaner, F.J. J. Am. Chem. Soc. 1974, 97, 3878.
8. McMillin, D.R.; Holwerda, R.A.; Gray, H.B. Proc. Natl. Acad. Sci. U.S.A. 1974, 71, 1338.
9. Guss, J.M.; Freeman, H.C.; J. Mol. Biol. 1983, 169, 521.
10. Adman, E.T.; Jensen, J.H. Israel J. Chem. 1981, 21, 8.
11. Penfield, K.W.; Gay, R.R.; Himmelwright, R.S.; Eickman, N.C.; Norris, V.A.; Freeman, H.C.; Solomon, E.I. J. Am. Chem. Soc. 1981, 103, 4382-4388.
12. Thamann, T.J.; Frank, P.; Willis, L.J.; Loehr, T.M.; Proc. Natl. Acad. Sci. U.S.A. 1982, 79, 6306-6400.
13. Scott, R.A.; Hahn, J.E.; Doniach, S.; Freeman, H.C.; Hodgson, K.O. J. Am. Chem. Soc. 1982, 104, 5364-5369.
14. Penfield, K.W.; Gewirth, A.A.; Solomon, E.I. J. Am. Chem. Soc. 1985, 107, 4519-4529.
15. Hitchman, M.A.; Cassidy, P.J. Inorg. Chem. 1978, 17, 1682.
16. Hitchman, M.A.; Cassidy, P.J. Inorg. Chem. 1979, 18, 1745.
17. Bates, C.A.; Moore, W.S.; Standley, K.J.; Stevens, K.W.H. Proc. Phys. Soc. 1962, 79, 73-83.
18. Roberts, J.E.; Brown, T.G.; Hoffman, B.M.; Peisach, J. J. Am. Chem. Soc. 1980, 102, 825-829.
19. Sharnoff, M. J. Chem. Phys. 1965, 42, 3383-3395.
20. Hahn, J.E.; Scott, R.A.; Hodgson, K.O.; Doniach, S.; Desjardins, S.R.; Solomon, E.I. Chem. Phys. Lett. 1982, 88, 595.
21. Hodgson, K.O.; Solomon, E.I. manuscript in preparation.
22. Ballhausen, C.J. "Introduction to Ligand Field Theory," McGraw-Hill: New York, 1961.
23. Goodgame, B.A.; Rayner, J.B. Adv. Inorg. Chem. Radiochem. 1970, 13, 135.
24. Cassidy, P.; Hitchman, M.A. Inorg. Chem. 1977, 16, 1568-1570.
25. McGarvey, B.R. In "Transition Metal Chemistry"; Carlin, R.L., Ed.; Marcel Dekker, Inc.; New York.
26. Maki, A.J.; McGarvey, B.R. J. Chem. Phys. 1958, 29, 31.
27. McGarvey, B.R. J. Phys. Chem. 1967, 71, 51-67.
28. Solomon, E.I. Comments on Inorganic Chemistry, 1984, 3, 225-320.
29. Fano, U.; Cooper, J. Rev. Mod. Phys. 1968, 40, 441.
30. Yeh, J.J. Atomic Data and Nuclear Data Tables, 1985, 32, 1.
31. Cohen, S.L.; Didziulis, S.V.; Solomon, E.I., to be published.
32. Desjardins, S.R.; Penfield, K.W., Cohen, S.L.; Musselman, R.L.; Solomon, E.I. J. Am. Chem. Soc., 1983, 105, 4590.
33. Goldman, J.; Tejeda, J.; Shevchic, N.J.; Cardona, M. Phys.Rev. B. 1974, 19, 4388.
34. Fano, U. Phys. Rev. 1966, 124, 1866.
35. Davis, L. Phys. Rev. B. 1982, 25, 2912.

36. Gelius, U.; Siegbahn, K. Far. Discuss. Chem. Soc. 1972, 54, 257-68.
37. Nefedov, V.I.; Sergushin, N.P.; Band, I.M.;Trzhaskovskaya, M.B. J. Electr. Spec. Relat. Phenom. 1973, 2, 383-403.
38. Nefedov, V.I.: Sergushin, N.P., Salyn, Y.V.; Band, I.M.; Trzhaskovskaya, M.B. J. Electr. Spec. Relat. Phenom. 1975, 7,175-185.
39. Kono, S.; Ishii, T., Sagawa, T.; Kobayashi, T. Phys. Rev. B. 1973, 8, 795.
40. Deconvolution of spectra (energy distribution of orbitals and widths of bands) accomplished by comparison of variable energy photoelectron spectra and SCF-Xα -SW ground state calculation.

RECEIVED November 8, 1985

INDEXES

Author Index

Subject Index

A

B

J

K

L

M

N

O

Production by Anne Riesberg
Indexing by Janet S. Dodd
Jacket design by Pamela Lewis

Elements typeset by Hot Type Ltd., Washington, DC
Printed and bound by Maple Press Co., York, PA